GLENCOE

MATH

YOUR COMMON CORE EDITION CCSS

AUTHORS

Carter • Cuevas • Day • Malloy • Kersaint • Luchin • McClain
Molix-Bailey • Price • Reynosa • Silbey • Vielhaber • Willard

Bothell, WA • Chicago, IL • Columbus, OH • New York, NY

connectED.mcgraw-hill.com

STEM McGraw-Hill is committed to providing instructional materials in Science, Technology, Engineering, and Mathematics (STEM) that give all students a solid foundation, one that prepares them for college and careers in the 21st century.

Send all inquiries to:
McGraw-Hill Education
8787 Orion Place
Columbus, OH 43240

ISBN: 978-0-07-661529-2 (*Volume 1*)
MHID: 0-07-661529-4

Printed in the United States of America.

18 19 20 21 LMN 21 20 19 18

Our mission is to provide educational resources that enable students to become the problem solvers of the 21st century and inspire them to explore careers within Science, Technology, Engineering, and Mathematics (STEM) related fields.

CONTENTS IN BRIEF

Units organized by CCSS domain

Glencoe Math is organized into units based on groups of related standards called domains. This year, you will study and understand the five domains shown below.

Richard Drury/Photodisc/Getty Images

Ratios and Proportional Relationships

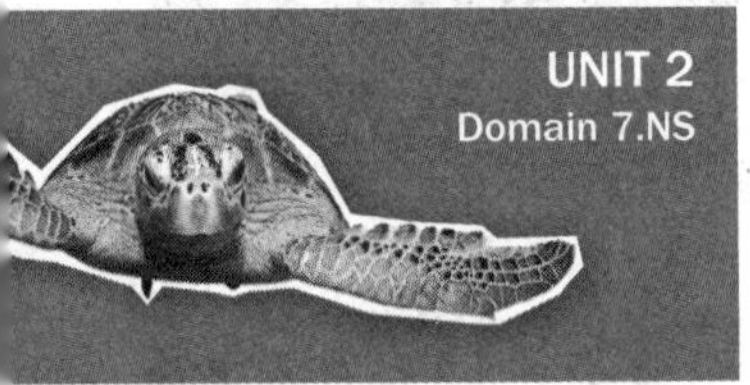

Gerald Nowak/Westend61/Photolibrary

The Number System

Jill Braaten/The McGraw-Hill Companies

Expressions and Equations

Tim Flach/Stone+/Getty Images

Geometry

Back in the Pack dog portraits/flickr RF/Getty Images

Statistics and Probability

it's all at **connectED.mcgraw-hill.com**

Go to the Student Center for your eBook, Resources, Homework, and Messages.

Glencoe Math: Student Center

Favorites | Glencoe Math: Student Center | Tools

Hello, Student | Home | ConnectED | Help | Logout

GLENCOE MATH CCSS

Search | Standards

Student Center

Home | Homework | Resources

Unit 4 > Chapter 6: Transformations

Lesson 4: Dilation

Open eBook

Unit 4: Geometry > Chapter 6: Transformations > Lesson 4: Dilation

Geometry CCSS

Homework

Due: Monday, October 7, 2011

Assignment Title

You have unread teacher comments.

More

Messages

Monday, October 7, 2011

Don't forget to study for the test on Friday.

More

The McGraw-Hill Companies

Terms of Use | Privacy Policy | Technical Support | Minimum System Requirements

Copyright© The McGraw-Hill Companies, Inc.

Write your Username ______ Password ______

Get your resources online to help you in class and at home.

Vocab

Find activities for building vocabulary.

Watch

Watch animations and videos.

Tutor

See a teacher illustrate examples and problems.

Tools

Explore concepts with virtual manipulatives.

Check

Self-assess your progress.

eHelp

Get targeted homework help.

Masters

Provides practice worksheets.

GO mobile

Scan this QR code with your smart phone* or visit mheonline.com/apps.

*May require quick response code reader app.

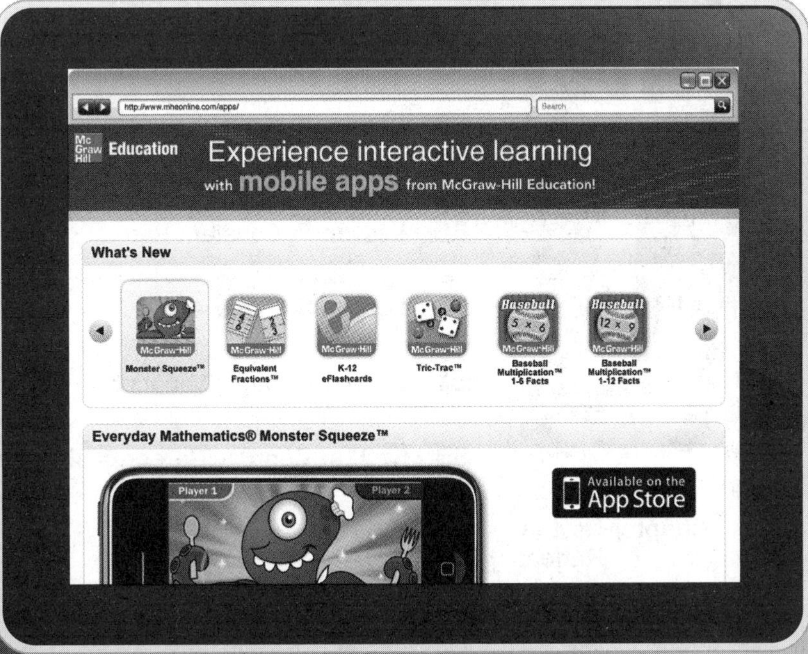

UNIT 1 Ratios and Proportional Relationships

UNIT PROJECT PREVIEW page 2

Chapter 1
Ratios and Proportional Reasoning

Essential Question

HOW can you show that two objects are proportional?

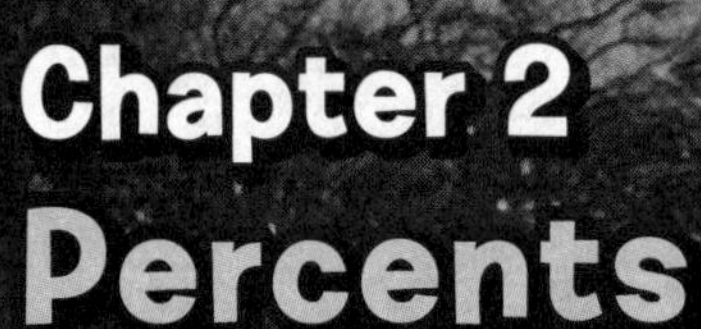

Chapter 2
Percents

Essential Question

HOW can percent help you understand situations involving money?

Become a Travel Expert

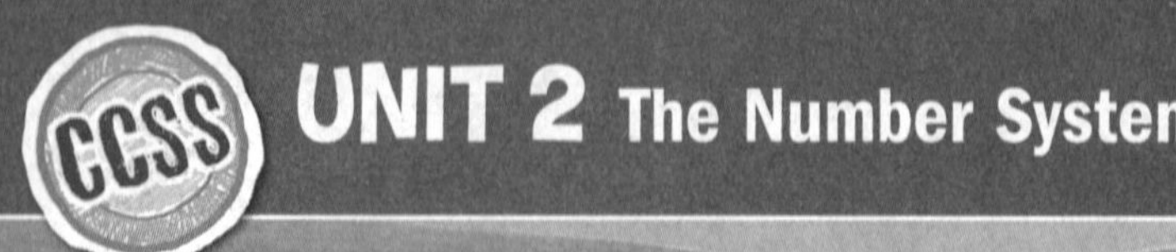

UNIT PROJECT PREVIEW
page 185

Chapter 3 Integers

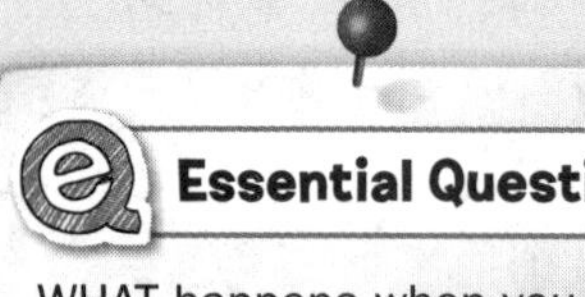

WHAT happens when you add, subtract, multiply, and divide integers?

p. 233

Chapter 4 Rational Numbers

Essential Question

WHAT happens when you add, subtract, multiply, and divide fractions?

Explore the Ocean Depths

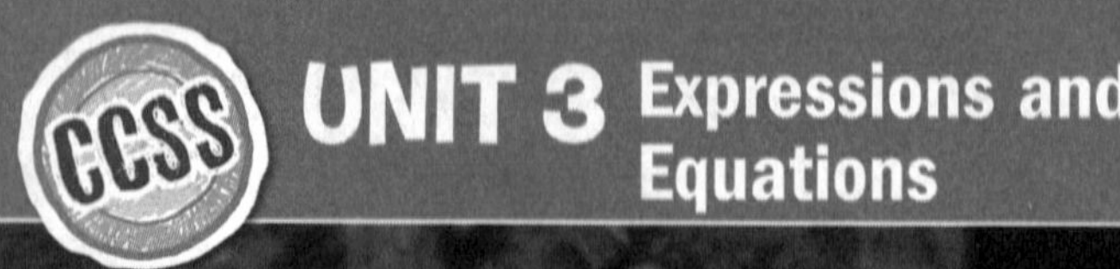

UNIT 3 Expressions and Equations

UNIT PROJECT PREVIEW page 344

Chapter 5 Expressions

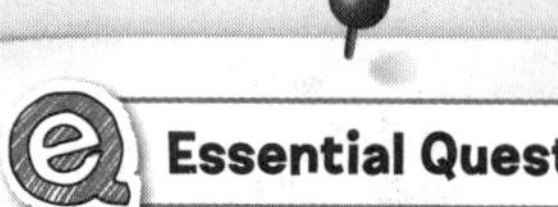

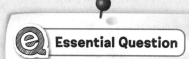

Essential Question

HOW can you use numbers and symbols to represent mathematical ideas?

p. 387

Chapter 6 Equations and Inequalities

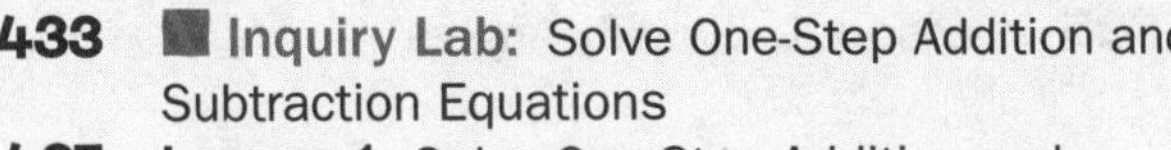

Real World p. 437

Essential Question

WHAT does it mean to say two quantities are equal?

Stand Up and Be Counted!

UNIT 4 Geometry

UNIT PROJECT PREVIEW
page 530

Chapter 7 Geometric Figures

Essential Question

HOW does geometry help us describe real-world objects?

Real World p. 535

Chapter 8
Measure Figures

Essential Question

HOW do measurements help you describe real-world objects?

p. 623

Turn Over a New Leaf

CCSS **UNIT 5** Statistics and Probability

UNIT PROJECT PREVIEW page 706

Chapter 9 Probability

Essential Question

HOW can you predict the outcome of future events?

Chapter 10
Statistics

HOW do you know which type of graph to use when displaying data?

Common Core State Standards for MATHEMATICS, Grade 7

Glencoe Math, Course 2, focuses on four critical areas: (1) developing understanding of and applying proportional relationships; (2) operations with rational numbers and working with expressions and linear equations; (3) solving problems involving scale drawings, geometric constructions, and surface area, and volume; and (4) drawing inferences about populations.

Content Standards

Domain 7.RP **Ratios and Proportional Relationships**

- Analyze proportional relationships and use them to solve real-world and mathematical problems.

Domain 7.NS **The Number System**

- Apply and extend previous understandings of operations with fractions to add, subtract, multiply, and divide rational numbers.

Domain 7.EE **Expressions and Equations**

- Use properties of operations to generate equivalent expressions.
- Solve real-life and mathematical problems using numerical and algebraic expressions and equations.

Domain 7.G **Geometry**

- Draw, construct and describe geometrical figures and describe the relationships between them.
- Solve real-life and mathematical problems involving angle measure, area, surface area, and volume.

Domain 7.SP **Statistics and Probability**

- Use random sampling to draw inferences about a population.
- Draw informal comparative inferences about two populations.
- Investigate chance processes and develop, use, and evaluate probability models.

Mathematical Practices

1. Make sense of problems and persevere in solving them.
2. Reason abstractly and quantitatively.
3. Construct viable arguments and critique the reasoning of others.
4. Model with mathematics.
5. Use appropriate tools strategically.
6. Attend to precision.
7. Look for and make use of structure.
8. Look for and express regularity in repeated reasoning.

Ratios and Proportional Relationships

Domain 7.RP

Analyze proportional relationships and use them to solve real-world and mathematical problems.

1. Compute unit rates associated with ratios of fractions, including ratios of lengths, areas and other quantities measured in like or different units.
2. Recognize and represent proportional relationships between quantities.
   a. Decide whether two quantities are in a proportional relationship, e.g., by testing for equivalent ratios in a table or graphing on a coordinate plane and observing whether the graph is a straight line through the origin.
   b. Identify the constant of proportionality (unit rate) in tables, graphs, equations, diagrams, and verbal descriptions of proportional relationships.
   c. Represent proportional relationships by equations.
   d. Explain what a point (x, y) on the graph of a proportional relationship means in terms of the situation, with special attention to the points (0, 0) and $(1, r)$ where r is the unit rate.
3. Use proportional relationships to solve multistep ratio and percent problems.

Related Unit Project: Unit 1
Become a Travel Expert

The Number System

Domain 7.NS

Apply and extend previous understandings of operations with fractions to add, subtract, multiply, and divide rational numbers.

1. Apply and extend previous understandings of addition and subtraction to add and subtract rational numbers; represent addition and subtraction on a horizontal or vertical number line diagram.
   a. Describe situations in which opposite quantities combine to make 0.
   b. Understand $p + q$ as the number located a distance $|q|$ from p, in the positive or negative direction depending on whether q is positive or negative. Show that a number and its opposite have a sum of 0 (are additive inverses). Interpret sums of rational numbers by describing real-world contexts.
   c. Understand subtraction of rational numbers as adding the additive inverse, $p - q = p + (-q)$. Show that the distance between two rational numbers on the number line is the absolute value of their difference, and apply this principle in real-world contexts.
   d. Apply properties of operations as strategies to add and subtract rational numbers.
2. Apply and extend previous understandings of multiplication and division and of fractions to multiply and divide rational numbers.
   a. Understand that multiplication is extended from fractions to rational numbers by requiring that operations continue to satisfy the properties of operations, particularly the distributive property, leading to products such as $(-1)(-1) = 1$ and the rules for multiplying signed numbers. Interpret products of rational numbers by describing real-world contexts.

Related Unit Project: Unit 2
Explore the Ocean Depths

b. Understand that integers can be divided, provided that the divisor is not zero, and every quotient of integers (with non-zero divisor) is a rational number. If p and q are integers, then $-(p/q) = (-p)/q = p/(-q)$. Interpret quotients of rational numbers by describing real-world contexts.

c. Apply properties of operations as strategies to multiply and divide rational numbers.

d. Convert a rational number to a decimal using long division; know that the decimal form of a rational number terminates in 0s or eventually repeats.

3. Solve real-world and mathematical problems involving the four operations with rational numbers.

Domain 7.EE

Expressions and Equations

Use properties of operations to generate equivalent expressions.

1. Apply properties of operations as strategies to add, subtract, factor, and expand linear expressions with rational coefficients.

2. Understand that rewriting an expression in different forms in a problem context can shed light on the problem and how the quantities in it are related.

Solve real-life and mathematical problems using numerical and algebraic expressions and equations.

Related Unit Project: Unit 3
Stand Up and Be Counted!

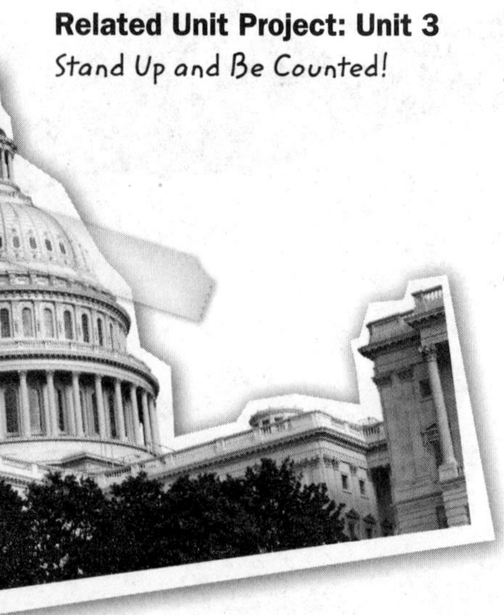

3. Solve multi-step real-life and mathematical problems posed with positive and negative rational numbers in any form (whole numbers, fractions, and decimals), using tools strategically. Apply properties of operations to calculate with numbers in any form; convert between forms as appropriate; and assess the reasonableness of answers using mental computation and estimation strategies.

4. Use variables to represent quantities in a real-world or mathematical problem, and construct simple equations and inequalities to solve problems by reasoning about the quantities.

a. Solve word problems leading to equations of the form $px + q = r$ and $p(x + q) = r$, where p, q, and r are specific rational numbers. Solve equations of these forms fluently. Compare an algebraic solution to an arithmetic solution, identifying the sequence of the operations used in each approach.

b. Solve word problems leading to inequalities of the form $px + q > r$ or $px + q < r$, where p, q, and r are specific rational numbers. Graph the solution set of the inequality and interpret it in the context of the problem.

Geometry

Domain 7.G

Draw, construct, and describe geometrical figures and describe the relationships between them.

1. Solve problems involving scale drawings of geometric figures, including computing actual lengths and areas from a scale drawing and reproducing a scale drawing at a different scale.
2. Draw (freehand, with ruler and protractor, and with technology) geometric shapes with given conditions. Focus on constructing triangles from three measures of angles or sides, noticing when the conditions determine a unique triangle, more than one triangle, or no triangle.
3. Describe the two-dimensional figures that result from slicing three-dimensional figures, as in plane sections of right rectangular prisms and right rectangular pyramids.

Related Unit Project: Unit 4
Turn Over a New Leaf

Solve real-life and mathematical problems involving angle measure, area, surface area, and volume.

4. Know the formulas for the area and circumference of a circle and use them to solve problems; give an informal derivation of the relationship between the circumference and area of a circle.
5. Use facts about supplementary, complementary, vertical, and adjacent angles in a multi-step problem to write and solve simple equations for an unknown angle in a figure.
6. Solve real-world and mathematical problems involving area, volume and surface area of two- and three-dimensional objects composed of triangles, quadrilaterals, polygons, cubes, and right prisms.

Statistics and Probability

Domain 7.SP

Use random sampling to draw inferences about a population.

1. Understand that statistics can be used to gain information about a population by examining a sample of the population; generalizations about a population from a sample are valid only if the sample is representative of that population. Understand that random sampling tends to produce representative samples and support valid inferences.
2. Use data from a random sample to draw inferences about a population with an unknown characteristic of interest. Generate multiple samples (or simulated samples) of the same size to gauge the variation in estimates or predictions.

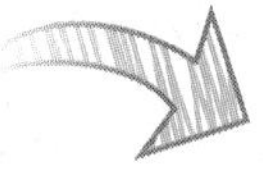

Draw informal comparative inferences about two populations.

3. Informally assess the degree of visual overlap of two numerical data distributions with similar variabilities, measuring the difference between the centers by expressing it as a multiple of a measure of variability.

4. Use measures of center and measures of variability for numerical data from random samples to draw informal comparative inferences about two populations.

Related Unit Project: Unit 5
Math Genes

Investigate chance processes and develop, use, and evaluate probability models.

5. Understand that the probability of a chance event is a number between 0 and 1 that expresses the likelihood of the event occurring. Larger numbers indicate greater likelihood. A probability near 0 indicates an unlikely event, a probability around 1/2 indicates an event that is neither unlikely nor likely, and a probability near 1 indicates a likely event.
6. Approximate the probability of a chance event by collecting data on the chance process that produces it and observing its long-run relative frequency, and predict the approximate relative frequency given the probability.

7. Develop a probability model and use it to find probabilities of events. Compare probabilities from a model to observed frequencies; if the agreement is not good, explain possible sources of the discrepancy.

    a. Develop a uniform probability model by assigning equal probability to all outcomes, and use the model to determine probabilities of events.

    b. Develop a probability model (which may not be uniform) by observing frequencies in data generated from a chance process.

8. Find probabilities of compound events using organized lists, tables, tree diagrams, and simulation.

    a. Understand that, just as with simple events, the probability of a compound event is the fraction of outcomes in the sample space for which the compound event occurs.

    b. Represent sample spaces for compound events using methods such as organized lists, tables and tree diagrams. For an event described in everyday language (e.g., "rolling double sixes"), identify the outcomes in the sample space which compose the event.

    c. Design and use a simulation to generate frequencies for compound events.

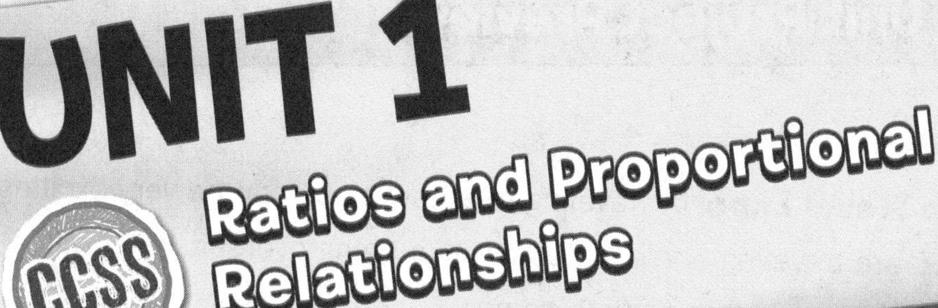

UNIT 1

Ratios and Proportional Relationships

CCSS

Essential Question

HOW can you use mathematics to describe change and model real-world situations?

Chapter 1
Ratios and Proportional Reasoning

Proportional relationships can be used to solve real-world problems. In this chapter, you will determine whether the relationship between two quantities is proportional. Then you will use proportions to solve multi-step problems.

Chapter 2
Percents

Proportional relationships can be used to solve percent problems. In this chapter, you will find percent of increase and decrease and use percents to solve problems involving sales tax, tips, markups and discounts, and simple interest.

Unit Project Preview

Become a Travel Expert Traveling to a new place can be very exciting. Whether you are traveling a few states away or going overseas, you will surely experience and learn something new.

When you plan a trip, one thing you should definitely consider is the amount of money the trip will cost. Carefully planning the budget for a trip will ensure that you have enough money and it might help you save a few bucks as well. At the end of Chapter 2, you'll complete a project about the many costs involved with travel.

Choose a city in the United States. Complete the table by estimating the cost of different parts of a one-week family vacation to your city.

My Trip to ________	
Item	**Cost ($)**
Round Trip Plane Ticket	
Hotel	
Rental Car	
Food	
Attractions	

Chapter 1

Ratios and Proportional Reasoning

Essential Question

HOW can you show that two objects are proportional?

Common Core State Standards

Content Standards
7.RP.1, 7.RP.2, 7.RP.2a, 7.RP.2b, 7.RP.2c, 7.RP.2d, 7.RP.3, 7.NS.3

Mathematical Practices
1, 2, 3, 4, 5, 6

Math in the Real World

Airplanes used for commercial flights travel at a speed of about 550 miles per hour.

Suppose an airplane travels 265 miles in one-half hour. Draw an arrow on the speedometer below to represent the speed of the airplane in miles per hour.

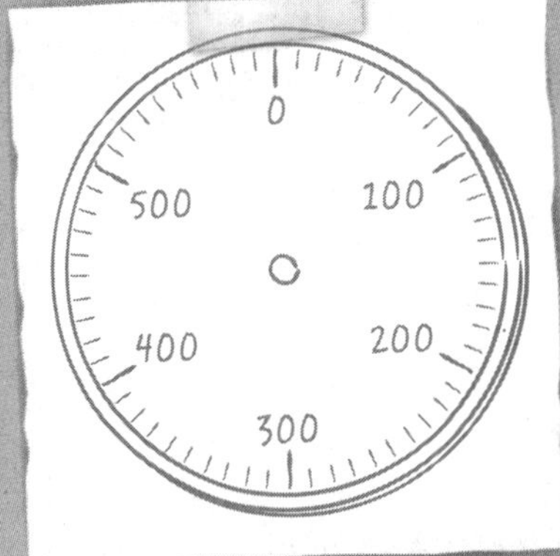

FOLDABLES® Study Organizer

Cut out the Foldable on page FL3 of this book.

Place your Foldable on page 92.

Use the Foldable throughout this chapter as you learn about proportional reasoning.

What Tools Do You Need?

Vocabulary

complex fraction
constant of proportionality
constant rate of change
constant of variation
coordinate plane
cross products
dimensional analysis
direct variation
equivalent ratios
nonproportional
proportion
proportional
ordered pair
origin
quadrants
rate
rate of change
slope
unit rate
unit ratio
x-axis
x-coordinate
y-axis
y-coordinate

Review Vocabulary

Functions A function is a relationship that assigns exactly one output value for each input value. The function rule is the operation performed on the input. Perform each indicated operation on the input 10. Then write each output in the organizer.

Input	Rule	Output
10	Add 2.	
	Subtract 3.	
	Multiply by 4.	
	Divide by 5.	

When Will You Use This?

Try the Quick Check below.
Or, take the Online Readiness Quiz.

Quick Review

Common Core Review 6.RP.1, 6.RP.3

Example 1

Write the ratio of wins to losses as a fraction in simplest form.

Madison Mavericks	
Team Statistics	
Wins	10
Losses	12
Ties	8

wins → $\frac{10}{12} = \frac{5}{6}$ ← losses

The ratio of wins to losses is $\frac{5}{6}$.

Example 2

Determine whether the ratios 250 miles in 4 hours and 500 miles in 8 hours are equivalent.

Compare the ratios by writing them in simplest form.

250 miles : 4 hours $= \frac{250}{4}$ or $\frac{125}{2}$

500 miles : 8 hours $= \frac{500}{8}$ or $\frac{125}{2}$

The ratios are equivalent because they simplify to the same fraction.

Quick Check

Ratios **Write each ratio as a fraction in simplest form.**

Seventh-Grade Field Trip	
Students	180
Adults	24
Buses	4

1. adults : students 24:180
2. students : buses 180:4
3. buses : people 4:204

Equivalent Ratios **Determine whether the ratios are equivalent. Explain.**

4. 20 nails for every 5 shingles
   12 nails for every 3 shingles
   yes

5. 12 out of 20 doctors agree
   15 out of 30 doctors agree
   3:5 so they are not same
   1:2

Which problems did you answer correctly in the Quick Check? Shade those exercise numbers below.

1 2 3 4 5

Inquiry Lab

Unit Rates

HOW can you use a bar diagram to solve a real-world problem involving ratios?

Content Standards
Preparation for 7.RP.1, 7.RP.2, and 7.RP.2b

Mathematical Practices
1, 3, 4

Money When Jeremy gets his allowance, he agrees to save part of it. His savings and expenses are in the ratio 7:5. If his daily allowance is \$3, find how much he saves each day.

Investigation

Step 1 Complete the bar diagram below by writing *savings, expenses,* and *\$3* in the correct boxes.

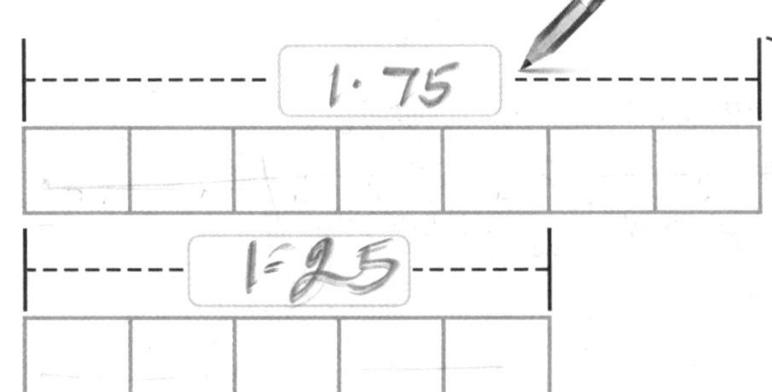

Total amount = 3
(Daily Allowance)

Step 2 Let x represent each part of a bar. Write and solve an equation to find the amount of money each bar represents.

$7x + 5x = 3$	Write the equation.
$12x = 3$	There are 12 parts in all.
$\frac{12x}{12} = \frac{3}{12}$	Division Property of Equality
$x = \frac{25}{100}$ or 0.25	Simplify.

$.25 \times 7 = 1.75$

Step 3 Determine the amount Jeremy saves each day. Since each part of the bar represents \$0.25, Jeremy's savings are represented by

7 × \$ .25 or \$1.75.

So, Jeremy saves \$ 1.75 each day.

Collaborate

Work with a partner to answer the following question.

1. The ratio of the number of boys to the number of girls on the swim team is 4:2. If there are 24 athletes on the swim team, how many more boys than girls are there? Use a bar diagram to solve.

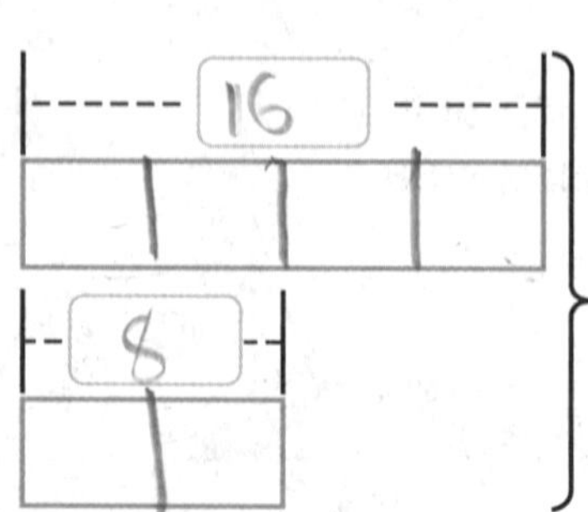

Total athletes = 24

$6x = 24$

$x = 4$

Analyze

Work with a partner to answer the following question.

2. **Reason Inductively** Suppose the swim team has 24 athletes, but the ratio of boys to girls on the swim team is 3 : 5. How would the bar diagram change? Instad of 16 it would be 9 and 15 would replace 8.

$8x = 24$

$x = 3$

Reflect

3. **Model with Mathematics** Write a real-world problem that could be represented by the bar diagram shown below. Then solve your problem.

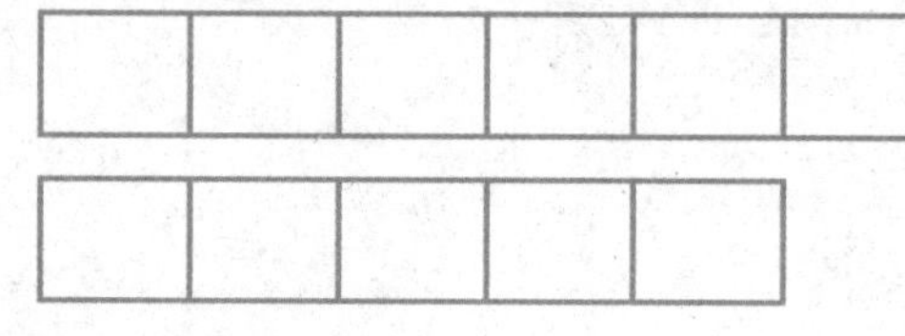

Total amount = 220

$22x = 220$

There are 6:5 chocolate chip cookies sold and snickerdoodles. If there are 220 in total how many more chocolate chips are there than snickerdoo

4. **Inquiry** HOW can you use a bar diagram to solve a real-world problem involving ratios?

To vislize how to solve a problem

Lesson 1
Rates

What You'll Learn

Scan the lesson. Predict two things you will learn about rates.

- ____________________
- ____________________

Essential Question

HOW can you show that two objects are proportional?

Vocabulary

rate
unit rate

Common Core State Standards

Content Standards
7.RP.2, 7.RP.2b

Mathematical Practices
1, 3, 4, 5

Real-World Link

Watch

Pulse Rate You can take a person's pulse by placing your middle and index finger on the underside of their wrist. Choose a partner and take their pulse for two minutes.

1. Record the results in the diagram below.

2. Use the results from Exercise 1 to complete the bar diagram and determine the number of beats per minute for your partner.

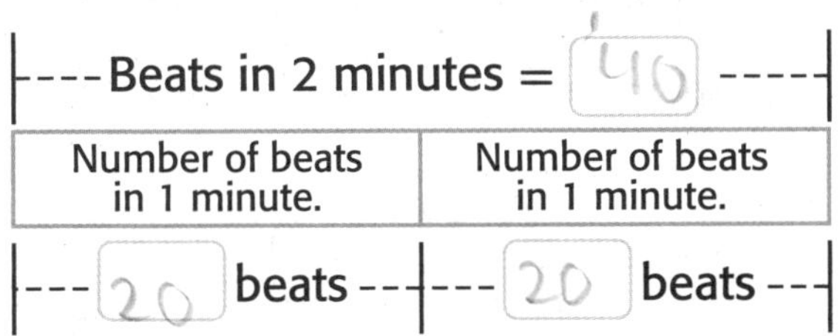

So, your partner's heart beats 20 times per minute.

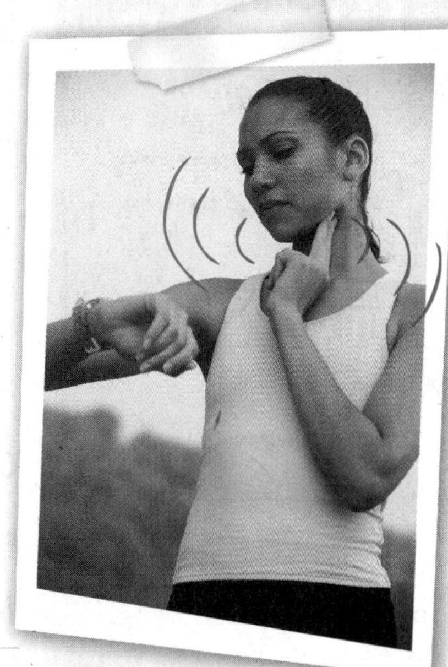

3. Use the results from Exercise 1 to determine the number of beats for $\frac{1}{2}$ minute for your partner.

Work Zone

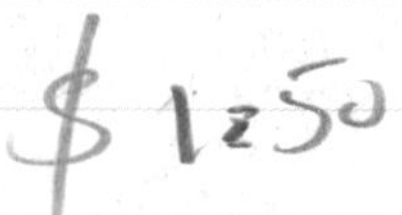

Find a Unit Rate

A ratio that compares two quantities with different kinds of units is called a **rate**. When you found each other's pulse, you were actually finding the heart *rate*.

$$\frac{\text{160 beats}}{\text{2 minutes}}$$

The units *beats* and *minutes* are different.

When a rate is simplified so that it has a denominator of 1 unit, it is called a **unit rate**.

$$\frac{\text{80 beats}}{\text{1 minute}}$$

The denominator is 1 unit.

The table below shows some common unit rates.

Rate	Unit Rate	Abbreviation	Name
$\frac{\text{number of miles}}{\text{1 hour}}$	miles per hour	mi/h or mph	average speed
$\frac{\text{number of miles}}{\text{1 gallon}}$	miles per gallon	mi/gal or mpg	gas mileage
$\frac{\text{number of dollars}}{\text{1 pound}}$	price per pound	dollars/lb	unit price

Example

Tutor

1. Adrienne biked 24 miles in 4 hours. If she biked at a constant speed, how many miles did she ride in one hour?

$$\text{24 miles in 4 hours} = \frac{24\text{ mi}}{4\text{ h}}$$ Write the rate as a fraction.

$$= \frac{24\text{ mi} \div 4}{4\text{ h} \div 4}$$ Divide the numerator and the denominator by 4.

$$= \frac{6\text{ mi}}{1\text{ h}}$$ Simplify.

Adrienne biked 6 miles in one hour.

STOP and Reflect

Circle the unit rate below that represents 18 cans for \$6.

$\frac{\text{9 cans}}{\$3}$ $\frac{\text{3 cans}}{\$1}$ $\frac{\text{4 cans}}{\$1}$

Got It? Do these problems to find out.

Find each unit rate. Round to the nearest hundredth if necessary.

a. \$300 for 6 hours

b. 220 miles on 8 gallons

Show your work.

a. \$50 per hour

b. 27.5 per gallons

Example

2. Find the unit price if it costs $2 for eight juice boxes.

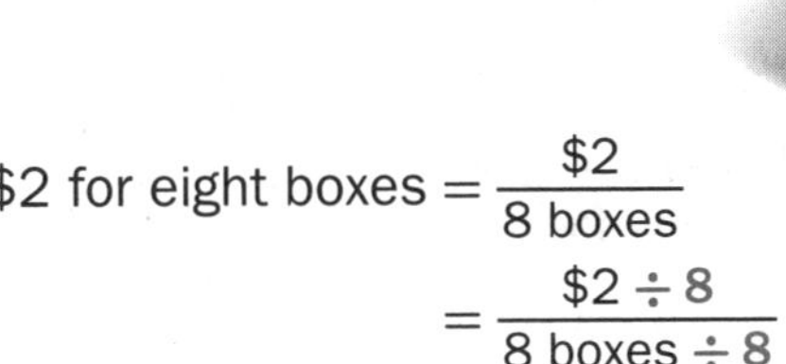

$$\$2 \text{ for eight boxes} = \frac{\$2}{8 \text{ boxes}}$$ Write the rate as a fraction.

$$= \frac{\$2 \div 8}{8 \text{ boxes} \div 8}$$ Divide the numerator and the denominator by 8.

$$= \frac{\$0.25}{1 \text{ box}}$$ Simplify.

The unit price is $0.25 per juice box.

Got It? Do this problem to find out.

c. Find the unit price if a 4-pack of mixed fruit sells for $2.12.

c. $0.53 per pack

Example

3. The prices of 3 different bags of dog food are given in the table. Which size bag has the lowest price per pound rounded to the nearest cent?

Dog Food Prices	
Bag Size (lb)	**Price ($)**
40	49.00
20	23.44
8	9.88

- 40-pound bag
  $49.00 ÷ 40 pounds ≈ $1.23 per pound
- 20-pound bag
  $23.44 ÷ 20 pounds ≈ $1.17 per pound
- 8-pound bag
  $9.88 ÷ 8 pounds ≈ $1.24 per pound

The 20-pound bag sells for the lowest price per pound.

Alternative Method

One 40-lb bag is equivalent to two 20-lb bags or five 8-lb bags. The cost for one 40-lb bag is $49, the cost for two 20-lb bags is about 2 x $23 or $46, and the cost for five 8-lb bags is about 5 x $10 or $50. So, the 20-lb bag has the lowest price per pound.

Got It? Do this problem to find out.

d. Tito wants to buy some peanut butter to donate to the local food pantry. Tito wants to buy as much peanut butter as possible. Which brand should he buy?

Peanut Butter Sales	
Brand	**Sale Price**
Nutty	12 ounces for $2.19
Grandma's	18 ounces for $2.79
Bee's	28 ounces for $4.69
Save-A-Lot	40 ounces for $6.60

d. Grandma

Example

4. Lexi painted 2 faces in 8 minutes at the Crafts Fair. At this rate, how many faces can she paint in 40 minutes?

Method 1 **Draw a Bar Diagram**

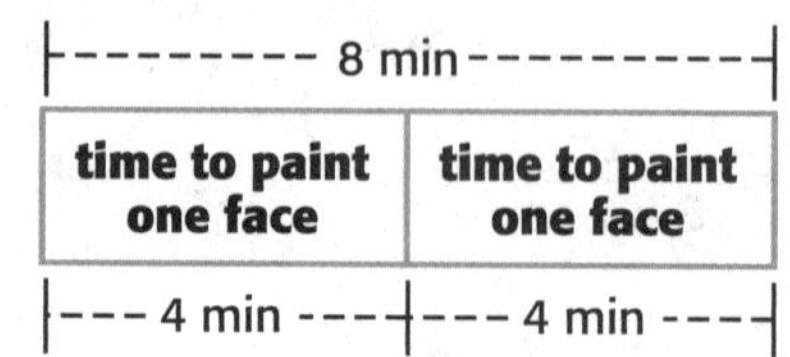

It takes 4 minutes to paint one face. In 40 minutes, Lexi can paint $40 \div 4$ or 10 faces.

Method 2 **Find a Unit Rate**

2 faces in 8 minutes $= \frac{2 \text{ faces} \div 8}{8 \text{ min} \div 8} = \frac{0.25 \text{ face}}{1 \text{ min}}$ Find the unit rate.

Multiply the unit rate by 40 minutes.

$\frac{0.25 \text{ face}}{1 \cancel{\text{min}}} \cdot 40 \cancel{\text{min}} = 10 \text{ faces}$ Divide out the common units.

Using either method, Lexi can paint 10 faces in 40 minutes.

Guided Practice

1. CD Express offers 4 CDs for $60. Music Place offers 6 CDs for $75. Which store offers the better buy? (Examples 1–3)

2. After 3.5 hours, Pasha had traveled 217 miles. If she travels at a constant speed, how far will she have traveled after 4 hours? (Example 4)

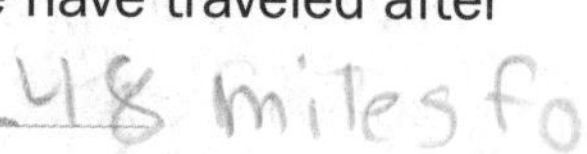

3. Write 5 pounds for $2.49 as a unit rate. Round to the nearest hundredth. (Example 2)

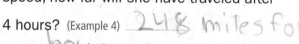

4. **Building on the Essential Question** Use an example to describe how a *rate* is a measure of one quantity per unit of another quantity.

Rate Yourself!

Are you ready to move on? Shade the section that applies.

For more help, go online to access a Personal Tutor.

Name ______________________ My Homework ______________________

Independent Practice

Go online for Step-by-Step Solutions  eHelp

Find each unit rate. Round to the nearest hundredth if necessary. (Examples 1 and 2)

1. 360 miles in 6 hours ______________

2. 6,840 customers in 45 days ______________

3. 45.5 meters in 13 seconds ______________

4. $7.40 for 5 pounds ______________

5. Estimate the unit rate if 12 pairs of socks sell for $5.79. (Examples 1 and 2)

____

6. CCSS **Justify Conclusions** The results of a swim meet are shown. Who swam the fastest? Explain your reasoning. (Example 3)

____

Name	Event	Time (s)
Tawni	50-m Freestyle	40.8
Pepita	100-m Butterfly	60.2
Susana	200-m Medley	112.4

7. Ben can type 153 words in 3 minutes. At this rate, how many words can he type in 10 minutes? (Example 4)

____

8. Kenji buys 3 yards of fabric for $7.47. Then he realizes that he needs 2 more yards. How much will the extra fabric cost? (Example 4)

____

9. The record for the Boston Marathon's wheelchair division is 1 hour, 18 minutes, and 27 seconds.

   a. The Boston Marathon is 26.2 miles long. What was the average speed of the record winner of the wheelchair division?

      Round to the nearest hundredth. ______________

   b. At this rate, about how long would it take this competitor to complete a 30-mile race? ______________

10. At Tire Depot, a pair of new tires sells for $216. The manager's special advertises the same tires selling at a rate of $380 for 4 tires. How much do you save per tire if you purchase the manager's special? yes 13

H.O.T. Problems Higher Order Thinking

11. CCSS **Use Math Tools** Find examples of grocery item prices in a newspaper, on television, or on the Internet. Compare unit prices of two different brands of the same item. Explain which item is the better buy.

you can bye 198 for 2 remotes for the same remote you can bye 4 remotes for 273.

12. CCSS **Find the Error** Seth is trying to find the unit price for a package of blank compact discs on sale at 10 for $5.49. Find his mistake and correct it.

0.549 for 1 disk

CCSS **Persevere with Problems Determine whether each statement is *sometimes, always,* or *never* true. Give an example or a counterexample.**

13. A ratio is a rate.

Somtimes

14. A rate is a ratio.

Always

Standardized Test Practice

15. The table shows the total distance traveled by a car driving at a constant rate of speed. How far will the car have traveled after 10 hours?

Ⓐ 520 miles

Ⓑ 585 miles

Ⓒ 650 miles

Ⓓ 715 miles

Time (h)	Distance (mi)
2	130
3.5	227.5
4	260
7	455

Name ______________________ My Homework ______________________

Extra Practice

Find each unit rate. Round to the nearest hundredth if necessary.

16. 150 people for 5 classes

30 people per class

Homework Help

$\frac{150 \text{ people} \div 5}{5 \text{ classes} \div 5} = \frac{30 \text{ people}}{1 \text{ class}}$

30 people per class

17. 815 Calories in 4 servings

203.75 Calories per serving

$\frac{815 \text{ Calories} \div 4}{4 \text{ servings} \div 4} = \frac{203.75 \text{ Calories}}{1 \text{ serving}}$

203.75 Calories per serving

18. \$1.12 for 8.2 ounces

0.136 for 1

19. 144 miles on 4.5 gallons

32 miles per gallon

20. CCSS **Justify Conclusions** A grocery store sells a 6-pack of bottled water for \$3.79, a 9-pack for \$4.50, and a 12-pack for \$6.89. Which package costs the least per bottle? Explain your reasoning.

.63 > .5

21. CCSS **Justify Conclusions** Dalila earns \$108.75 for working 15 hours as a holiday helper wrapping gifts. At this rate, how much money will she earn if she works 18 hours the next week? Explain.

7.25 × 18 = 130.5

22. CCSS **Use Math Tools** Use the graph that shows the average number of heartbeats for an active adult brown bear and a hibernating brown bear.

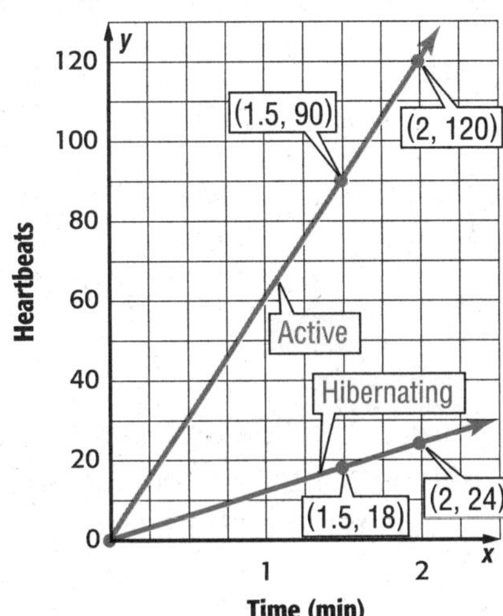

a. What does the point (2, 120) represent on the graph?

The most number of heart beats.

b. What does the ratio of the *y*-coordinate to the *x*-coordinate for each pair of points on the graph represent?

Y : X

c. Use the graph to find the bear's average heart rate when it is active and when it is hibernating.

(2, 24) and (2, 120)

Standardized Test Practice

23. Mrs. Ross needs to buy dish soap. There are four different sized containers.

Dish Soap Prices	
Brand	**Price**
Lots of Suds	$0.98 for 8 ounces
Bright Wash	$1.29 for 12 ounces
Spotless Soap	$3.14 for 30 ounces
Lemon Bright	$3.45 for 32 ounces

Which brand costs the least per ounce?

Ⓐ Lots of Suds Ⓒ Spotless Soap
Ⓑ Bright Wash Ⓓ Lemon Bright

24. **Short Response** Bonita spent $2.00 for 20 pencils, Jamal spent $1.50 for 10 pencils, and Hasina spent $2.10 for 15 pencils. List the students from least to greatest according to unit price paid.

Jamal

25. The Jimenez family took a four-day road trip. They traveled 300 miles in 5 hours on Sunday, 200 miles in 3 hours on Monday, 150 miles in 2.5 hours on Tuesday, and 250 miles in 6 hours on Wednesday. On which day did they average the greatest miles per hour?

Ⓕ Sunday Ⓗ Tuesday
Ⓖ Monday Ⓘ Wednesday

26. **Short Response** Suppose that 1 euro is worth $1.25. In Europe, a book costs 19 euros. In Los Angeles, the same book costs $22.50. In which location is the book less expensive?

Los Angeles

Common Core Review

Solve. Write in simplest form. 5.NF.4

27. $\frac{1}{2} \times \frac{4}{7} = \frac{4}{14}$

28. $\frac{2}{3} \times \frac{1}{6} = \frac{2}{18}$

29. $\frac{1}{4} \div \frac{3}{8} = \frac{3}{32}$

30. Lenora is following the recipe at the right. How many batches of the recipe can she make if she has 5 cups of vegetable oil? 5.NF.4

16

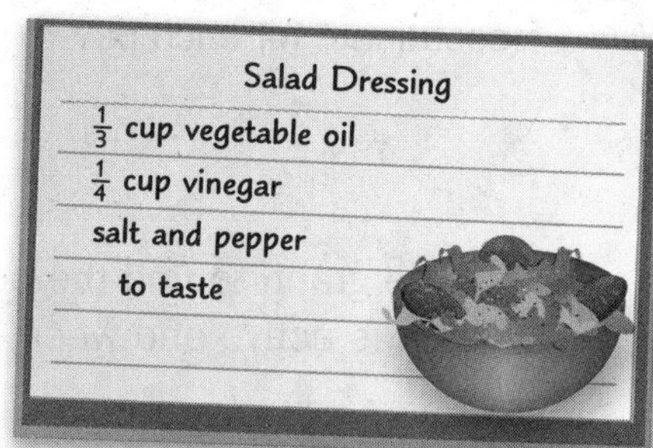

Salad Dressing
1/3 cup vegetable oil
1/4 cup vinegar
salt and pepper
to taste

Lesson 2

Complex Fractions and Unit Rates

What You'll Learn

List two headings you would use to make an outline of the lesson.

- ______________________________
- ______________________________

Essential Question

HOW can you show that two objects are proportional?

Vocabulary

complex fraction

Common Core State Standards

Content Standards
7.RP.1, 7.NS.3

Mathematical Practices
1, 3, 4, 6

Real-World Link

Speed Skating Dana is skating laps to train for a speed skating competition. She can skate 1 lap in 40 seconds.

1. Write a ratio in simplest form comparing Dana's time to her number of laps.

2. Suppose Dana skates for 20 seconds. How many laps will she skate?

____

3. Write the ratio of Dana's time from Exercise 2 to her number of laps.

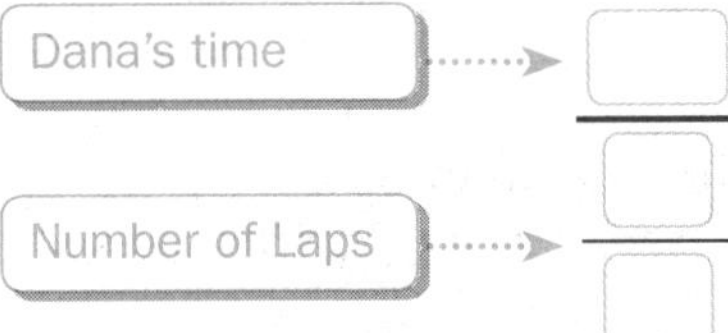

4. How could you simplify the ratio you wrote in Exercise 3?

____

Work Zone

Simplify a Complex Fraction

Fractions like $\frac{20}{\frac{1}{2}}$ are called complex fractions. **Complex fractions** are fractions with a numerator, denominator, or both that are also fractions. Complex fractions are simplified when both the numerator and denominator are integers.

Examples

1. Simplify $\frac{\frac{1}{4}}{2}$.

Recall that a fraction can also be written as a division problem.

$\frac{\frac{1}{4}}{2} = \frac{1}{4} \div 2$ — Write the complex fraction as a division problem.

$= \frac{1}{4} \times \frac{1}{2}$ — Multiply by the reciprocal of 2, which is $\frac{1}{2}$.

$= \frac{1}{8}$ — Simplify.

So, $\frac{\frac{1}{4}}{2}$ is equal to $\frac{1}{8}$.

Divide Fractions

To divide by a whole number, first write it as a fraction with a denominator of 1. Then multiply by the reciprocal.

So, $\frac{\frac{1}{4}}{2}$ can be written as $\frac{1}{4} \div \frac{2}{1}$.

2. Simplify $\frac{1}{\frac{1}{2}}$.

Write the fraction as a division problem.

$\frac{1}{\frac{1}{2}} = 1 \div \frac{1}{2}$ — Write the complex fraction as a division problem.

$= \frac{1}{1} \times \frac{2}{1}$ — Multiply by the reciprocal of $\frac{1}{2}$, which is $\frac{2}{1}$.

$= \frac{2}{1}$ or 2 — Simplify.

So, $\frac{1}{\frac{1}{2}}$ is equal to 2.

Got It? Do these problems to find out.

a. $\frac{2}{\frac{2}{3}}$ $\frac{2}{3} \times \frac{1}{2} = \frac{2}{6}$

b. $\frac{6}{\frac{1}{3}}$ $\frac{1}{3} \times \frac{1}{6} = \frac{1}{18}$

c. $\frac{\frac{2}{3}}{7}$ $\frac{7}{1} \times \frac{3}{2} = \frac{21}{2}$

d. $\frac{\frac{2}{4}}{2}$ $\frac{2}{4} \times \frac{2}{1} = \frac{4}{4}$

Show your work.

a. $\frac{1}{3}$

b. $\frac{1}{18}$

c. $1\frac{1}{2}$

d. $\frac{4}{4}$ or 1

Find Unit Rates

When the fractions of a complex fractions represent different units, you can find the unit rate.

Examples

3. Josiah can jog $1\frac{1}{3}$ miles in $\frac{1}{4}$ hour. Find his average speed in miles per hour.

Write a rate that compares the number of miles to hours.

$$\frac{1\frac{1}{3}\text{ mi}}{\frac{1}{4}\text{ h}} = 1\frac{1}{3} \div \frac{1}{4}$$ Write the complex fraction as a division problem.

$$= \frac{4}{3} \div \frac{1}{4}$$ Write the mixed number as an improper fraction.

$$= \frac{4}{3} \times \frac{4}{1}$$ Multiply by the reciprocal of $\frac{1}{4}$, which is $\frac{4}{1}$.

$$= \frac{16}{3} \text{ or } 5\frac{1}{3}$$ Simplify.

So, Josiah jogs at an average speed of $5\frac{1}{3}$ miles per hour.

4. Tia is painting her house. She paints $34\frac{1}{2}$ square feet in $\frac{3}{4}$ hour. At this rate, how many square feet can she paint each hour?

Write a ratio that compares the number of square feet to hours.

$$\frac{34\frac{1}{2}\text{ ft}^2}{\frac{3}{4}\text{ h}} = 34\frac{1}{2} \div \frac{3}{4}$$ Write the complex fraction as a division problem.

$$= \frac{69}{2} \div \frac{3}{4}$$ Write the mixed number as an improper fraction.

$$= \frac{69}{2} \times \frac{4}{3}$$ Multiply by the reciprocal of $\frac{3}{4}$, which is $\frac{4}{3}$.

$$= \frac{276}{6} \text{ or } 46$$ Simplify.

So, Tia can paint 46 square feet per hour.

Got It? Do these problems to find out.

e. Mr. Ito is spreading mulch in his yard. He spreads $4\frac{2}{3}$ square yards in 2 hours. How many square yards can he mulch per hour?

f. Aubrey can walk $4\frac{1}{2}$ miles in $1\frac{1}{2}$ hours. Find her average speed in miles per hour.

e. $2\frac{1}{3}$

f. 9

Example

5. On Javier's soccer team, about $33\frac{1}{3}$% of the players have scored a goal. Write $33\frac{1}{3}$% as a fraction in simplest form.

$33\frac{1}{3}\% = \dfrac{33\frac{1}{3}}{100}$	Definition of percent
$= 33\frac{1}{3} \div 100$	Write the complex fraction as a division problem.
$= \dfrac{100}{3} \div 100$	Write $33\frac{1}{3}$ as an improper fraction.
$= \dfrac{\overset{1}{\cancel{100}}}{3} \times \dfrac{1}{\underset{1}{\cancel{100}}}$	Multiply by the reciprocal of 100, which is $\frac{1}{100}$.
$= \dfrac{1}{3}$	Simplify.

So, about $\frac{1}{3}$ of Javier's team has scored a goal.

Guided Practice

Simplify. (Examples 1 and 2)

1. $\dfrac{18}{\frac{3}{4}} =$ $1\frac{1}{3}$

2. $\dfrac{\frac{3}{6}}{4} =$ 

3. $\dfrac{\frac{1}{3}}{\frac{1}{4}} =$ $1\frac{1}{3}$

4. Pep Club members are making spirit buttons. They make 490 spirit buttons in $3\frac{1}{2}$ hours. Find the number of buttons the Pep Club makes per hour. (Examples 3 and 4) 1715

5. A county sales tax is $6\frac{2}{3}$%. Write the percent as a fraction in simplest form. (Example 5) $\frac{66.5}{100}$

6. **Building on the Essential Question** What is a complex fraction? A fraction with a fraction for a numerator of denominter

Rate Yourself!

How confident are you about simplifying complex fractions? Check the box that applies.

For more help, go online to access a Personal Tutor.

Name ______________________ My Homework ______________________

Independent Practice

Go online for Step-by-Step Solutions 

Simplify. (Examples 1 and 2)

1. $\dfrac{\frac{1}{2}}{3} =$ ______

2. $\dfrac{\frac{2}{3}}{11} =$ ______

3. $\dfrac{\frac{8}{9}}{6} =$ ______

4. $\dfrac{\frac{2}{5}}{9} =$ ______

5. $\dfrac{\frac{4}{5}}{10} =$ ______

6. $\dfrac{\frac{1}{4}}{\frac{7}{10}} =$ ______

7. Mary is making pillows for her Life Skills class. She bought $2\frac{1}{2}$ yards of fabric. Her total cost was $15. What was the cost per yard? (Examples 3 and 4)

______

8. Doug entered a canoe race. He rowed $3\frac{1}{2}$ miles in $\frac{1}{2}$ hour. What is his average speed in miles per hour? (Examples 3 and 4)

______

9. Monica reads $7\frac{1}{2}$ pages of a mystery book in 9 minutes. What is her average reading rate in pages per minute? (Examples 3 and 4) ______

Write each percent as a fraction in simplest form. (Example 5)

10. $56\frac{1}{4}\% =$ ______

11. $15\frac{3}{5}\% =$ ______

12. $13\frac{1}{3}\% =$ ______

13. A bank is offering home loans at an interest rate of $5\frac{1}{2}\%$. Write the percent as a fraction in simplest form. (Example 5) ______

14. **Be Precise** Karl measured the wingspan of the butterfly and the moth shown below. How many times larger is the moth than the butterfly?

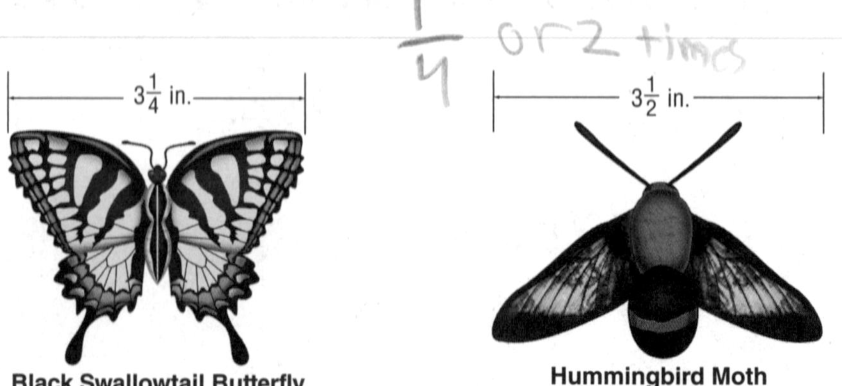

Black Swallowtail Butterfly

Hummingbird Moth

H.O.T. Problems Higher Order Thinking

15. **Construct an Argument** Explain how complex fractions can be used to solve problems involving ratios.

16. **Reason Inductively** Write three different complex fractions that simplify to $\frac{1}{4}$.

17. **Persevere with Problems** Use mental math to find the value of $\frac{15}{124} \cdot \frac{230}{30} \div \frac{230}{124}$.

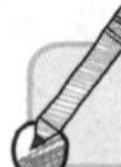

Standardized Test Practice

18. Which statement explains how to use the model to simplify the complex fraction?

$$\frac{\frac{2}{3}}{\frac{1}{12}}$$

$\frac{1}{12}$	$\frac{1}{12}$	$\frac{1}{12}$	$\frac{1}{12}$	$\frac{1}{12}$	$\frac{1}{12}$	$\frac{1}{12}$	$\frac{1}{12}$	$\frac{1}{12}$	$\frac{1}{12}$	$\frac{1}{12}$	$\frac{1}{12}$

Ⓐ Count the twelfths that fit within $\frac{2}{3}$ of the figure.

Ⓑ Remove $\frac{2}{3}$ of the twelfths, and count those remaining.

Ⓒ Count the number of thirds in the figure. Multiply this number by 12.

Ⓓ Count the number of rectangles in the figure. Divide this number by 3.

Name ______________________ My Homework ______________________

Extra Practice

Simplify.

19. $\dfrac{1}{\frac{1}{4}} = 4$

$\dfrac{1}{\frac{1}{4}} = 1 \div \dfrac{1}{4}$

$= \dfrac{1}{1} \times \dfrac{4}{1}$

$= \dfrac{4}{1}$ or 4

Homework Help

20. $\dfrac{12}{\frac{3}{5}} =$ ______

21. $\dfrac{\frac{9}{10}}{9} =$ ______

22. $\dfrac{\frac{1}{2}}{\frac{1}{4}} =$ ______

23. $\dfrac{\frac{1}{12}}{\frac{5}{6}} =$ ______

24. $\dfrac{\frac{5}{6}}{\frac{5}{9}} =$ ______

25. Mrs. Frasier is making costumes for the school play. Each costume requires 0.75 yard of fabric. She bought 6 yards of fabric. How many costumes can Mrs. Frasier make?

______

26. A lawn company advertises that they can spread 7,500 square feet of grass seed in $2\frac{1}{2}$ hours. Find the number of square feet of grass seed that can be spread per hour.

______

Write each percent as a fraction in simplest form.

27. $2\frac{2}{5}\% =$ ______

28. $7\frac{3}{4}\% =$ ______

29. $8\frac{1}{3}\% =$ ______

30. CCSS **Justify Conclusions** The value of a certain stock increased by $1\frac{1}{4}\%$. Explain how to write $1\frac{1}{4}\%$ as a fraction in simplest form. ______

______

______

Standardized Test Practice

31. Debra can run $20\frac{1}{2}$ miles in $2\frac{1}{4}$ hours. How many miles per hour can she run?

Ⓐ $46\frac{1}{8}$ miles per hour

Ⓑ $22\frac{3}{4}$ miles per hour

Ⓒ $18\frac{1}{4}$ miles per hour

Ⓓ $9\frac{1}{9}$ miles per hour

32. Which of the following is equivalent to $\frac{1}{2}$?

Ⓕ $\frac{\frac{1}{4}}{\frac{1}{2}}$

Ⓗ $\frac{\frac{1}{4}}{\frac{1}{4}}$

Ⓖ $\frac{\frac{1}{2}}{\frac{1}{2}}$

Ⓘ $\frac{\frac{1}{8}}{\frac{1}{2}}$

33. Tina wants to give away 6 bundles of thyme from her herb garden. If she has $\frac{1}{2}$ pound of thyme, how much will each bundle weigh?

Ⓐ $\frac{1}{2}$ lb

Ⓑ 3 lb

Ⓒ $\frac{1}{12}$ lb

Ⓓ 12 lb

34. **Short Response** Write $32\frac{1}{8}\%$ as a fraction in simplest form.

Common Core Review

Fill in each box with the equivalent customary measurement. 5.MD.1

35. 2 feet = ☐ inches

36. 5 tons = ☐ pounds

37. 8 gallons = ☐ quarts

Fill in each box with the equivalent metric measurement. 5.MD.1

38. 1 meter = ☐ centimeters

39. 1 liter = ☐ milliliters

40. 1 kilogram = ☐ grams

Lesson 3

Convert Unit Rates

What You'll Learn

Scan the lesson. Write the definitions of unit ratio and dimensional analysis.

- ____________________
- ____________________

Essential Question

HOW can you show that two objects are proportional?

Vocabulary

unit ratio
dimensional analysis

Common Core State Standards

Content Standards
7.RP.2, 7.RP.3

Mathematical Practices
1, 3, 4, 5

Real-World Link

Animals Squirrels, chipmunks, and rabbits are capable of running at fast speeds. The table shows the top running speeds of these animals.

Animal	Speed (mph)
Squirrel	10
Chipmunk	15
Cottontail Rabbit	30

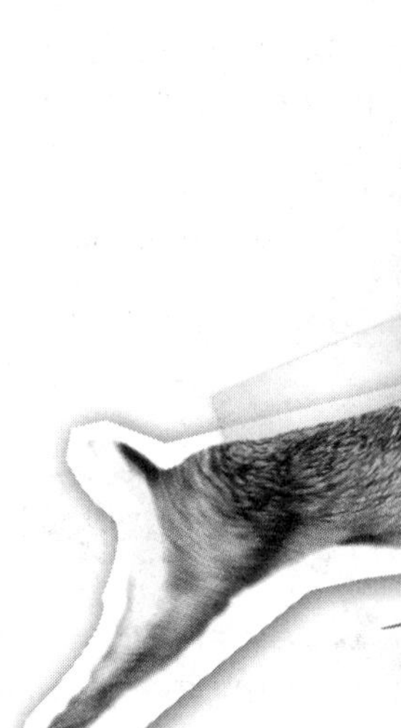

1. How many feet are in 1 mile? 10 miles?

   1 mile = ________ feet

   10 miles = ________ feet

2. How many seconds are in 1 minute? 1 hour?

   1 minute = ________ seconds

   1 hour = ________ seconds

3. How could you determine the number of feet per second a squirrel can run?

   ____

4. Complete the following statement. Round to the nearest tenth.

   10 miles per hour ≈ ☐ feet per second

Work Zone

Convert Rates

The relationships among some commonly used customary and metric units of measure are shown in the tables below.

Customary Units of Measure	
Smaller	**Larger**
12 inches	1 foot
16 ounces	1 pound
8 pints	1 gallon
3 feet	1 yard
5,280 feet	1 mile

Metric Units of Measure	
Smaller	**Larger**
100 centimeters	1 meter
1,000 grams	1 kilogram
1,000 milliliters	1 liter
10 millimeters	1 centimeter
1,000 milligrams	1 gram

Each of the relationships in the tables can be written as a **unit ratio**. Like a unit rate, a unit ratio is one in which the denominator is 1 unit. Below are three examples of unit ratios.

$$\frac{\text{12 inches}}{\text{1 foot}} \qquad \frac{\text{16 ounces}}{\text{1 pound}} \qquad \frac{\text{100 centimeters}}{\text{1 meter}}$$

The numerator and denominator of each of the unit ratios shown are equal. So, the value of each ratio is 1.

You can convert one rate to an equivalent rate by multiplying by a unit ratio or its reciprocal. When you convert rates, you include the units in your computation.

The process of including units of measure as factors when you compute is called **dimensional analysis**.

$$\frac{10 \text{ ft}}{1 \text{ s}} = \frac{10 \cancel{\text{ft}}}{1 \text{ s}} \cdot \frac{12 \text{ in.}}{1 \cancel{\text{ft}}} = \frac{10 \cdot 12 \text{ in.}}{1 \text{ s} \cdot 1} = \frac{120 \text{ in.}}{1 \text{ s}}$$

Example

1. A remote control car travels at a rate of 10 feet per second. How many inches per second is this?

$$\frac{10 \text{ ft}}{1 \text{ s}} = \frac{10 \text{ ft}}{1 \text{ s}} \cdot \frac{12 \text{ in.}}{1 \text{ ft}}$$ Use 1 foot = 12 inches. Multiply by $\frac{12 \text{ in.}}{1 \text{ ft}}$.

$$= \frac{10 \cancel{\text{ft}}}{1 \text{ s}} \cdot \frac{12 \text{ in.}}{1 \cancel{\text{ft}}}$$ Divide out common units.

$$= \frac{10 \cdot 12 \text{ in.}}{1 \text{ s} \cdot 1}$$ Simplify.

$$= \frac{120 \text{ in.}}{1 \text{ s}}$$ Simplify.

So, 10 feet per second equals 120 inches per second.

Examples

2. A swordfish can swim at a rate of 60 miles per hour. How many feet per hour is this?

You can use 1 mile = 5,280 feet to convert the rates.

$$\frac{60 \text{ mi}}{1 \text{ h}} = \frac{60 \text{ mi}}{1 \text{ h}} \cdot \frac{5{,}280 \text{ ft}}{1 \text{ mi}} \quad \text{Multiply by } \frac{5{,}280 \text{ ft}}{1 \text{ mi}}.$$

$$= \frac{60 \not{\text{mi}}}{1 \text{ h}} \cdot \frac{5{,}280 \text{ ft}}{1 \not{\text{mi}}} \quad \text{Divide out common units.}$$

$$= \frac{60 \cdot 5{,}280 \text{ ft}}{1 \cdot 1 \text{ h}} \quad \text{Simplify.}$$

$$= \frac{316{,}800 \text{ ft}}{1 \text{ h}} \quad \text{Simplify.}$$

A swordfish can swim at a rate of 316,800 feet per hour.

3. Marvin walks at a speed of 7 feet per second. How many feet per hour is this?

You can use 60 seconds = 1 minute and you can use 60 minutes = 1 hour to convert the rates.

$$\frac{7 \text{ ft}}{1 \text{ s}} = \frac{7 \text{ ft}}{1 \text{ s}} \cdot \frac{60 \text{ s}}{1 \text{ min}} \cdot \frac{60 \text{ min}}{1 \text{ h}} \quad \text{Multiply by } \frac{60 \text{ s}}{1 \text{ min}} \text{ and } \frac{60 \text{ min}}{1 \text{ h}}.$$

$$= \frac{7 \text{ ft}}{1 \not{\text{s}}} \cdot \frac{60 \not{\text{s}}}{1 \not{\text{min}}} \cdot \frac{60 \not{\text{min}}}{1 \text{ h}} \quad \text{Divide out common units.}$$

$$= \frac{7 \cdot 60 \cdot 60 \text{ ft}}{1 \cdot 1 \cdot 1 \text{ h}} \quad \text{Simplify.}$$

$$= \frac{25{,}200 \text{ ft}}{1 \text{ h}} \quad \text{Simplify.}$$

Marvin walks 25,200 feet in 1 hour.

and Reflect

To convert meters per hour to kilometers per hour, circle the relationship you need to know.

100 cm = 1 m
60 s = 1 min
1,000 m = 1 km

Got It? Do these problems to find out.

a. A gull can fly at a speed of 22 miles per hour. About how many feet per hour can the gull fly?

b. An AMTRAK train travels at 125 miles per hour. Convert the speed to miles per minute. Round to the nearest tenth.

a. ______________

b. ______________

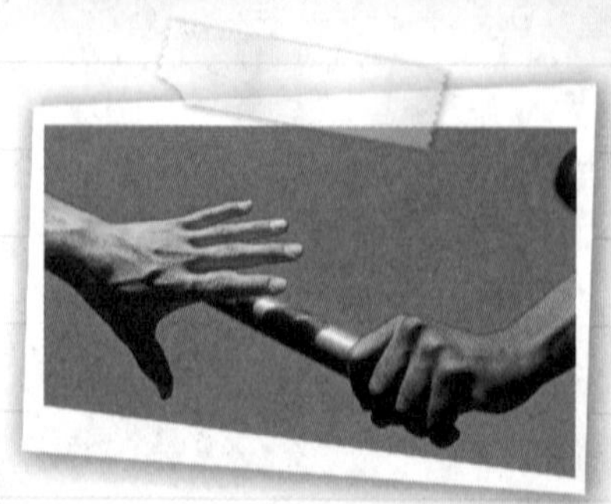

Example

4. The average speed of one team in a relay race is about 10 miles per hour. What is this speed in feet per second?

We can use 1 mile = 5,280 feet, 1 hour = 60 minutes, and 1 minute = 60 seconds to convert the rates.

$$\frac{10 \text{ mi}}{1 \text{ h}} = \frac{10 \text{ mi}}{1 \text{ h}} \cdot \frac{5{,}280 \text{ ft}}{1 \text{ mi}} \cdot \frac{1 \text{ h}}{60 \text{ min}} \cdot \frac{1 \text{ min}}{60 \text{ s}}$$ Multiply by distance and time unit ratios.

$$= \frac{10 \cancel{\text{mi}}}{1 \cancel{\text{h}}} \cdot \frac{5{,}280 \text{ ft}}{1 \cancel{\text{mi}}} \cdot \frac{1 \cancel{\text{h}}}{60 \cancel{\text{min}}} \cdot \frac{1 \cancel{\text{min}}}{60 \text{ s}}$$ Divide out common units.

$$= \frac{10 \cdot 5{,}280 \cdot 1 \cdot 1 \text{ ft}}{1 \cdot 1 \cdot 60 \cdot 60 \text{ s}}$$ Simplify.

$$= \frac{52{,}800 \text{ ft}}{3{,}600 \text{ s}}$$ Simplify.

$$\approx \frac{14.7 \text{ ft}}{1 \text{ s}}$$ Simplify.

The relay team runs at an average speed of about 14.7 feet per second.

Guided Practice

1. Water weighs about 8.34 pounds per gallon. About how many ounces per gallon is the weight of the water? (Examples 1 and 2) ________

2. A skydiver is falling at about 176 feet per second. How many feet per minute is he falling? (Example 3) ________

3. Lorenzo rides his bike at a rate of 5 yards per second. About how many miles per hour can Lorenzo ride his bike? (Example 4)

4. **Building on the Essential Question** Explain why the ratio $\frac{3 \text{ feet}}{1 \text{ yard}}$ has a value of one.

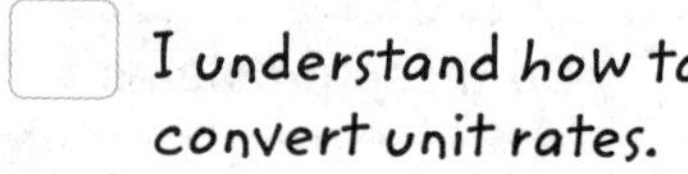

Rate Yourself!

☐ I understand how to convert unit rates.

Great! You're ready to move on!

☐ I still have questions about converting unit rates.

No Problem! Go online to access a Personal Tutor.

Independent Practice

Go online for Step-by-Step Solutions

1 A go-kart's top speed is 607,200 feet per hour. What is the speed in miles per hour? (Examples 1 and 2)

2. The fastest a human has ever run is 27 miles per hour. How many miles per minute did the human run? (Example 3)

3 A peregrine falcon can fly 322 kilometers per hour. How many meters per hour can the falcon fly? (Example 3)

4. A pipe is leaking at 1.5 cups per day. About how many gallons per week is the pipe leaking? (***Hint***: 1 gallon = 16 cups) (Example 4)

5. Charlie runs at a speed of 3 yards per second. About how many miles per hour does Charlie run? (Example 4)

6. CCSS **Model with Mathematics** Refer to the graphic novel frame below. Seth traveled 1 mile in 57.1 seconds. About how fast does Seth travel in miles per hour?

Watch Replay it online!

SKREEECH!
3
I can't believe how fast I was going.

7. The speed at which a certain computer can access the Internet is 2 megabytes per second. How fast is this in megabytes per hour?

8. CCSS **Use Math Tools** The approximate metric measurement of length is given for a U.S. customary unit of length. Use your estimation skills to complete the graphic organizer below. Fill in each blank with *foot, yard, inch,* or *mile*.

Metric		Customary
2.54 centimeters	→	1
0.30 meter	→	1
0.91 meter	→	1
1.61 kilometers	→	1

H.O.T. Problems Higher Order Thinking

9. CCSS **Model with Mathematics** Give an example of a unit rate used in a real-world situation.

10. CCSS **Reason Inductively** When you convert 100 feet per second to inches per second, will there be more or less than 100 inches. Explain.

11. CCSS **Persevere with Problems** Use the information in Exercise 8 to convert 7 meters per minute to yards per hour. Round to the nearest tenth.

Standardized Test Practice

12. A salt truck drops 39 kilograms of salt per minute. How many grams of salt does the truck drop per second?

Ⓐ 600 Ⓒ 650

Ⓑ 625 Ⓓ 6,000

Name ______________________ My Homework ______________________

Extra Practice

13. 20 mi/h = [1,760] ft/min

$\frac{20 \text{ mi}}{1 \text{ h}} \cdot \frac{5,280 \text{ ft}}{1 \text{ mi}} \cdot \frac{1 \text{ h}}{60 \text{ min}} =$

Homework Help

$\frac{105,600 \text{ ft}}{60 \text{ min}} = 1,760 \text{ ft/min}$

14. 16 cm/min = [9.6] m/h

$\frac{16 \text{ cm}}{1 \text{ min}} \cdot \frac{1 \text{ m}}{100 \text{ cm}} \cdot \frac{60 \text{ min}}{1 \text{ h}} =$

$\frac{960 \text{ m}}{100 \text{ h}} = 9.6 \text{ m/h}$

15. 45 mi/h = [ ] ft/s

16. 26 cm/s = [ ] m/min

17. 24 mi/h = [ ] ft/s

18. 105.6 L/h = [ ] L/min

19. The table shows the speed and number of wing beats per second for various flying insects.

a. What is the speed of a housefly in feet per second? Round to the nearest hundredth.

____

b. How many times does a dragonfly's wing beat per minute?

____

c. About how many miles can a bumblebee travel in one minute?

____

d. How many times can a honeybee beat its wings in one hour?

____

Flying Insects		
Insect	Speed (miles per hour)	Wing Beats per Second
Housefly	4.4	190
Honeybee	5.7	250
Dragonfly	15.6	38
Hornet	12.8	100
Bumblebee	6.4	130

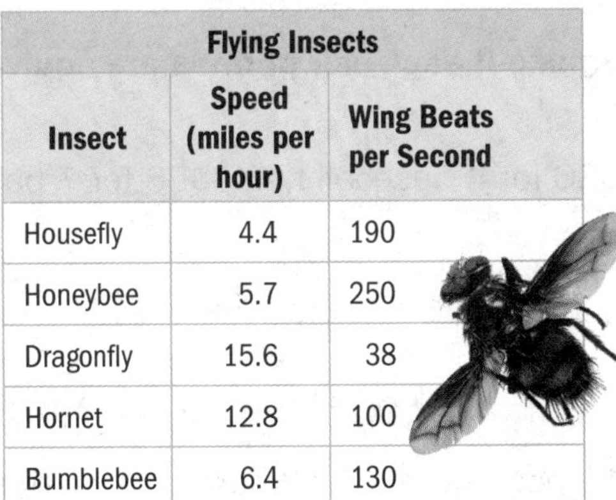

Standardized Test Practice

20. Thirty-five miles per hour is the same rate as which of the following?

Ⓐ 150 feet per minute

Ⓑ 1,500 feet per minute

Ⓒ 2,200 feet per minute

Ⓓ 3,080 feet per minute

21. A boat is traveling at an average speed of 15 meters per second. How many kilometers per second is the boat traveling?

Ⓕ 1.5

Ⓖ 0.15

Ⓗ 0.015

Ⓘ 1,500

22. Short Response An oil tanker empties at 3.5 gallons per minute. Convert this rate to cups per second. Round to the nearest tenth. Show the steps you used.

Common Core Review

Determine if each pair of rates are equivalent. Explain your reasoning.
6.RP.3b

23. $36 for 4 baseball hats; $56 for 7 baseball hats

24. 12 posters for 36 students; 21 posters for 63 students

25. An employer pays $22 for 2 hours. Use the ratio table to determine how much she charges for 5 hours. 6.RP.3a

Payment	$22		
Hours	2		5

Lesson 4

Proportional and Nonproportional Relationships

What You'll Learn

Scan the lesson. Write the definitions of proportional and nonproportional.

- proportional ______________________
- nonporportional ______________________

Essential Question

HOW can you show that two objects are proportional?

Vocabulary

proportional
nonproportional
equivalent ratios

Common Core State Standards

Content Standards
7.RP.2, 7.RP.2a, 7.RP.2b

Mathematical Practices
1, 3, 4

Real-World Link

Pizza Party Ms. Cochran is planning a year-end pizza party for her students. Ace Pizza offers free delivery and charges $8 per medium pizza.

1. Complete the table to determine the cost for different numbers of pizzas ordered.

Cost ($)	8				
Pizza	1	2	3	4	5

2. For each number of pizzas, fill in the boxes to write the relationship of the cost and number of pizzas as a ratio in simplest form.

$\frac{16}{2} = \frac{\square}{1}$ $\quad \frac{24}{3} = \frac{\square}{\square}$

$\frac{32}{\square} = \frac{\square}{\square}$ $\quad \frac{\square}{5} = \frac{\square}{\square}$

3. What do you notice about the simplified ratios?

____

____

Work Zone

Identify Proportional Relationships

Two quantities are **proportional** if they have a constant ratio or unit rate. For relationships in which this ratio is not constant, the two quantities are **nonproportional**.

In the pizza example on the previous page, the cost of an order is *proportional* to the number of pizzas ordered.

$$\frac{\text{cost of order}}{\text{pizzas ordered}} = \frac{8}{1} = \frac{16}{2} = \frac{24}{3} = \frac{32}{4} = \frac{40}{5} \text{ or \$8 per pizza}$$

All of the ratios above are **equivalent ratios** because they all have the same value.

Example

1. Andrew earns $18 per hour for mowing lawns. Is the amount of money he earns proportional to the number of hours he spends mowing? Explain.

Find the amount of money he earns for working a different number of hours. Make a table to show these amounts.

Earnings ($)	18	36	54	72
Time (h)	1	2	3	4

For each number of hours worked, write the relationship of the amount he earned and hour as a ratio in simplest form.

$$\frac{\text{amount earned}}{\text{number of hours}} \rightarrow \frac{18}{1} \text{ or } 18 \quad \frac{36}{2} \text{ or } 18 \quad \frac{54}{3} \text{ or } 18 \quad \frac{72}{4} \text{ or } 18$$

All of the ratios between the two quantities can be simplified to 18.

The amount of money he earns is proportional to the number of hours he spends mowing.

Got It? **Do this problem to find out.**

a. At Lakeview Middle School, there are 2 homeroom teachers assigned to every 48 students. Is the number of students at this school proportional to the number of teachers? Explain your reasoning.

a. ____________

Examples

2. Uptown Tickets charges $7 per baseball game ticket plus a $3 processing fee per order. Is the cost of an order proportional to the number of tickets ordered? Explain.

Cost ($)	10	17	24	31
Tickets Ordered	1	2	3	4

For each number of tickets, write the relationship of the cost and number of tickets as a ratio in simplest form.

$\frac{\text{cost of order}}{\text{tickets ordered}} \rightarrow \frac{10}{1}$ or 10 $\quad \frac{17}{2}$ or 8.5 $\quad \frac{24}{3}$ or 8 $\quad \frac{31}{4}$ or 7.75

Since the ratios of the two quantities are not the same, the cost of an order is *not* proportional to the number of tickets ordered.

3. You can use the recipe shown to make a fruit punch. Is the amount of sugar used proportional to the amount of mix used? Explain.

Find the amount of sugar and mix needed for different numbers of batches. Make a table to help you solve.

Cups of Sugar	$\frac{1}{2}$	1	$1\frac{1}{2}$	2
Envelopes of Mix	1	2	3	4

For each number of cups of sugar, write the relationship of the cups and number of envelopes of mix as a ratio in simplest form.

$\frac{\text{cups of sugar}}{\text{envelopes of mix}} \rightarrow \frac{\frac{1}{2}}{1}$ or 0.5 $\quad \frac{1}{2}$ or 0.5 $\quad \frac{1\frac{1}{2}}{3}$ or 0.5 $\quad \frac{2}{4}$ or 0.5

All of the ratios between the two quantities can be simplified to 0.5. The amount of mix used is proportional to the amount of sugar used.

Got It? Do this problem to find out.

b. At the beginning of the year, Isabel had $120 in the bank. Each week, she deposits another $20. Is her account balance proportional to the number of weeks of deposits? Use the table below. Explain your reasoning.

Time (wk)	1	2	3
Balance ($)			

b. ______________

Example

4. The tables shown represent the number of pages Martin and Gabriel read over time. Which situation represents a proportional relationship between the time spent reading and the number of pages read? Explain.

Pages Martin Read	2	4	6
Time (min)	5	10	15

Pages Gabriel Read	3	4	7
Time (min)	5	10	15

Write the ratios for each time period in simplest form.

$\frac{\text{pages}}{\text{minutes}} \rightarrow \frac{2}{5}, \frac{4}{10} \text{ or } \frac{2}{5}, \frac{6}{15} \text{ or } \frac{2}{5}$ $\quad \frac{3}{5}, \frac{4}{10} \text{ or } \frac{2}{5}, \frac{7}{15}$

All of the ratios between Martin's quantities are $\frac{2}{5}$. So, Martin's reading rate represents a proportional relationship.

Guided Practice

For Exercises 1 and 2, use a table to solve. Then explain your reasoning.

1. The Vista Marina rents boats for $25 per hour. In addition to the rental fee, there is a $12 charge for fuel. Is the number of hours you can rent the boat proportional to the total cost? Explain. (Examples 1–3)

Rental Time (h)			
Cost ($)			

2. Which situation represents a proportional relationship between the hours worked and amount earned for Matt and Jane? Explain. (Example 4)

Matt's Earnings ($)	12	20	31
Time (h)	1	2	3

Jane's Earnings ($)	12	24	36
Time (h)	1	2	3

3. **Building on the Essential Question** Explain what makes two quantities proportional.

Rate Yourself!

How confident are you about determining proportional relationships? Shade the ring on the target.

For more help, go online to access a Personal Tutor.

FOLDABLES *Time to update your Foldable!*

Name ______________________ My Homework ______________________

Independent Practice

Go online for Step-by-Step Solutions 

For Exercises 1 and 2, use a table to solve. Then explain your reasoning. (Examples 1 and 2)

1. An adult elephant drinks about 225 liters of water each day. Is the number of days the water supply lasts proportional to the number of liters of water the elephant drinks?

Time (days)	1	2	3	4
Water (L)				

2. An elevator *ascends*, or goes up, at a rate of 750 feet per minute. Is the height to which the elevator ascends proportional to the number of minutes it takes to get there? (Examples 1–3)

Time (min)	1	2	3	4
Height (ft)				

3. Which situation represents a proportional relationship between the number of laps run by each student and their time? (Example 4)

Desmond's Time (s)	146	292	584
Laps	2	4	8

Maria's Time (s)	150	320	580
Laps	2	4	6

Copy and Solve **Use a table to help you solve. Then explain your reasoning. Show your work on a separate piece of paper.**

4. Plant A is 18 inches tall after one week, 36 inches tall after two weeks, 56 inches tall after three weeks. Plant B is 18 inches tall after one week, 36 inches tall after two weeks, 54 inches tall after three weeks. Which situation represents a proportional relationship between the plants' height and number of weeks? (Example 4)

5. Determine whether the measures for the figure shown are proportional.
   a. the length of a side and the perimeter

   b. the length of a side and the area

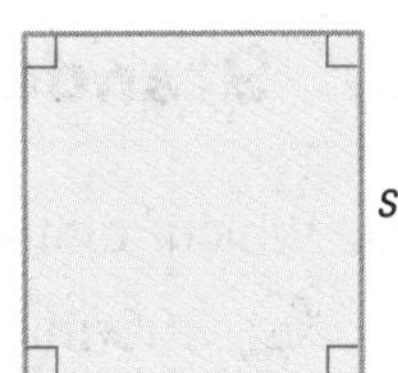

6. **Justify Conclusions** MegaMart collects a sales tax equal to $\frac{1}{16}$ of the retail price of each purchase. The tax is sent to the state government.

a. Is the amount of tax collected proportional to the cost of an item before tax is added? Explain.

Retail Price ($)	16	32	48	64
Tax Collected ($)				

b. Is the amount of tax collected proportional to the cost of an item after tax has been added? Explain.

Retail Price ($)	16	32	48	
Tax Collected ($)				
Cost Including Tax ($)				

H.O.T. Problems Higher Order Thinking

7. **Find the Error** Blake ran laps around the gym. His times are shown in the table. Blake is trying to decide whether the number of laps is proportional to the time. Find his mistake and correct it.

Time (min)	1	2	3	4
Laps	4	6	8	10

It is proportional because the number of laps always increases by 2.

8. **Persevere with Problems** Determine whether the cost for ordering multiple items that will be delivered is *sometimes, always,* or *never* proportional. Explain your reasoning.

Standardized Test Practice

9. Which relationship has a unit rate of 60 miles per hour?

Ⓐ 300 miles in 6 hours Ⓒ 240 miles in 6 hours

Ⓑ 300 miles in 5 hours Ⓓ 240 miles in 5 hours

Name ________________ My Homework ________________

Extra Practice

For Exercises 10-12, use a table to solve. Then explain your reasoning.

10. A vine grows 7.5 feet every 5 days. Is the length of the vine on the last day proportional to the number of days of growth?

Homework Help

Yes; the length to time ratios are all equal to 1.5 ft per day.

Time (days)	5	10	15	20
Length (ft)	7.5	15	22.5	30

11. STEM To convert a temperature in degrees Celsius to degrees Fahrenheit, multiply the Celsius temperature by $\frac{9}{5}$ and then add 32°. Is a temperature in degrees Celsius proportional to its equivalent temperature in degrees Fahrenheit?

Degrees Celsius	0	10	20	30
Degrees Fahrenheit				

12. On Saturday, Querida gave away 416 coupons for a free appetizer at a local restaurant. The next day, she gave away about 52 coupons an hour.

a. Is the number of coupons Querida gave away on Sunday proportional to the number of hours she worked that day?

Hours Worked on Sunday	1	2	3	4
Coupons Given Away on Sunday				

b. Is the total number of coupons Querida gave away on Saturday and Sunday proportional to the number of hours she worked on Sunday?

Hours Worked on Sunday	1	2	3	4
Coupons Given Away on Weekend				

13. CCSS **Justify Conclusions** The fee for ride tickets at a carnival is shown in the table at the right.

Tickets	5	10	15	20
Fee ($)	5	9.50	13.50	16

a. Is the fee for ride tickets proportional to the number of tickets? Explain your reasoning.

b. Can you determine the fee for 30 ride tickets? Explain.

Standardized Test Practice

14. Mr. Martinez is comparing the price of oranges from several different markets. Which market's pricing guide is based on a constant unit price?

Ⓐ

Number of Oranges	5	10	15	20
Total Cost ($)	3.50	6.00	8.50	11.00

Ⓑ

Number of Oranges	5	10	15	20
Total Cost ($)	3.50	6.50	9.50	12.50

Ⓒ

Number of Oranges	5	10	15	20
Total Cost ($)	3.00	5.00	7.00	9.00

Ⓓ

Number of Oranges	5	10	15	20
Total Cost ($)	3.00	6.00	9.00	12.00

15. Short Response The middle school is planning a family movie night where popcorn will be served. The constant relationship between the number of people n and the number of cups of popcorn p is shown in the table. How many people can be served with 519 cups of popcorn?

n	30	60	120	
p	90	180	360	519

Common Core Review

Find the value of each expression if $x = 12$. 6.EE.2

16. $3x$ ______

17. $2x - 4$ ______

18. $5x + 30$ ______

19. $3x - 2x$ ______

20. $x - 12$ ______

21. $\frac{x}{4}$ ______

Make a table to solve the situation. 6.RP.3a

22. Brianna downloads 9 songs each month onto her MP3 player. Show the total number of songs downloaded after 1, 2, 3, and 4 months.

Month				
Number of Songs				

Problem-Solving Investigation

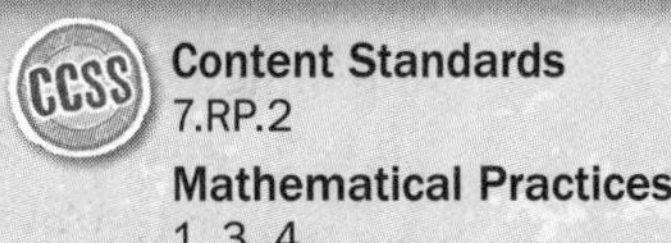

CCSS **Content Standards**
7.RP.2
Mathematical Practices
1, 3, 4

Case #1 'Round and 'Round

The Forte family visited the Mall of America in Minneapolis. The Ferris wheel in the mall's amusement park is about 22.5 meters tall.

What is the approximate height of the Mall of America Ferris wheel in feet if 1 foot is about 0.3 meter?

In mathematics, there is a *four-step problem-solving plan* you can use to help you solve any problem. The four steps are *Understand, Plan, Solve,* and *Check*.

Understand *What are the facts?*

- The Mall of America Ferris wheel is about 22.5 meters tall.
- You need to find the height of the Ferris wheel in feet.

Plan *What is your strategy to solve this problem?*

To solve the problem, write an expression that converts meters to feet. Then divide out common units.

Solve *How can you apply the strategy?*

One foot is about 0.3 meter. Convert 22.5 meters to feet.

$22.5 \text{ meters} \cdot \frac{1 \text{ foot}}{0.3 \text{ meter}} \approx \frac{22.5}{0.3}$ or ☐ feet

So, the Ferris wheel is about 75 feet tall.

Check *Does the answer make sense?*

There is a little more than 3 feet in a meter.

Since 3 • 22.5 is 67.5 and 75 feet is a little more than 67.5 feet, the answer is reasonable.

Analyze the Strategy

CCSS **Reason Inductively** Explain in your own words how the four-step plan helps you solve real-world problems.

Case #2 Cool Treats

Mr. Martino's class learned the average American consumes about 23 quarts of ice cream every year. The class also learned the average American in the north-central United States consumes about 19 quarts more.

How much ice cream in gallons is consumed every year by the average American in the north-central United States?

1 Understand

Read the problem. What are you being asked to find?

I need to find ____

____.

Fill in each box with the information you know.

The average American consumes about ☐ quarts of ice cream.

The average American in the north-central United States consumes about ☐ quarts more.

2 Plan

Choose two operations to solve the problem.

I will ____.

3 Solve

How will you use the operations?

I will ____.

Find total quarts.

☐ + ☐ = ☐

Convert to gallons.

☐ quarts $\cdot \dfrac{\text{1 gallon}}{\square \text{ quarts}}$ = ________ gallons

The average American in the north-central United States consumes about ☐ gallons of ice cream each year.

4 Check

Use information from the problem to check your solution.

____.

Collaborate Work with a small group to solve the following cases. Show your work on a separate piece of paper.

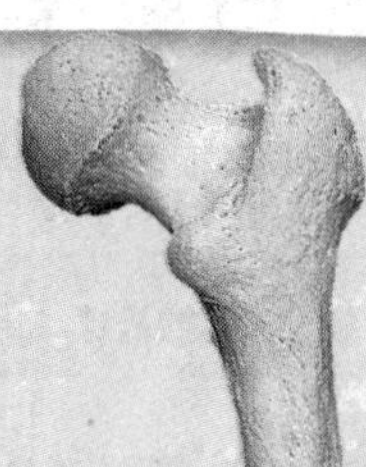

Case #3 Financial Literacy

Terry opened a savings account in December with $150 and deposited $30 each month beginning in January.

What is the value of Terry's account at the end of July?

Case #4 STEM

About how many centimeters longer is the average femur than the average tibia? (Hint: 1 inch ≈ 2.54 centimeters)

Bones in a Human Leg	
Bone	**Length (in.)**
Femur (upper leg)	19.88
Tibia (inner lower leg)	16.94
Fibula (outer lower leg)	15.94

Case #5 Patterns

Numbers that can be represented by a triangular arrangement of dots are called *triangular numbers*. The first four triangular numbers are shown.

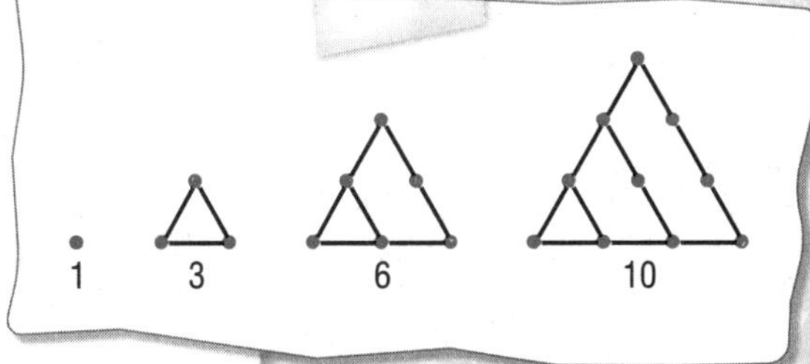

Describe the pattern in the first four numbers. Then list the next three triangular numbers.

Circle a strategy below to solve the problem.

- Draw a diagram.
- Solve a simpler problem.
- Guess, check, and revise.
- Make a table.

Case #6 School

The Boosters expect 500 people at the annual awards banquet.

If each table seats 8 people, how many tables are needed?

Mid-Chapter Check

Vocabulary Check

1. CCSS **Be Precise** Define *complex fraction*. Give two examples of a complex fraction. (Lesson 2)

2. Fill in the blank in the sentence below with the correct term. (Lesson 1)

   When a rate is simplified so that it has a denominator of 1 unit, it is called a(n) ______________ rate.

Skills Check and Problem Solving

Find each unit rate. Round to the nearest hundredth if necessary. (Lesson 1)

3. 750 yards in 25 minutes ______________

4. \$420 for 15 tickets 

Simplify. (Lesson 2)

5. $\dfrac{9}{\frac{1}{3}} =$ ______________

6. $\dfrac{\frac{1}{2}}{4} =$ ______________

7. $\dfrac{\frac{1}{6}}{1\frac{3}{8}} =$ ______________

8. A tourist information center charges \$10 per hour to rent a bicycle. Is the rental charge proportional to the number of hours you rent the bicycle? Justify your response. (Lesson 4)

9. **Standardized Test Practice** Which of the following is the same as 2,088 feet per minute? (Lesson 3)

   Ⓐ 696 meters per minute
   Ⓑ 696 yards per minute
   Ⓒ 696 feet per minute
   Ⓓ 696 yards per second

Lesson 5

Graph Proportional Relationships

What You'll Learn

Scan the lesson. Predict two things you will learn about graphing proportional relationships.

-
-

Essential Question

HOW can you show that two objects are proportional?

Vocabulary

coordinate plane
quadrants
ordered pair
x-coordinate
y-coordinate
y-axis
origin
x-axis

Common Core State Standards

Content Standards
7.RP.2, 7.RP.2a

Mathematical Practices
1, 2, 3, 4

Vocabulary Start-Up

Maps have grids to locate cities. The **coordinate plane** is a type of grid that is formed when two number lines intersect at their zero points. The number lines separate the coordinate plane into four regions called **quadrants**.

An **ordered pair** is a pair of numbers, such as (1, 2), used to locate or graph points on the coordinate plane.

The **x-coordinate** corresponds to a number on the *x*-axis. → (1, 2) ← The **y-coordinate** corresponds to a number on the *y*-axis.

Label the coordinate plane with the terms *ordered pair, x-coordinate,* and *y-coordinate*.

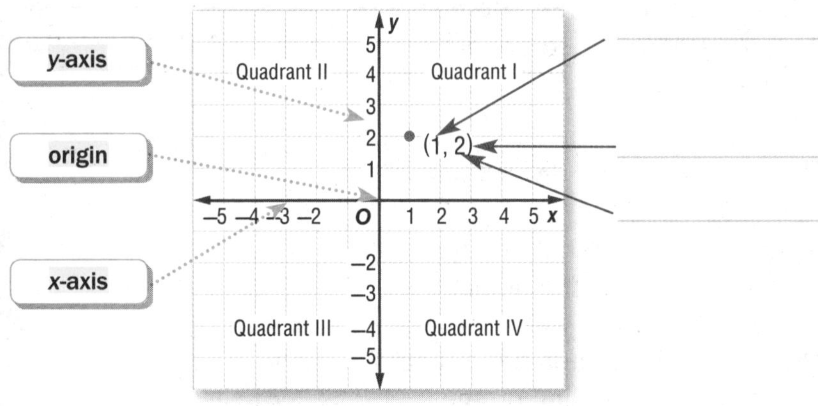

Graph points (2, 3) and (–3, –2) above. Connect the three points on the coordinate plane. Describe the graph.

Work Zone

Identify Proportional Relationships

Another way to determine whether two quantities are proportional is to graph the quantities on the coordinate plane. If the graph of the two quantities is a straight line through the origin, then the two quantities are proportional.

Example

Linear Relationships
Relationships that have straight-line graphs are called linear relationships.

1. The slowest mammal on Earth is the tree sloth. It moves at a speed of 6 feet per minute. Determine whether the number of feet the sloth moves is proportional to the number of minutes it moves by graphing on the coordinate plane. Explain your reasoning.

Step 1 Make a table to find the number of feet walked for 0, 1, 2, 3, and 4 minutes.

Time (min)	0	1	2	3	4
Distance (ft)	0	6	12	18	24

Step 2 Graph the ordered pairs (time, distance) on the coordinate plane. Then connect the ordered pairs.

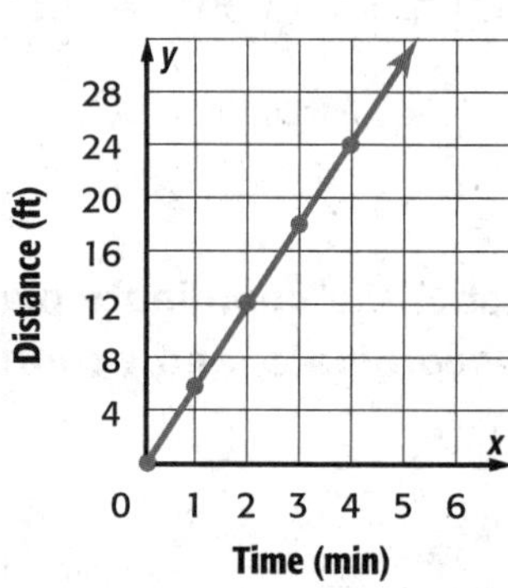

The line passes through the origin and is a straight line. So, the number of feet traveled is proportional to the number of minutes.

Got It? **Do this problem to find out.**

a. James earns $5 an hour babysitting. Determine whether the amount of money James earns babysitting is proportional to the number of hours he babysits by graphing on the coordinate plane. Explain your reasoning in the work zone.

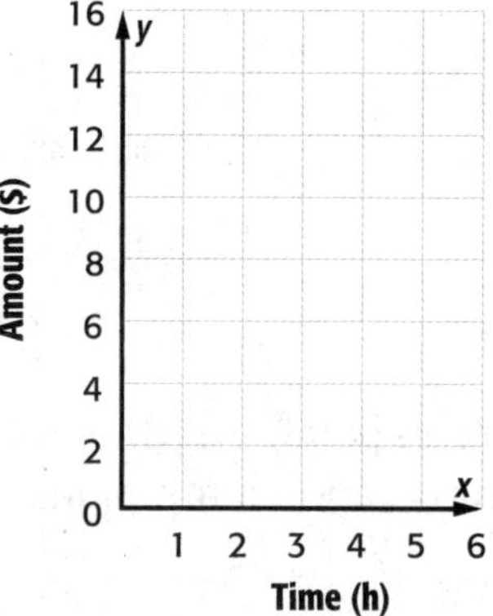

a. ______________

Example

2. The cost of renting video games from Games Inc. is shown in the table. Determine whether the cost is proportional to the number of games rented by graphing on the coordinate plane. Explain your reasoning.

Video Game Rental Rates	
Number of Games	**Cost ($)**
1	3
2	5
3	7
4	9

Step 1 Write the two quantities as ordered pairs (number of games, cost).

The ordered pairs are (1, 3), (2, 5), (3, 7), and (4, 9).

Step 2 Graph the ordered pairs on the coordinate plane. Then connect the ordered pairs and extend the line to the *y*-axis.

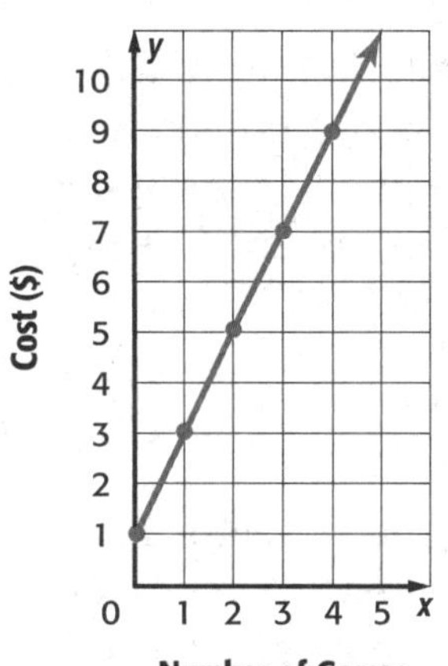

The line does not pass through the origin. So, the cost of the video games is not proportional to the number of games rented.

Check The ratios are not constant. $\frac{1}{3} \neq \frac{2}{5}$ ✓

Quick Review
When drawing a graph, include a title and labels for the horizontal and vertical axes.

Got It? Do this problem to find out.

b. The table shows the number of Calories an athlete burned per minute of exercise. Determine whether the number of Calories burned is proportional to the number of minutes by graphing on the coordinate plane. Explain your reasoning in the Work Zone.

Calories Burned	
Number of Minutes	**Number of Calories**
0	0
1	4
2	8
3	13

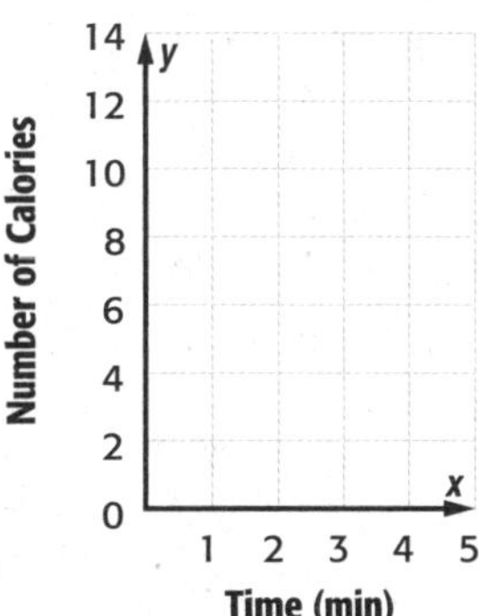

b. ____________

Example

3. Which batting cage represents a proportional relationship between the number of pitches thrown and the cost? Explain.

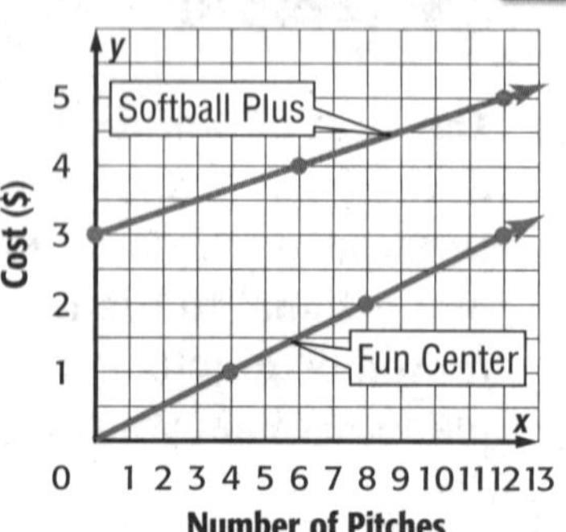

The graph for Softball Plus is a straight line, but it does not pass through the origin. So, the relationship is not proportional.

The graph for the Fun Center is a straight line through the origin. So, the relationship between the number of the pitches thrown and the cost is proportional.

Guided Practice

1. The cost of 3-D movie tickets is $12 for 1 ticket, $24 for 2 tickets, and $36 for 3 tickets. Determine whether the cost is proportional to the number of tickets by graphing on the coordinate plane. Explain your reasoning. (Examples 1 and 2)

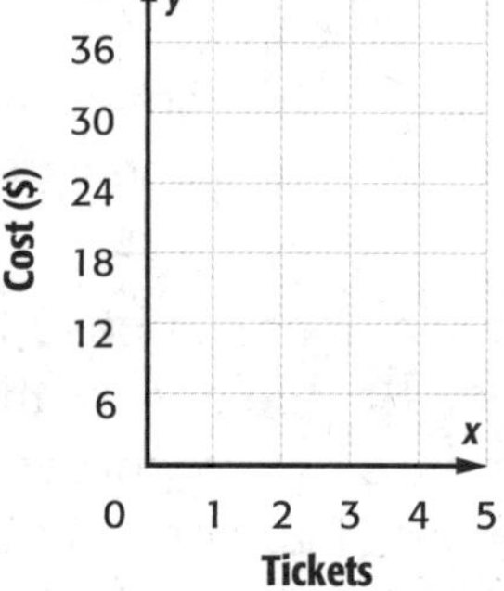

2. The number of books two stores sell after 1, 2, and 3 days is shown. Which book sale represents a proportional relationship between time and books? Explain. (Example 3)

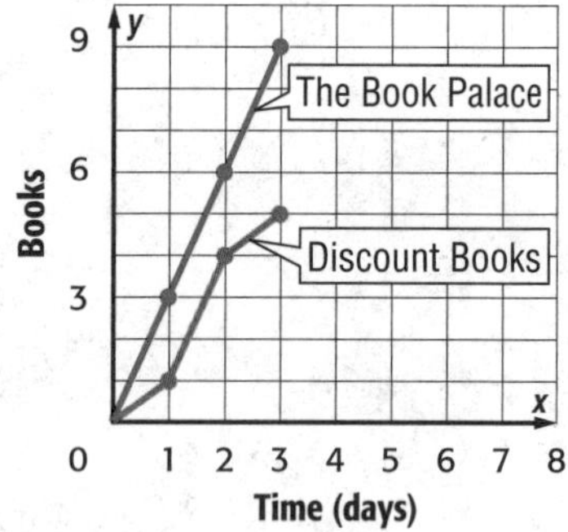

3. **Building on the Essential Question** How does graphing relationships help you determine whether the relationship is proportional or not?

Rate Yourself!

How confident are you about identifying proportional relationships by graphing? Check the box that applies.

For more help, go online to access a Personal Tutor.

FOLDABLES *Time to update your Foldable!*

Independent Practice

Go online for Step-by-Step Solutions

Model with Mathematics Determine whether the relationship between the two quantities shown in each table are proportional by graphing on the coordinate plane. Explain your reasoning. (Examples 1 and 2)

1.

Savings Account	
Week	Account Balance ($)
1	125
2	150
3	175

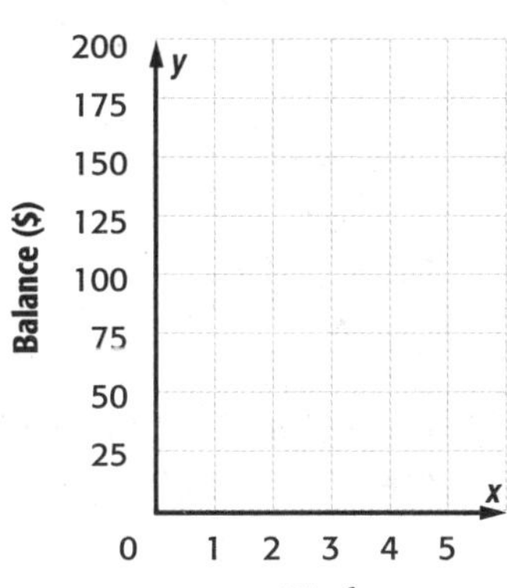

2.

Calories in Fruit Cups	
Servings	Calories
1	70
3	210
5	350

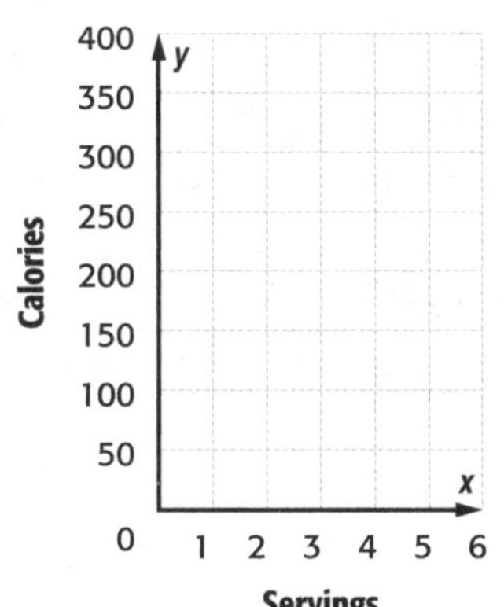

3. The height of two plants is recorded after 1, 2, and 3 weeks as shown in the graph at the right. Which plants' growth represents a proportional relationship between time and height? Explain. (Example 3)

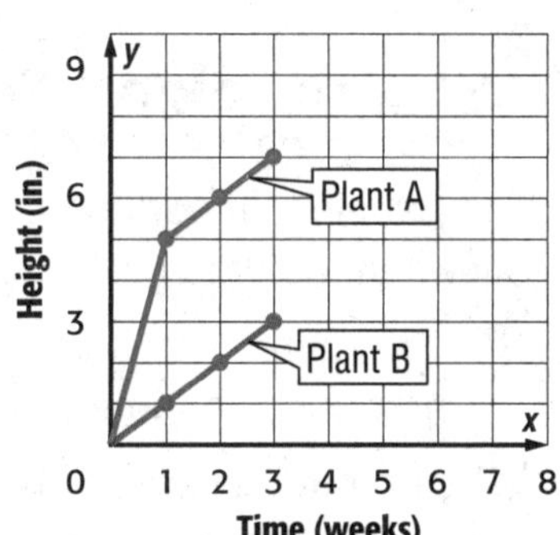

4. The perimeter of a square is 4 times as great as the length of any of its sides. Determine if the perimeter of a square is proportional to its side length. Explain.

5. A health club charges $35 a month for membership fees. Determine whether the cost of membership is proportional to the number of months. Explain your reasoning.

H.O.T. Problems Higher Order Thinking

6. CCSS **Reason Abstractly** Describe some data that when graphed would represent a proportional relationship. Explain your reasoning.

7. CCSS **Persevere with Problems** The greenhouse temperatures at certain times are shown in the table. The greenhouse maintains temperatures between 65°F and 85°F. Suppose the temperature increases at a constant rate. Create a graph of the time and temperatures at each hour from 1:00 P.M. to 8:00 P.M. Is the relationship proportional? Explain.

Time	Temperature (°F)
1:00 P.M.	66
6:00 P.M.	78.5
8:00 P.M.	83.5

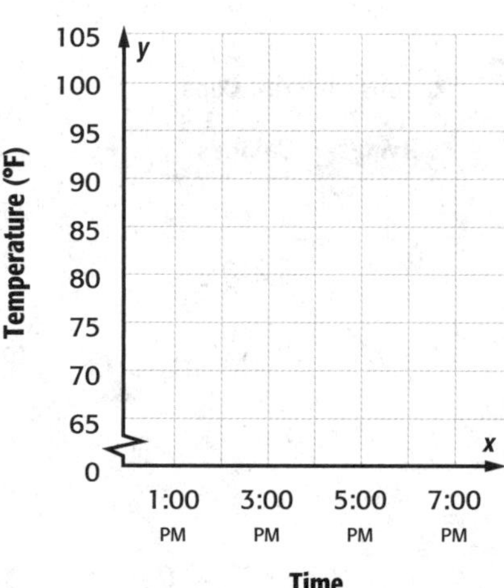

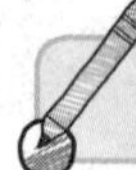

Standardized Test Practice

8. The Calories burned for exercising various number of minutes are shown in the graph. Which statement about the graph is *not* true?

Ⓐ The number of Calories burned is proportional to the number of minutes spent exercising.

Ⓑ The number of Calories burned is *not* proportional to the number of minutes spent exercising.

Ⓒ If the line were extended, it would pass through the origin.

Ⓓ The line is straight.

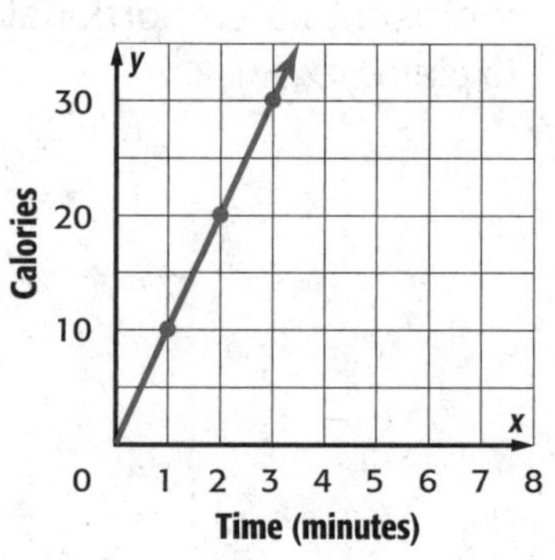

Extra Practice

Determine whether the relationship between the two quantities shown in each table are proportional by graphing on the coordinate plane. Explain your reasoning.

9.

Cooling Water	
Time (min)	Temperature (°F)
5	95
10	90
15	85

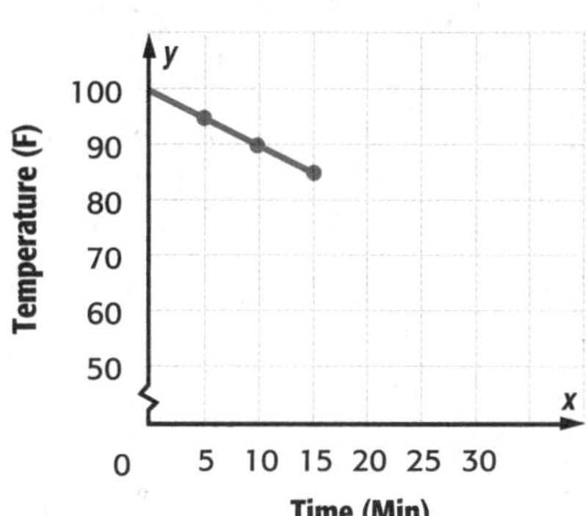

Homework Help

Not proportional; The graph does not pass through the origin.

10.

Pizza Recipe	
Number of Pizzas	Cheese (oz)
1	8
4	32
7	56

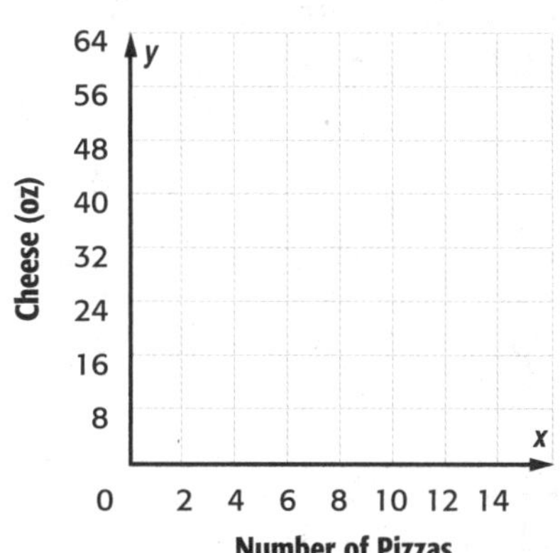

Copy and Solve **Determine if each situation represents a proportional relationship. Graph on a separate piece of paper. Write an explanation for each situation.**

11. CCSS **Justify Conclusions** An airplane is flying at an altitude of 4,000 feet and descends at a rate of 200 feet per minute. Determine whether the altitude is proportional to number of minutes. Explain your reasoning.

12. Frank and Allie purchased cell phone plans through different providers. Their costs for several minutes are shown. Graph each plan to determine whose plan is proportional to the number of minutes the phone is used. Explain your reasoning.

Cell Phone Plans		
Time (min)	Frank's Cost ($)	Allie's Cost ($)
0	0	4.00
3	1.50	4.50
6	3.00	5.00

Standardized Test Practice

13. **Short Response** Determine whether the relationship between the number of heartbeats and the time shown in the graph is proportional. Explain your reasoning.

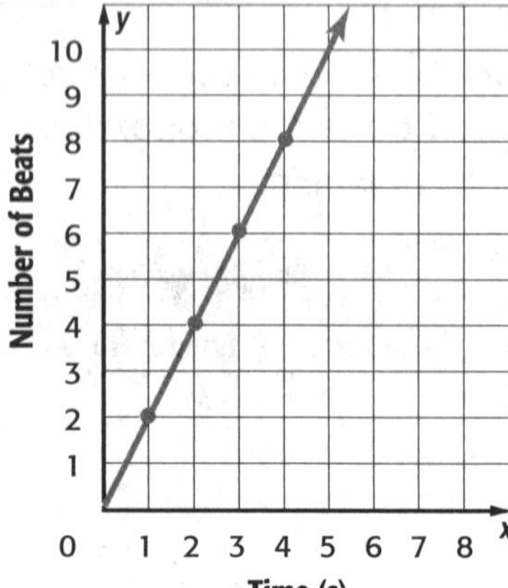

14. Refer to the graph in Exercise 13. Which of the following ordered pairs represent the unit rate?

Ⓐ (0, 0)

Ⓑ (1, 2)

Ⓒ (2, 4)

Ⓓ (3, 6)

15. **Short Response** The distance Charlie and Samora travel after 1, 2, and 3 hours of jogging is shown. Which exercise routine represents a proportional relationship between time and distance? Explain.

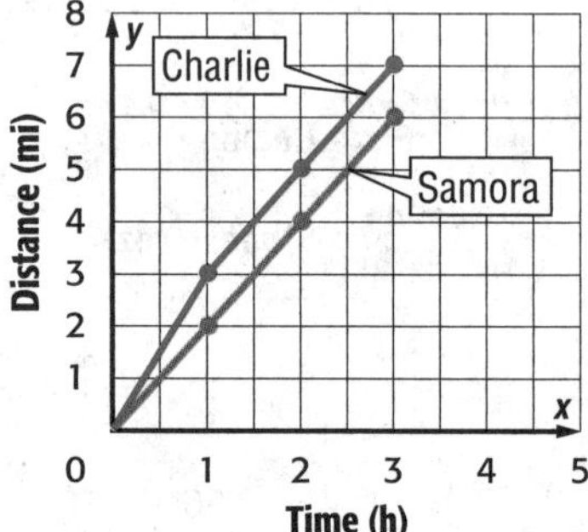

Common Core Review

Write each ratio as a fraction in simplest form. 6.RP.1

16. A class has 10 boys and 15 girls. What is the ratio of boys to girls? ______

17. A car dealership has 55 cars and 11 vans. What is the ratio of cars to vans? ______

18. A drawer has 4 red shirts and 8 green shirts. What is the ratio of red shirts to the total number of shirts? ______

19. A store sells 13 coffees and 65 hot chocolates. What is the ratio of coffees to hot chocolates? ______

Inquiry Lab

Proportional and Nonproportional Relationships

HOW are proportional and nonproportional linear relationships alike? HOW are they different?

Content Standards
7.RP.2, 7.RP.2a

Mathematical Practices
1, 3, 4

Albert and Bianca joined an online discussion group. Each student posted four comments. The number of replies to each of their comments is shown in the table. Determine if each data set represents a proportional relationship.

Investigation

Step 1 Arrange centimeter cubes to model the number of replies per comment as shown in the diagram below.

Student	Albert				Bianca			
Comment Number	1	2	3	4	1	2	3	4
Number of Replies								

Step 2 Complete each table. Then graph the data on the coordinate plane. You may wish to use a different color pencil for each data set.

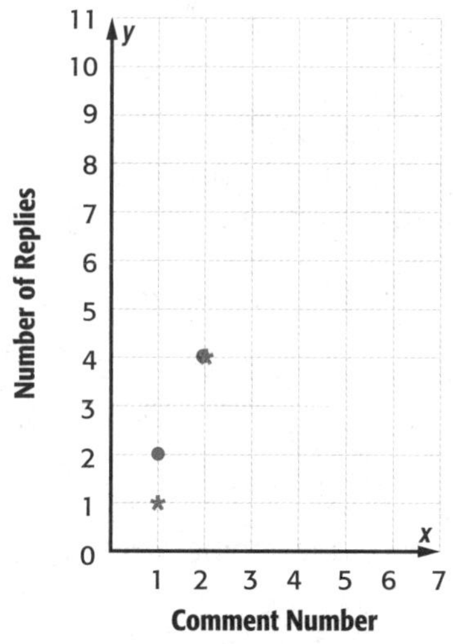

Albert's Comments

Comment Number (x)	Number of Replies (y)
1	2
2	4
3	
4	

Bianca's Comments

Comment Number (x)	Number of Replies (y)
1	1
2	4
3	
4	

Work with a partner to answer the following questions.

1. CCSS **Justify Conclusions** Does Albert's graph represent a proportional relationship? Does Bianca's? Explain.

2. How can you use constant ratios to determine if a relationship is proportional?

Work with a partner to complete the table. Describe the type of relationship shown by each set of ordered pairs. The first one is already done for you.

	Ordered Pairs	Type of Relationship
	(0, 5), (1, 7), (2, 9), (3, 11), (4, 13)	linear and nonproportional
3.	(0, 0), (1, 3), (2, 6), (3, 9), (4, 12)	
4.	(0, 0), (1, 1), (2, 4), (3, 9), (4, 16)	

5. CCSS **Model with Mathematics** Describe a real-world situation that represents a proportional relationship. Then explain how you could change your situation so that it represents a nonproportional relationship.

6. Inquiry HOW are proportional and nonproportional linear relationships alike? HOW are they different?

Lesson 6

Solve Proportional Relationships

What You'll Learn

Scan the lesson. Write the definitions of equivalent ratios and proportion.

- equivalent ratios
- proportion

Essential Question

HOW can you show that two objects are proportional?

Vocabulary

proportion
cross product

Common Core State Standards

Content Standards
7.RP.2, 7.RP.2b, 7.RP.2c, 7.RP.3

Mathematical Practices
1, 2, 3, 4

Real-World Link

Fruit Smoothies Katie and some friends want to buy fruit smoothies. They go to a health food store that advertises a sale of 2 fruit smoothies for $5.

1. Fill in the boxes to write a ratio that compares the cost to the number of fruit smoothies.

$$\frac{\$\square}{\square \text{ smoothies}}$$

2. Suppose Katie and her friends buy 6 fruit smoothies. Complete the ratio that compares the cost to the number of fruit smoothies.

$$\frac{\$\square}{6 \text{ smoothies}}$$

3. Is the cost proportional to the number of fruit smoothies for two and six smoothies? Explain.

Key Concept Write and Solve Proportions

Work Zone

Words A **proportion** is an equation stating that two ratios or rates are equivalent.

Numbers	Algebra
$\frac{6}{8} = \frac{3}{4}$	$\frac{a}{b} = \frac{c}{d}, b \neq 0, d \neq 0$

Consider the following proportion.

$$\frac{a}{b} = \frac{c}{d}$$

$$\frac{a}{\cancel{b}_1} \cdot \cancel{b}^1 d = \frac{c}{\cancel{d}_1} \cdot b\cancel{d}^1$$ Multiply each side by bd and divide out common factors.

$$ad = bc$$ Simplify.

The products ad and bc are called the **cross products** of this proportion. The cross products of any proportion are equal.

$$\frac{6}{8} = \frac{3}{4} \rightarrow 8 \cdot 3 = 24, \quad 6 \cdot 4 = 24$$

Real World

Example

1. After 2 hours, the air temperature had risen 7°F. Write and solve a proportion to find the amount of time it will take at this rate for the temperature to rise an additional 13°F.

Write a proportion. Let t represent the time in hours.

temperature → $\frac{7}{2} = \frac{13}{t}$ ← temperature
time → ← time

$7 \cdot t = 2 \cdot 13$ Find the cross products.

$7t = 26$ Multiply

$\frac{7t}{7} = \frac{26}{7}$ Divide each side by 7.

$t \approx 3.7$ Simplify.

It will take about 3.7 hours to rise an additional 13°F.

Got It? **Do these problems to find out.**

Show your work.

Solve each proportion.

a. $\frac{x}{4} = \frac{9}{10}$ **b.** $\frac{2}{34} = \frac{5}{y}$ **c.** $\frac{7}{3} = \frac{n}{21}$

a. ______

b. ______

c. ______

Example

2. If the ratio of Type O to non-Type O donors at a blood drive was 37:43, how many donors would be Type O, out of 300 donors?

Type O donors → / total donors → $\frac{37}{37 + 43}$ or $\frac{37}{80}$

Write a proportion. Let t represent the number of Type O donors.

Type O donors → / total donors → $\frac{37}{80} = \frac{t}{300}$ ← Type O donors / ← total donors

$37 \cdot 300 = 80t$ Find the cross products.

$11{,}100 = 80t$ Multiply.

$\frac{11{,}100}{80} = \frac{80t}{80}$ Divide each side by 80.

$138.75 = t$ Simplify.

There would be about 139 Type O donors.

Got It? **Do this problem to find out.**

d. The ratio of 7th grade students to 8th grade students in a soccer league is 17:23. If there are 200 students in all, how many are in the 7th grade?

d. ____________

Use Unit Rate

You can also use the unit rate to write an equation expressing the relationship between two proportional quantities.

Examples

3. **Olivia bought 6 containers of yogurt for $7.68. Write an equation relating the cost *c* to the number of yogurts *y*. How much would Olivia pay for 10 yogurts at this same rate?**

Find the unit rate between cost and containers of yogurt.

$\frac{\text{cost in dollars}}{\text{containers of yogurt}} = \frac{7.68}{6}$ or $1.28 per container

The cost is $1.28 times the number of containers of yogurt.

$c = 1.28y$ Let c represent the cost. Let y represent the number of yogurts.

$= 1.28(10)$ Replace y with 10.

$= 12.80$ Multiply.

The cost for 10 containers of yogurt is $12.80.

Consider the proportion $\frac{6}{b} = \frac{c}{7}$. What is true about bc?

4. Jaycee bought 8 gallons of gas for $31.12. Write an equation relating the cost c to the number of gallons g of gas. How much would Jaycee pay for 11 gallons at this same rate?

Find the unit rate between cost and gallons.

$\frac{\text{cost in dollars}}{\text{gasoline in gallons}} = \frac{31.12}{8}$ or $3.89 per gallon

The cost is $3.89 times the number of gallons.

$c = 3.89g$ — Let c represent the cost. Let g represent the number of gallons.

$= 3.89(11)$ — Replace g with 11.

$= 42.79$ — Multiply.

The cost for 11 gallons of gas is $42.79.

Got It? **Do this problem to find out.**

e. Olivia typed 2 pages in 15 minutes. Write an equation relating the number of minutes *m* to the number of pages *p* typed. How long will it take her to type 10 pages at this rate?

e. ___________

Guided Practice

Solve each proportion. (Examples 1 and 2)

1. $\frac{k}{7} = \frac{32}{56}$ $k =$ ________

2. $\frac{3.2}{9} = \frac{n}{36}$ $n =$ ________

3. $\frac{41}{x} = \frac{5}{2}$ $x =$ ________

4. Trina earns $28.50 tutoring for 3 hours. Write an equation relating her earnings *m* to the number of hours *h* she tutors. Assuming the situation is proportional, how much would Trina earn tutoring for 2 hours? for 4.5 hours? (Examples 3 and 4)

5. **Building on the Essential Question** How do you solve a proportion?

Rate Yourself!

How confident are you about solving proportions? Check the box that applies.

For more help, go online to access a Personal Tutor.

FOLDABLES Time to update your Foldable!

Name ______________________ My Homework ______________________

Independent Practice

Go online for Step-by-Step Solutions 

Solve each proportion. (Examples 1 and 2)

1. $\frac{1.5}{6} = \frac{10}{p}$ $p =$ ______

2. $\frac{44}{p} = \frac{11}{5}$ $p =$ ______

3. $\frac{2}{w} = \frac{0.4}{0.7}$ $w =$ ______

Assume the situations are proportional. Write and solve by using a proportion. (Examples 1 and 2)

4. Evarado paid $1.12 for a dozen eggs at his local grocery store. Determine the cost of 3 eggs.

5. Sheila mixed 3 ounces of blue paint with 2 ounces of yellow paint. She decided to create 20 ounces of the same mixture. How many ounces of yellow paint does Sheila need for the new mixture?

Assume the situations are proportional. Use the unit rate to write an equation, then solve. (Examples 3 and 4)

6. A car can travel 476 miles on 14 gallons of gas. Write an equation relating the distance d to the number of gallons g. How many gallons of gas does this car need to travel 578 miles.

7. Mrs. Baker paid $2.50 for 5 pounds of bananas. Write an equation relating the cost c to the number of pounds p of bananas. How much would Mrs. Baker pay for 8 pounds of bananas?

8. A woman who is 64 inches tall has a shoulder width of 16 inches. Write an equation relating the height h to the width w. Find the height of a woman who has a shoulder width of 18.5 inches.

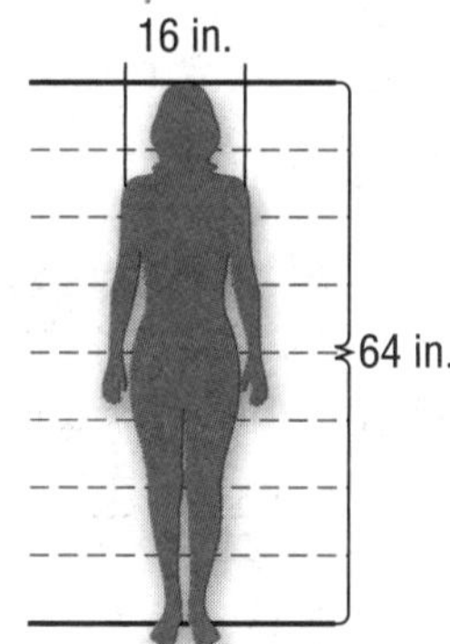

9. At an amusement park, 360 visitors rode the roller coaster in 3 hours. Write and solve a proportion to find the number of visitors at this rate who will ride the roller coaster in 7 hours. (Examples 3 and 4)

10. CCSS **Reason Abstractly** Use the table to write a proportion relating the weights on two planets. Then find the missing weight. Round to the nearest tenth.

Weights on Different Planets Earth Weight = 120 pounds	
Mercury	45.6 pounds
Venus	109.2 pounds
Uranus	96 pounds
Jupiter	304.8 pounds

a. Earth: 90 pounds; Venus: ______ pounds

b. Mercury: 55 pounds; Earth: ______ pounds

c. Jupiter: 350 pounds; Uranus: ______ pounds

d. Venus: 115 pounds; Mercury: ______ pounds

H.O.T. Problems Higher Order Thinking

11. CCSS **Justify Conclusions** A powdered drink mix calls for a ratio of powder to water of 1 : 8. If there are 32 cups of powder, how many total cups of water are needed? Explain your reasoning.

CCSS **Persevere with Problems** **Solve each equation.**

12. $\frac{2}{3} = \frac{18}{x + 5}$ ______

13. $\frac{x - 4}{10} = \frac{7}{5}$ ______

14. $\frac{4.5}{17 - x} = \frac{3}{8}$ ______

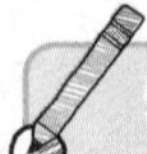

Standardized Test Practice

15. In which proportion does x have a value of 4?

Ⓐ $\frac{x}{21} = \frac{12}{7}$

Ⓑ $\frac{12}{21} = \frac{x}{7}$

Ⓒ $\frac{5}{2} = \frac{1}{x}$

Ⓓ $\frac{1}{x} = \frac{5}{200}$

Name ______________________ My Homework ______________________

Extra Practice

Solve each proportion.

16. $\frac{x}{13} = \frac{18}{39}$ $x =$ 6

$x \cdot 39 = 13 \cdot 18$

Homework Help

$39x = 234$

$\frac{39x}{39} = \frac{234}{39}$

$x = 6$

17. $\frac{6}{25} = \frac{d}{30}$ $d =$ ______

18. $\frac{2.5}{6} = \frac{h}{9}$ $h =$ ______

Assume the situations are proportional. Write and solve by using a proportion.

19. For every person who has the flu, there are 6 people who have only flu-like symptoms. If a doctor sees 40 patients, determine approximately how many patients you would expect to have only flu-like symptoms.

______

20. For every left-handed person, there are about 4 right-handed people. If there are 30 students in a class, predict the number of students who are right-handed.

______

21. Jeremiah is saving money from a tutoring job. After the first three weeks, he saved $135. Assume the situation is proportional. Use the unit rate to write an equation relating the amount saved *s* to the number of weeks *w* worked. At this rate, how much will Jeremiah save after eight weeks?

______

22. CCSS **Make a Prediction** A speed limit of 100 kilometers per hour (kph) is approximately equal to 62 miles per hour (mph). Write an equation relating kilometers per hour *k* to miles per hour *m*. Then predict the following measures. Round to the nearest tenth.

a. a speed limit in mph for a speed limit of 75 kph

______

b. a speed limit in kph for a speed limit of 20 mph

______

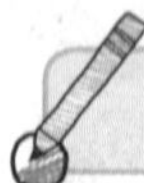

Standardized Test Practice

23. A recipe for making 3 dozen muffins requires 1.5 cups of flour. At this rate, how many cups of flour are required to make 5 dozen muffins?

Ⓐ 2 cups
Ⓑ 2.5 cups
Ⓒ 3 cups
Ⓓ 3.5 cups

24. An amusement park line is moving about 4 feet every 15 minutes. At this rate, approximately how long will it take for a person at the back of the 50-foot line to reach the front of the line?

Ⓕ 1 hour
Ⓖ 3 hours
Ⓗ 5 hours
Ⓘ 13 hours

25. **Short Response** Crystal's mother kept a record of Crystal's height at different ages. She recorded the information in a table.

Age (yr)	Height (in.)
0 (birth)	19
1	25
2	30
5	42
10	55
12	60

Is the relationship between Crystal's age and her height proportional? Explain.

Common Core Review

26. The table shows the cost to have various numbers of pizzas delivered from Papa's Slice of Italy pizzeria. Is the relationship between the cost and the number of pizzas proportional? Explain. 7.RP.2a

Number of Pizzas	Cost ($)
1	12.50
2	20
3	27.50
4	35

27. Brenna charges $15, $30, $45, and $60 for babysitting 1, 2, 3, and 4 hours, respectively. Is the relationship between the amount charged and the number of hours proportional? If so, find the unit rate. If not, explain why not. 7.RP.2a

Find each unit rate. 6.RP.3b

28. 50 miles on 2.5 gallons

29. 2,500 kilobytes in 5 minutes

Inquiry Lab

Rate of Change

HOW is unit rate related to rate of change?

CCSS Content Standards 7.RP.2, 7.RP.2b

Mathematical Practices 1, 3

Pets Happy Hound is a doggie daycare where people drop off their dogs while they are at work. It costs $3 for 1 hour, $6 for 2 hours, and $9 for 3 hours of doggie daycare. Farah takes her dog to Happy Hound several days a week. Farah wants to determine if the number of hours of daycare is related to the cost.

Investigation

Step 1 Assume the pattern in the table continues. Complete the table shown.

Happy Hound Doggie Daycare	
Number of Hours	Cost ($)
1	3
2	6
3	9
4	
5	

Step 2 The cost depends on the number of hours. So, the cost is the output y, and the number of hours is the ________. Graph the data on the coordinate plane below.

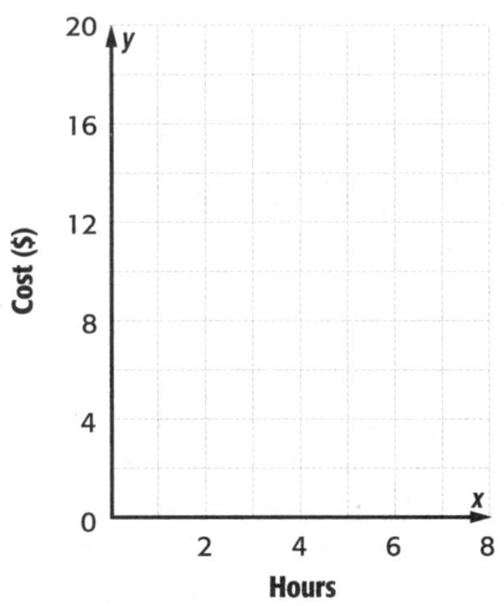

Refer to the Investigation. Work with a partner.

1. CCSS **Justify Conclusions** Is the graph linear? Explain.

2. What is the cost per hour, or unit rate, charged by Happy Hound?

3. CCSS **Justify Conclusions** Is the relationship proportional? Explain.

4. Use the graph to examine any two consecutive points. By how much does y change? By how much does x change?

5. The first two ordered pairs on the graph are (1, 3) and (2, 6). You can find the *rate of change* by writing the ratio of the change in y to the change in x.

   Find the rate of change shown in the graph. ______

Work with a partner to answer the following question.

6. Pampered Pooch charges $5 for 1 hour of doggie daycare, $10 for 2 hours, and $15 for 3 hours.

   a. What is the unit rate? ______

   b. What is the rate of change? ______

   c. CCSS **Reason Inductively** How do the rates of change for doggie daycare at Pampered Pooch and Happy Hound compare?

7. Inquiry HOW is unit rate related to rate of change?

Lesson 7

Constant Rate of Change

What You'll Learn

Scan the text on the following two pages. Write two facts you learned about constant rate of change.

- ______________________________

- ______________________________

Essential Question

HOW can you show that two objects are proportional?

Vocabulary

rate of change
constant rate of change

Common Core State Standards

Content Standards
7.RP.2, 7.RP.2b, 7.RP.2d

Mathematical Practices
1, 3, 4

Vocabulary Start-Up

A **rate of change** is a rate that describes how one quantity changes in relation to another. In a linear relationship, the rate of change between any two quantities is the same. A linear relationship has a **constant rate of change**.

Real-World Link

A computer programmer charges customers per line of code written. Fill in the blanks with the amount of change between consecutive numbers.

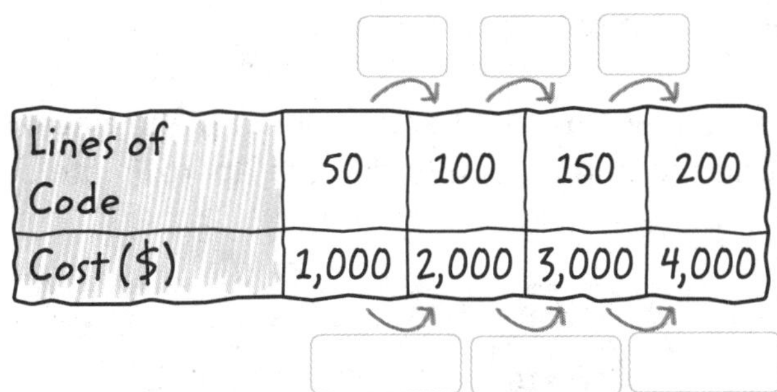

Lines of Code	50	100	150	200
Cost ($)	1,000	2,000	3,000	4,000

Label the diagram below with the terms *change in lines*, *change in dollars*, and *constant rate of change*.

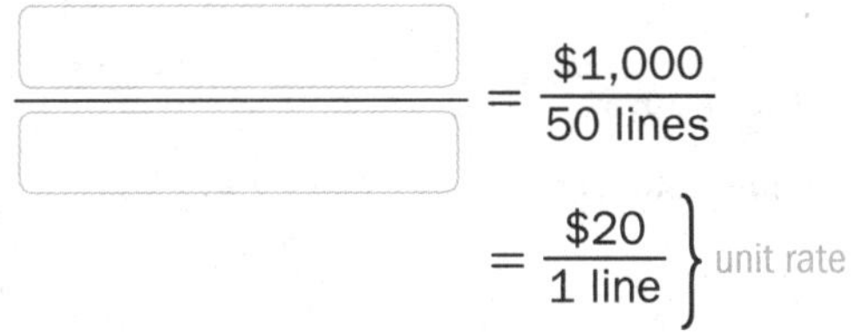

$$\frac{\square}{\square} = \frac{\$1,000}{50 \text{ lines}}$$

$$= \frac{\$20}{1 \text{ line}} \Big\} \text{ unit rate}$$

The ______________________________ is $20 per line of programming code.

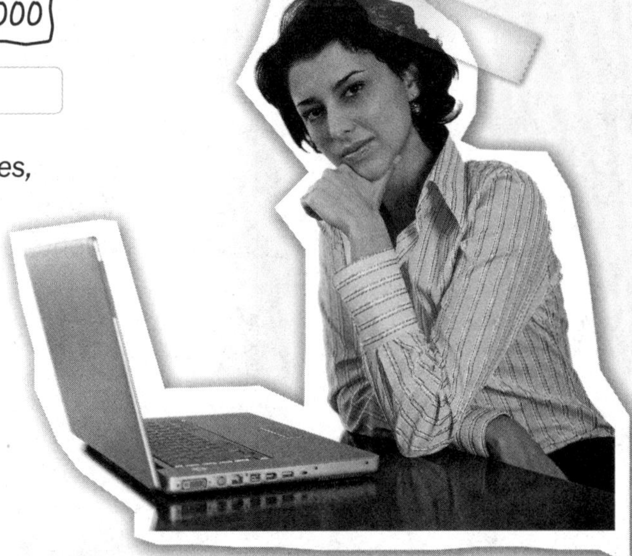

Work Zone

Use a Table

You can use a table to find a constant rate of change.

Example

1. The table shows the amount of money a booster club makes washing cars for a fundraiser. Use the information to find the constant rate of change in dollars per car.

Cars Washed			
	Number	Money ($)	
	5	40	
+5	10	80	+40
+5	15	120	+40
+5	20	160	+40

Find the unit rate to determine the constant rate of change.

$$\frac{\text{change in money}}{\text{change in cars}} = \frac{40 \text{ dollars}}{5 \text{ cars}}$$ The money earned increases by $40 for every 5 cars.

$$= \frac{8 \text{ dollars}}{1 \text{ car}}$$ Write as a unit rate.

So, the number of dollars earned increases by $8 for every car washed.

Unit Rate
A rate of change is usually expressed as a unit rate.

Got It? Do these problems to find out.

a. The table shows the number of miles a plane traveled while in flight. Use the information to find the approximate constant rate of change in miles per minute.

Time (min)	30	60	90	120
Distance (mi)	290	580	870	1,160

b. The table shows the number of students that buses can transport. Use the table to find the constant rate of change in students per school bus.

Number of Buses	2	3	4	5
Number of Students	144	216	288	360

a. ________

b. ________

Use a Graph

You can also use a graph to find a constant rate of change and to analyze points on the graph.

Examples

2. The graph represents the distance traveled while driving on a highway. Find the constant rate of change.

To find the rate of change, pick any two points on the line, such as (0, 0) and (1, 60).

$$\frac{\text{change in miles}}{\text{change in hours}} = \frac{(60 - 0)\text{ miles}}{(1 - 0)\text{ hours}}$$

$$= \frac{60\text{ miles}}{1\text{ hour}}$$

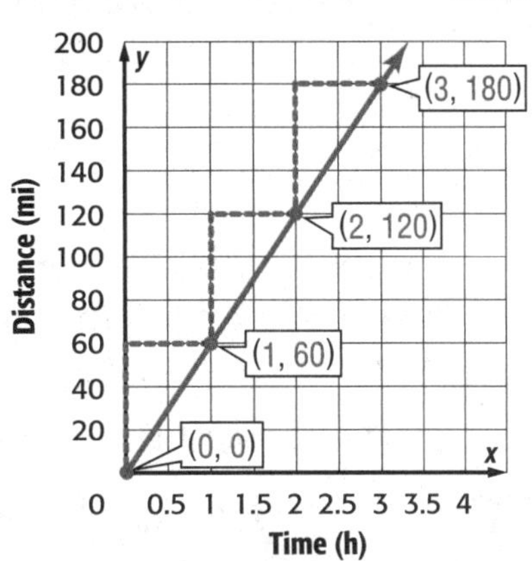

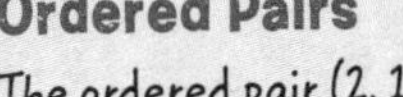

Ordered Pairs
The ordered pair (2, 120) represents traveling 120 miles in 2 hours.

3. Explain what the points (0, 0) and (1, 60) represent.

The point (0, 0) represents traveling zero miles in zero hours. The point (1, 60) represents traveling 60 miles in 1 hour. Notice that this is the constant rate of change.

Got It? Do these problems to find out.

c. Use the graph to find the constant rate of change in miles per hour while driving in the city.

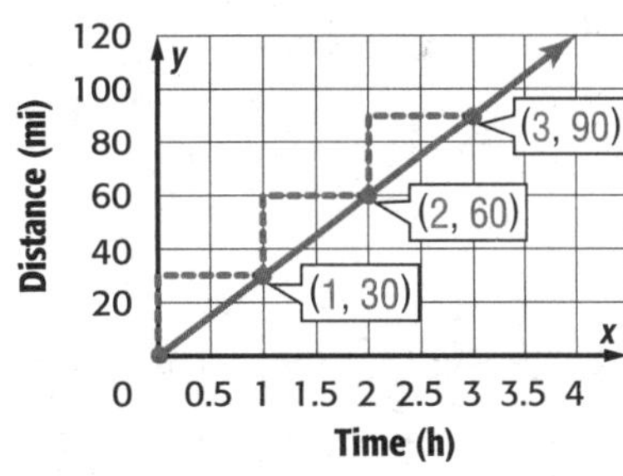

d. On the lines below, explain what the points (0, 0) and (1, 30) represent.

c. ________

Example

4. The table and graph below show the hourly charge to rent a bicycle at two different stores. Which store charges more per bicycle? Explain.

Pedals Rentals	
Time (hour)	**Cost ($)**
2	24
3	36
4	48

+1, +1 (Time); +12, +12 (Cost)

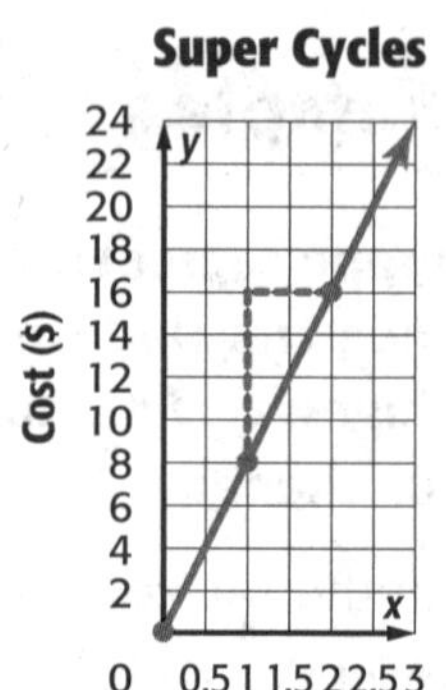

The cost at Pedals Rentals increases by $12 every hour. The cost at Super Cycles increases by $8 every hour.

So, Pedals Rentals charges more per hour to rent a bicycle.

Guided Practice

1. The table and graph below show the amount of money Mi-Ling and Daniel save each week. Who saves more each week? Explain. (Examples 1, 2, and 4)

Mi-Ling's Savings	
Time (weeks)	**Savings ($)**
2	$30
3	$45
4	$60

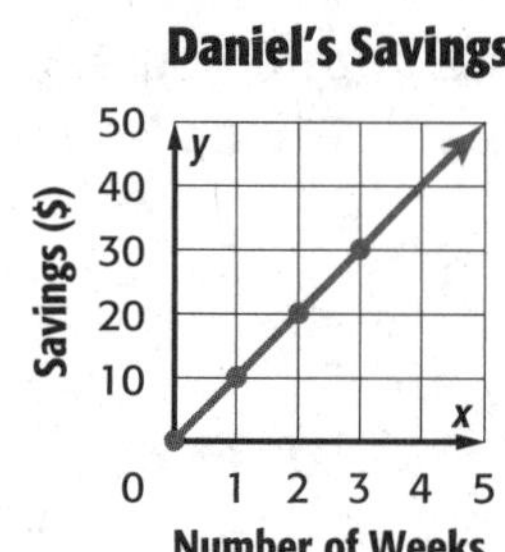

2. Refer to the graph in Exercise 1. Explain what the points (0, 0) and (1, 10) represent. (Example 3)

3. **Building on the Essential Question** How can you find the unit rate on a graph that goes through the origin?

Rate Yourself!

Are you ready to move on? Shade the section that applies.

For more help, go online to access a Personal Tutor.

Name ______________________ My Homework ______________________

Independent Practice

Go online for Step-by-Step Solutions

Find the constant rate of change for the table. (Example 1)

1

Time (s)	Distance (m)
1	6
2	12
3	18
4	24

____

2.

Items	Cost ($)
2	18
4	36
6	54
8	72

____

3 The graph shows the cost of purchasing T-shirts. Find the constant rate of change for the graph. Then explain what points (0, 0) and (1, 9) represent. (Examples 2 and 3)

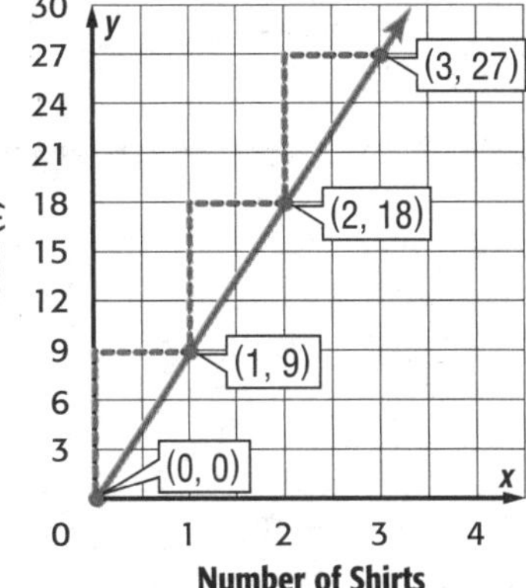

____

____

____

____

4. The Guzman and Hashimoto families each took a 4-hour road trip. The distances traveled by each family are shown in the table and graph below. Which family averaged fewer miles per hour? Explain. (Example 4)

Guzman's Road Trip	
Time (hours)	**Distance (miles)**
2	90
3	135
4	180

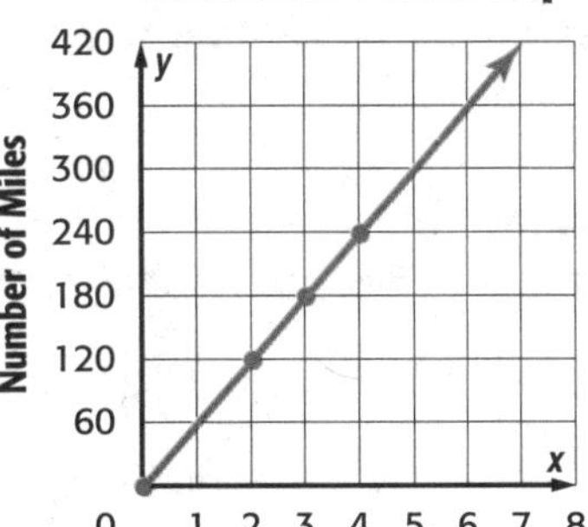

____

____

____

5. At 1:00 P.M., the water level in a pool is 13 inches. At 1:30 P.M., the water level is 18 inches. At 2:30 P.M., the water level is 28 inches. What is the constant rate of change?

____

6 **CCSS Model with Mathematics** Refer to the lap times for Exercises **a** and **b**.

a. How long does it take Seth to race 1 mile? Write the constant rate of change in miles per second. Round to the nearest hundredth. ______________

b. Graph the distance y and time x on the coordinate plane at the right. Graph the distance on the y-axis and the time on the x-axis.

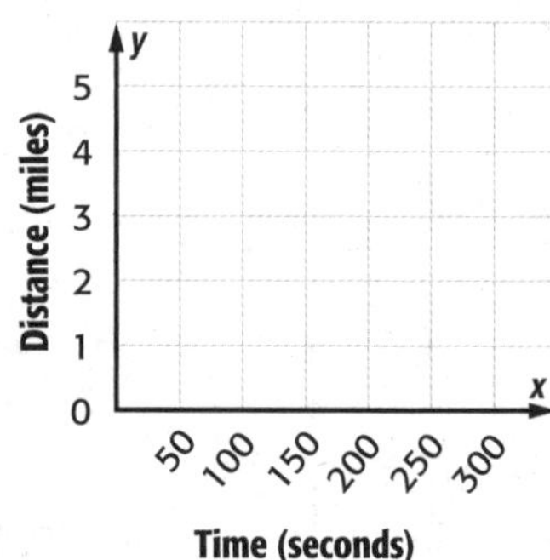

H.O.T. Problems

7. **CCSS Model with Mathematics** Make a table where the constant rate of change is 6 inches for every foot.

Feet	Inches

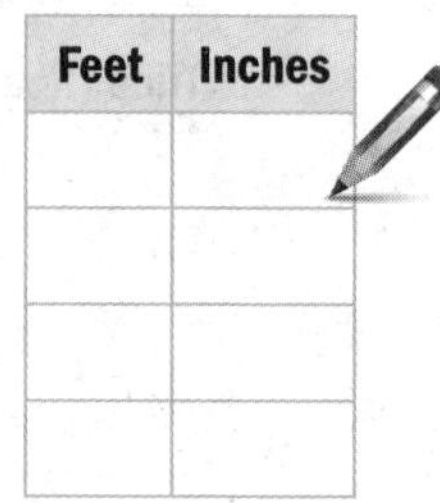

8. **CCSS Persevere with Problems** The constant rate of change for the relationship shown in the table is \$8 per hour. Find the missing values.

$x =$ ______ $y =$ ______ $z =$ ______

Time (h)	1	2	3
Earnings (\$)	x	y	z

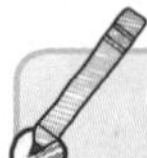

Standardized Test Practice

9. The information in the table represents a constant rate of change. Find the missing value.

Ⓐ 30
Ⓑ 90
Ⓒ 105
Ⓓ 120

Number of Packages	2	4	7
Number of Raisins	30	60	x

Name ______________________ My Homework ______________________

Extra Practice

Find the constant rate of change for each table.

10.

Time (h)	0	1	2	3
Wage ($)	0	9	18	27

$9 per hour

$\frac{\text{change in wages}}{\text{change in hours}} = \frac{\$9}{1 \text{ hour}}$

Homework Help

11.

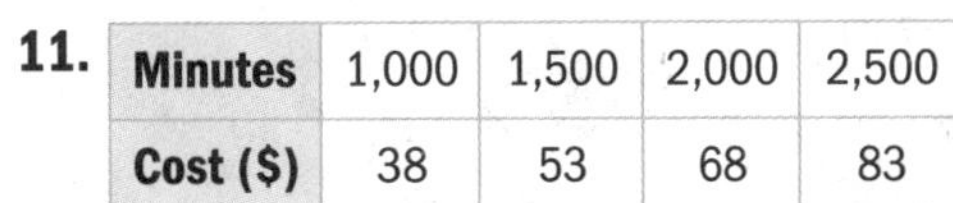

Minutes	1,000	1,500	2,000	2,500
Cost ($)	38	53	68	83

12. Use the graph to find the constant rate of change. Then, explain what the points (0, 0) and (6, 72) represent.

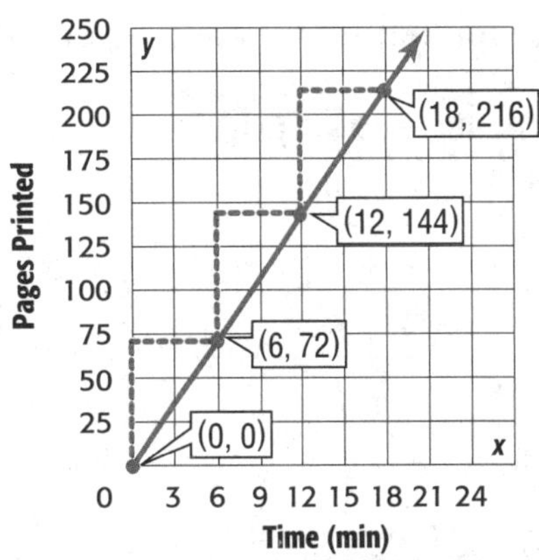

13. CCSS **Justify Conclusions** Ramona and Josh earn money by babysitting. The amounts earned for one evening are shown in the table and graph. Who charged more per hour? Explain.

Ramona's Earnings	
Time (hours)	**Earnings ($)**
2	18
3	27
4	36

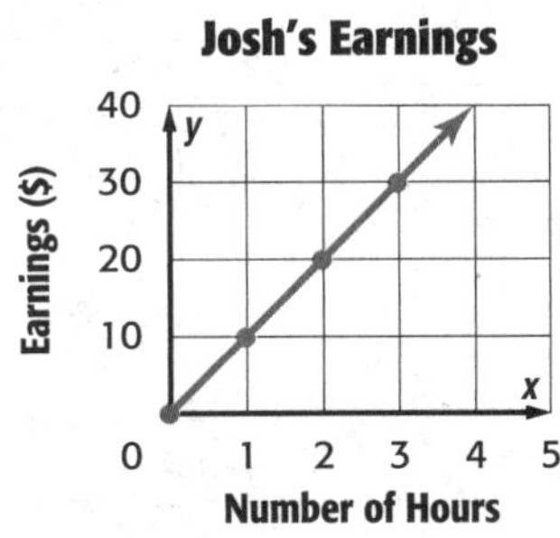

14. The cost of 1 movie ticket is $7.50. The cost of 2 movie tickets is $15. Based on this constant rate of change, what is the cost of 4 movie tickets? ________

Standardized Test Practice

15. Use the information in the table to find the constant rate of change.

Number of Apples	3	7	11
Number of Seeds	30	70	110

Ⓐ $\frac{10}{1}$

Ⓑ $\frac{1}{10}$

Ⓒ $\frac{40}{4}$

Ⓓ $\frac{4}{40}$

16. **Short Response** Reggie started a running program to prepare for track season. Every day for 60 days, he ran a half hour in the morning and a half hour in the evening. He averaged 6.5 miles per hour. At this rate, what is the total number of miles Reggie ran over the 60-day period? ______

Common Core Review

Write the output for each given input in the tables below. 5.OA.3

17.

Input	Add 4	Output
1	1 + 4	
2	2 + 4	
3	3 + 4	
4	4 + 4	

18.

Input	Subtract 5	Output
30	30 − 5	
40	40 − 5	
50	50 − 5	
60	60 − 5	

19.

Input	Multiply by 2	Output
1	1 × 2	
2	2 × 2	
3	3 × 2	
4	4 × 2	

20.

Input	Divide by 3	Output
3	3 ÷ 3	
6	6 ÷ 3	
9	9 ÷ 3	
12	12 ÷ 3	

Write the rule shown in each table. 5.OA.3

21.

Input		Output
4	?	10
5	?	11
6	?	12
7	?	13

22.

Input		Output
2	?	10
4	?	20
6	?	30
8	?	40

Lesson 8
Slope

What You'll Learn

Scan the lesson. Predict two things you will learn about slope.

-

-

Essential Question

HOW can you show that two objects are proportional?

Vocabulary

slope

Common Core State Standards

Content Standards
7.RP.2, 7.RP.2b

Mathematical Practices
1, 3, 4

Real-World Link

Recycling Hero Comics prints on recycled paper. The table shows the total number of pounds of recycled paper that has been used each day during the month.

Day of Month	Total Recycled (lbs)
3	36
5	60
6	72
7	84
12	144

1. Graph the ordered pairs on the coordinate plane.

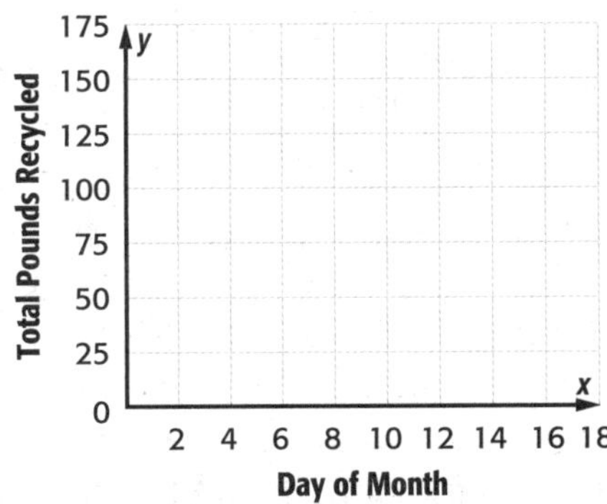

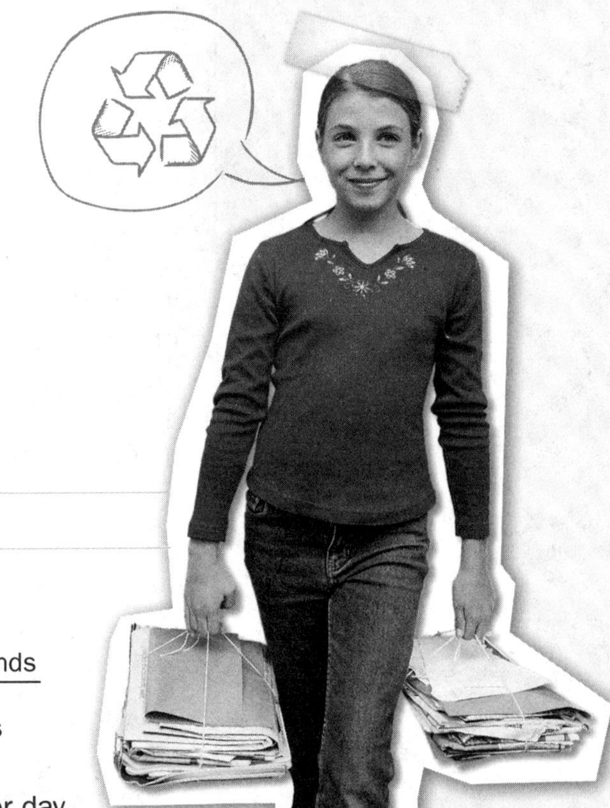

2. Explain why the graph is linear.

3. Use two points to find the constant rate of change.

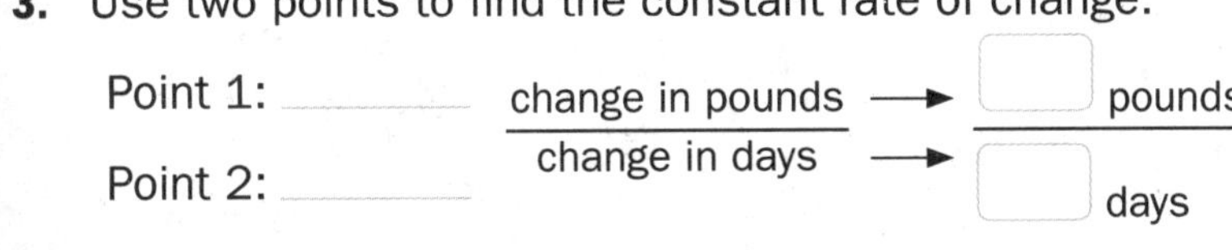

So, the constant rate of change is $\frac{24}{2}$ or ☐ pounds per day.

Key Concept

Slope

Slope is the rate of change between any two points on a line.

Work Zone

$$\text{slope} = \frac{\text{change in } y}{\text{change in } x} \begin{array}{l} \leftarrow \text{vertical change} \\ \leftarrow \text{horizontal change} \end{array}$$

$$= \frac{2}{1} \text{ or } 2$$

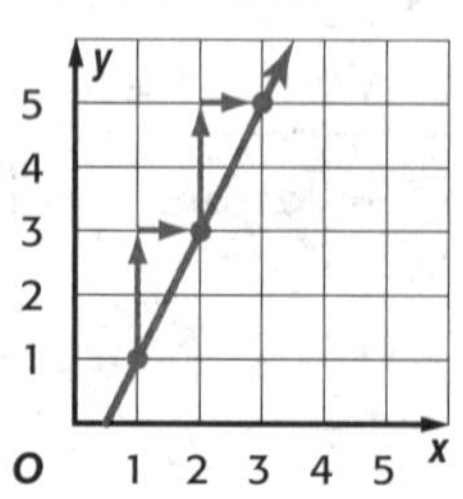

In a linear relationship, the vertical change (change in *y*-value) per unit of horizontal change (change in *x*-value) is always the same. This ratio is called the **slope** of the function. The constant rate of change, or unit rate, is the same as the slope of the related linear relationship.

The slope tells how steep the line is. The vertical change is sometimes called "rise" while the horizontal change is called "run." You can say that slope $= \frac{\text{rise}}{\text{run}}$.

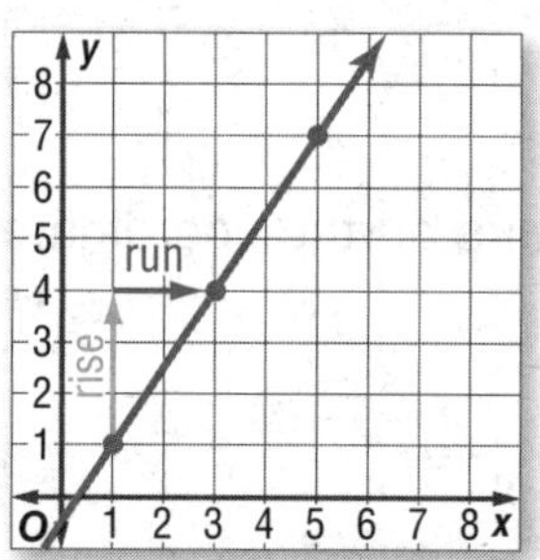

Count the number of units that make up the rise of the line in the graph shown above. Write this number for the numerator of the fraction below. Count the number of units that make up the run of the line. Write this number for the denominator of the fraction below.

$$\frac{\text{rise}}{\text{run}} = \frac{\square}{\square}$$

So, the slope of the line is $\frac{3}{2}$.

Example

1. The table below shows the relationship between the number of seconds *y* it takes to hear thunder after a lightning strike and the miles *x* you are from the lightning. Graph the data and find the slope. Explain what the slope represents.

Miles (*x*)	0	1	2	3	4	5
Seconds (*y*)	0	5	10	15	20	25

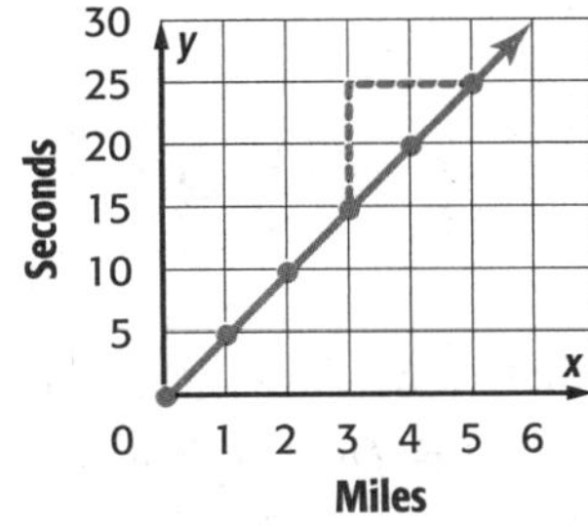

$$\text{slope} = \frac{\text{change in } y}{\text{change in } x}$$ Definition of slope

$$= \frac{25 - 15}{5 - 3}$$ Use (3, 15) and (5, 25).

$$= \frac{10}{2}$$ ← seconds, ← miles

$$= \frac{5}{1}$$ Simplify.

So, for every 5 seconds between a lightning flash and the sound of thunder, there is 1 mile between you and the lightning strike.

Got It? **Do this problem to find out.**

a. Graph the data about plant height for a science fair project. Then find the slope of the line. Explain what the slope represents in the work zone.

Week	Plant Height (cm)
1	1.5
2	3
3	4.5
4	6

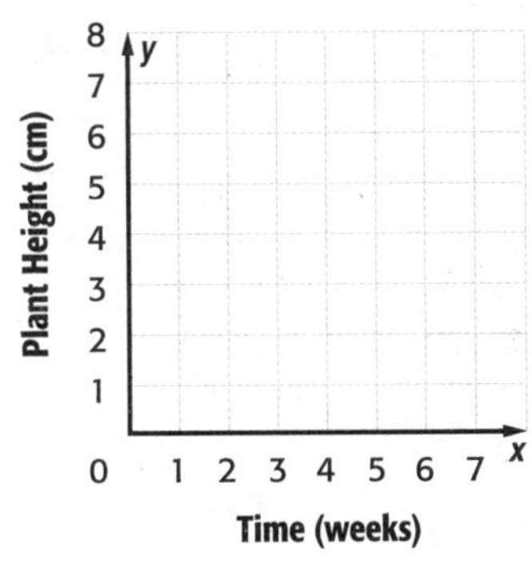

Slope In everyday language, slope means inclination or slant.

In math language, slope means the ratio of vertical change per unit of horizontal change; the steepness of a line.

a. ___________

Example

2. Renaldo opened a savings account. Each week he deposits $300. Draw a graph of the the account balance versus time. Find the numerical value of the slope and interpret it in words.

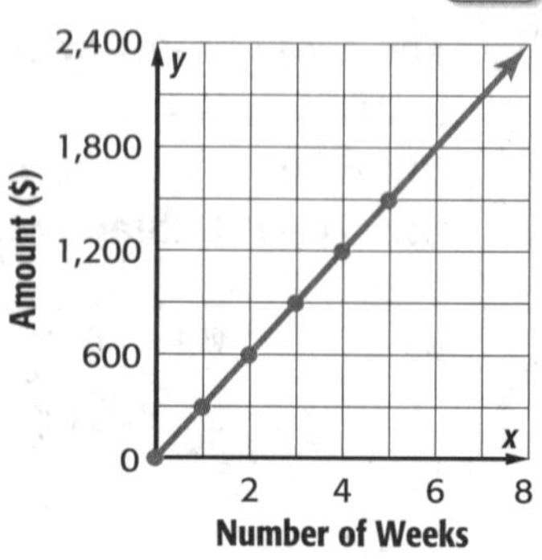

The slope of the line is the rate at which the account balance rises, or $\frac{\$300}{1 \text{ week}}$.

Got It? Do this problem to find out.

b. Jessica has a balance of $45 on her cell phone account. She adds $10 each week for the next four weeks. In the work zone, graph the account balance versus time. Find the numerical value of the slope and interpret it in words.

Show your work.

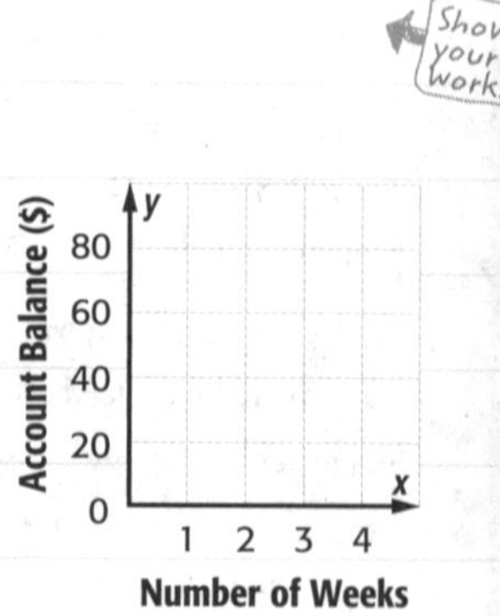

b. ______________

Guided Practice

1. The table at the right shows the number of small packs of fruit snacks *y* per box *x*. Graph the data. Then find the slope of the line. Explain what the slope represents. (Examples 1 and 2)

Boxes, *x*	3	5	7
Fruit Snacks, *y*	12	20	28

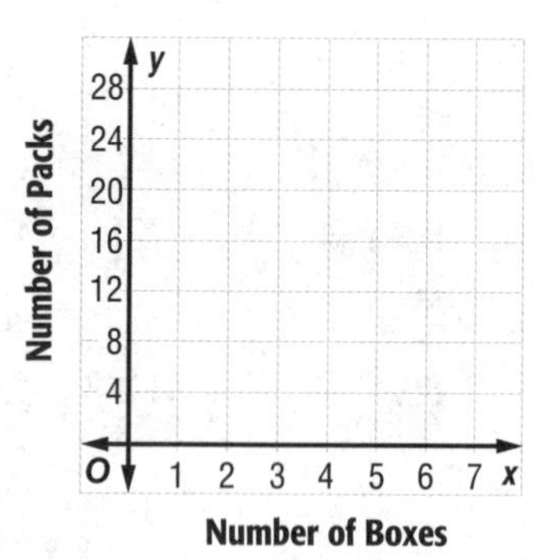

2. **Building on the Essential Question** How is rate of change related to slope? ______________

Rate Yourself!

How well do you understand slope? Circle the image.

Clear | Somewhat Clear | Not So Clear

For more help, go online to access a Personal Tutor.

Independent Practice

Go online for Step-by-Step Solutions

1 The table shows the number of pages Adriano read in x hours. Graph the data. Then find the slope of the line. Explain what the slope represents. (Example 1)

Time (h)	1	2	3	4
Number of pages	50	100	150	200

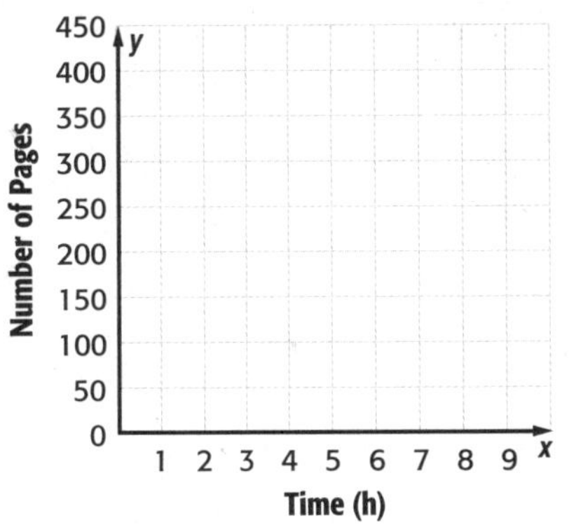

2. Graph the data. Find the numerical value of the slope and interpret it in words. (Example 2)

Number of Yards	1	2	3
Number of Feet	3	6	9

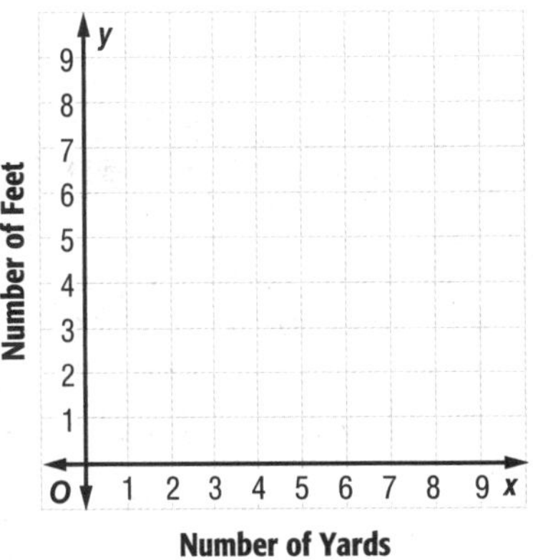

3 The graph shows the average speed of two cars on the highway.

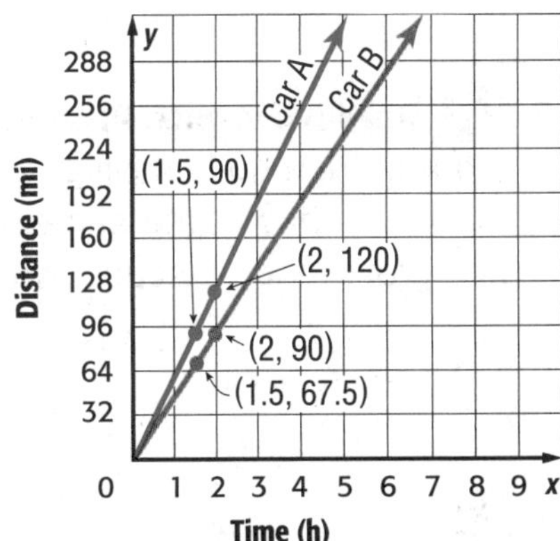

a. What does (2, 120) represent?

b. What does (1.5, 67.5) represent?

c. What does the ratio of the y-coordinate to the x-coordinate for each pair of points on the graph represent?

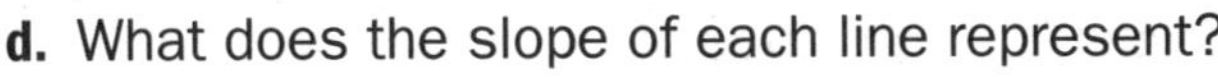

d. What does the slope of each line represent?

e. Which car is traveling faster? How can you tell from the graph?

4. **Multiple Representations** Complete the graphic organizer on slope.

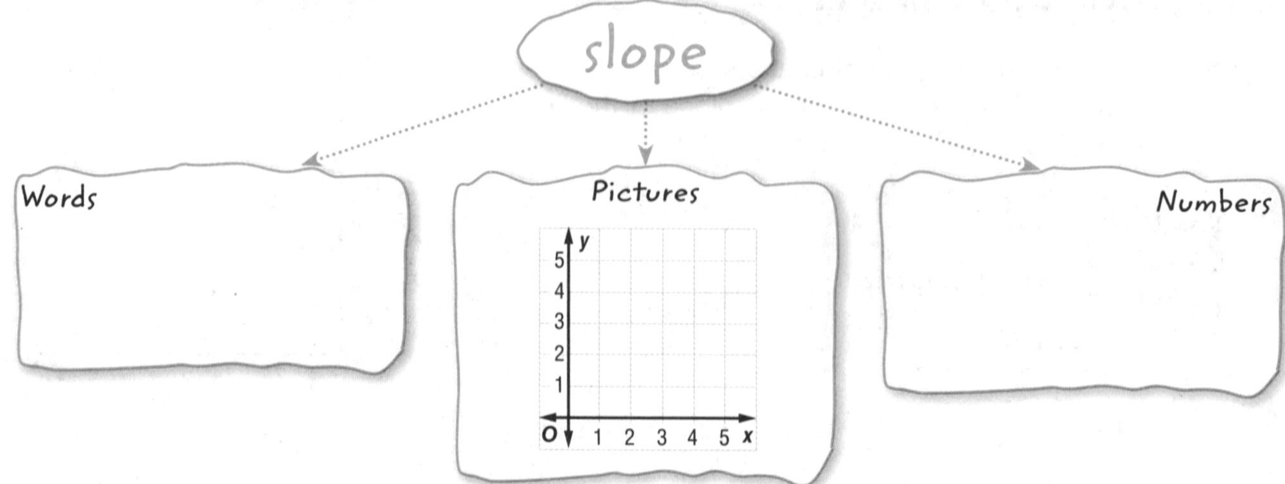

H.O.T. Problems Higher Order Thinking

5. **Find The Error** Marisol is finding the slope of the line containing the points (3, 7) and (5, 10). Find her mistake and correct it.

The slope between the two points (3, 7) and (5, 10) is found like this:
$\text{slope} = \frac{\text{rise}}{\text{run}} = \frac{5-3}{10-7}$
$= \frac{2}{3}$

6. **Persevere with Problems** Kaya is saving money at a rate of $30 per month. Edgardo is saving money at a rate of $35 per month. They both started saving at the same time. If you were to create a table of values and graph each function, what would be the slope of each graph?

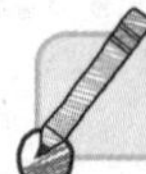

Standardized Test Practice

7. The table shows the number of packs and the number of sticks of chewing gum. If the data were graphed, what would be the slope of the line?

Packs	Sticks of Gum
1	5
2	10
3	15
4	20

Ⓐ (1, 5)

Ⓑ (5, 1)

Ⓒ $\frac{1}{5}$

Ⓓ $\frac{5}{1}$

Extra Practice

8. **Justify Conclusions** The table to the right shows the number of markers per box. Graph the data. Then find the slope of the line. Explain what the slope represents.

Boxes	1	2	3	4
Markers	8	16	24	32

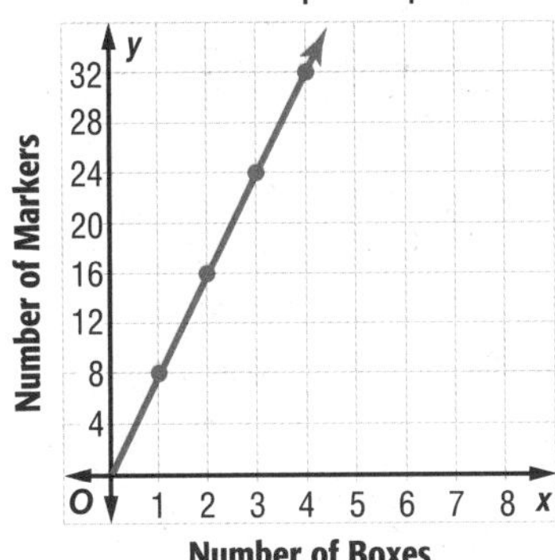

Use (1, 8) and (2, 16).

$$\text{slope} = \frac{\text{change in } y}{\text{change in } x} = \frac{16 - 8}{2 - 1} = \frac{8}{1}$$

So, there are 8 markers in every 1 box.

9. The table shows the cost to rent a paddle boat from two businesses.

Paddle Boat Rentals		
Number of Hours	Water Wheels Cost ($)	Fun in the Sun Cost ($)
1	20	25
2	40	50
3	60	75
4	80	100

a. What does (1, 20) represent?

______

______

b. What does (2, 50) represent?

______

______

Copy and Solve **For Exercises 10–13, draw a graph on a separate sheet of grid paper to find each slope. Then record each slope and interpret its meaning.**

10. The table shows the amount Maggie earns for various numbers of hours she babysits. Graph the data. Then find the slope of the line. Explain what the slope represents.

Number of Hours	Earnings ($)
1	8
2	16
3	24
4	32

11. Joshua swims 25 meters in 1 minute. Draw a graph of meters swam versus time. Find the value of the slope and interpret it in words.

12. The Jackson family rents 6 movies each month. Draw a graph of movies rented versus time. Find the value of the slope and interpret it in words.

13. Zack completes 20 homework problems in 1 hour. Draw a graph of homework problems versus time. Find the value of the slope and interpret it in words.

Standardized Test Practice

14. **Short Response** Find the slope of the line below that shows the distance Jairo traveled while jogging.

____

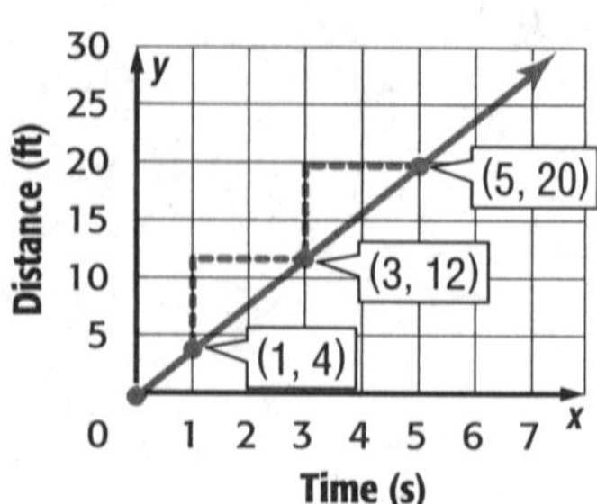

15. Line *RS* represents a bike ramp.

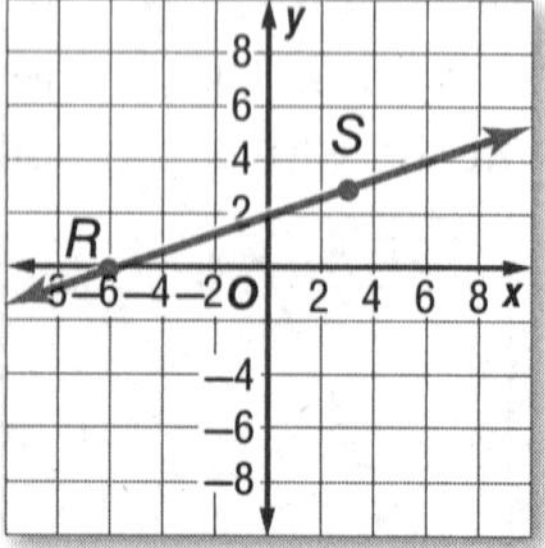

What is the slope of the ramp?

Ⓐ (1, 3)

Ⓑ (3, 1)

Ⓒ $\frac{1}{3}$

Ⓓ $\frac{3}{1}$

16. **Short Response** Two weeks ago, Audrey earned $84 for 7 hours of work. This week, she earned $132 for 11 hours of work. Find the numerical value of the slope of the line that would represent Audrey's earnings.

____

Common Core Review

Determine if each situation is proportional. Explain your reasoning. 7.RP.2

17. Taxi cab passengers are charged $2.50 upon entering a cab. They are then charged $1.00 for every mile traveled.

____

18. A restaurant charges $5 for one sandwich, $9.90 for two sandwiches, and $14.25 for three sandwiches.

____

____

19.

Tickets Purchased	1	2	3	4
Cost ($)	7.50	15	22.50	30

____

____

20.

Cups of Flour	3	6	9	12
Cups of Sugar	2	4	6	8

____

____

Lesson 9

Direct Variation

What You'll Learn

Scan the text on the following two pages. Write the definitions of direct variation and constant of proportionality.

- direct variation ______________________
- constant of proportionality ______________________

Essential Question

HOW can you show that two objects are proportional?

Vocabulary

direct variation
constant of variation
constant of proportionality

Common Core State Standards

Content Standards
7.RP.2, 7.RP.2a, 7.RP.2b

Mathematical Practices
1, 2, 3, 4

Real-World Link

Speed The distance d a car travels after t hours can be represented by $d = 65t$. The table and graph also represent the situation.

Time (hours)	Distance (miles)
2	130
3	195
4	260

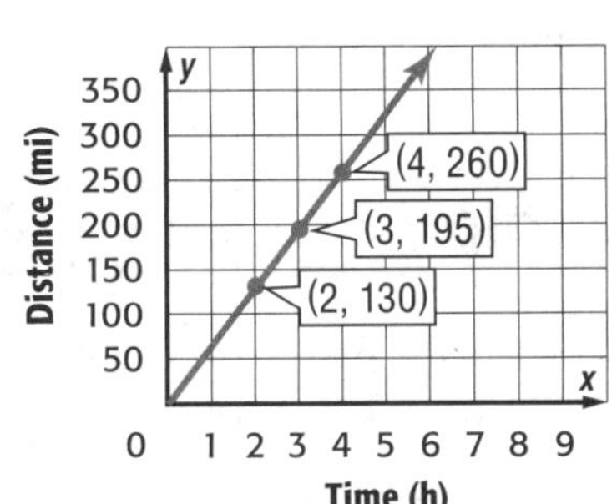

1. Fill in the blanks to find the constant ratio.

$$\frac{\text{distance traveled}}{\text{driving time}} = \frac{130}{2} = \frac{195}{\square} = \frac{\square}{4}$$

The constant ratio is ☐ miles per hour.

2. The constant rate of change, or slope, of the line is $\frac{\text{change in miles}}{\text{change in time}}$, which is equal to $\frac{195 - 130}{3 - 2}$ or ☐ miles per hour.

3. Write a sentence that compares the constant rate of change and the constant ratio.

____

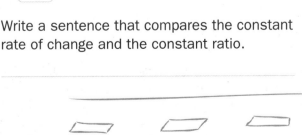

Work Zone

Key Concept Direct Variation

Words A direct variation is a relationship in which the ratio of y to x is a constant, k. We say y varies directly with x.

Symbols $\frac{y}{x} = k$ or $y = kx$, where $k \neq 0$

Example $y = 2x$

Model

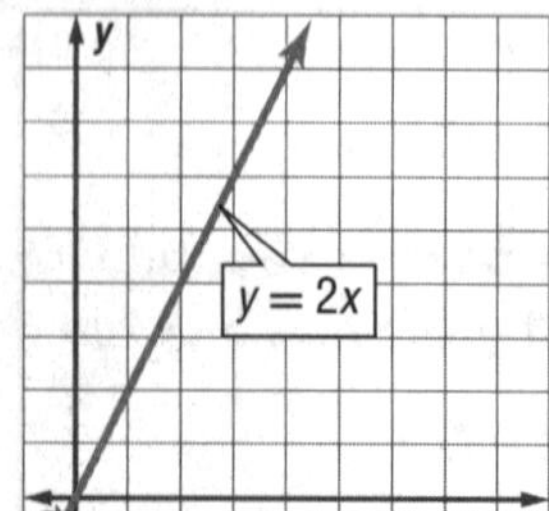

When two variable quantities have a constant ratio, their relationship is called a **direct variation**. The constant ratio is called the **constant of variation**. The constant of variation is also known as the **constant of proportionality**.

In a direct variation equation, the constant rate of change, or slope, is assigned a special variable, k.

Real World Example

1. The height of the water as a pool is being filled is shown in the graph. Determine the rate in inches per minute.

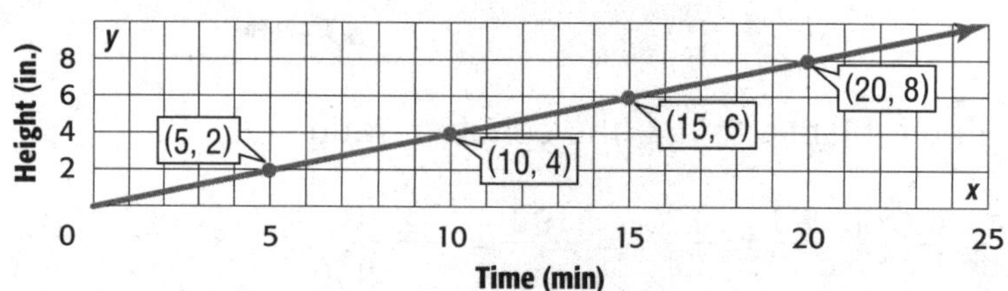

Direct Variation
When a relationship varies directly, the graph of the function will always go through the origin, (0, 0). Also, the unit rate r is located at $(1, r)$.

Since the graph of the data forms a line, the rate of change is constant. Use the graph to find the constant of proportionality.

$\frac{\text{height}}{\text{time}} \rightarrow \frac{2}{5}$ or $\frac{0.4}{1}$ $\quad \frac{4}{10}$ or $\frac{0.4}{1}$ $\quad \frac{6}{15}$ or $\frac{0.4}{1}$ $\quad \frac{8}{20}$ or $\frac{0.4}{1}$

The pool fills at a rate of 0.4 inch every minute.

Got It? Do this problem to find out.

Show your work.

a. Two minutes after a diver enters the water, he has descended 52 feet. After 5 minutes, he has descended 130 feet. At what rate is the scuba diver descending?

a. ____________

Example

2. The equation $y = 10x$ represents the amount of money y Julio earns for x hours of work. Identify the constant of proportionality. Explain what it represents in this situation.

$y = kx$

$y = 10x$

Compare the equation to $y = kx$, where k is the constant of proportionality.

The constant of proportionality is 10. So, Julio earns $10 for every hour that he works.

Got It? **Do this problem to find out.**

b. The distance y traveled in miles by the Chang family in x hours is represented by the equation $y = 55x$. Identify the constant of proportionality. Then explain what it represents.

b. ____________

Determine Direct Variation

Not all situations with a constant rate of change are proportional relationships. Likewise, not all linear functions are direct variations.

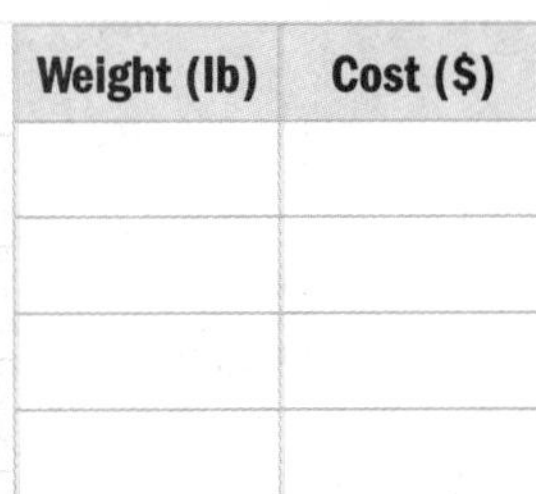

Weight (lb)	Cost ($)

Example

2. Pizzas cost $8 each plus a $3 delivery charge. Show the cost of 1, 2, 3, and 4 pizzas. Is there a direct variation?

Number of Pizzas	1	2	3	4
Cost ($)	$11	$19	$27	$35

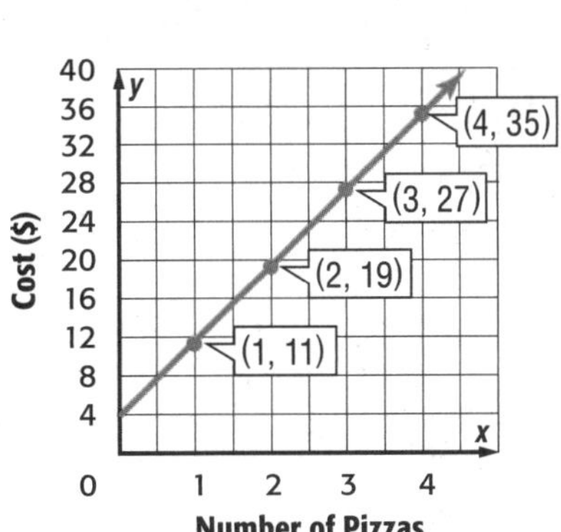

$\frac{\text{cost}}{\text{number of pizzas}} \rightarrow \frac{11}{1}, \frac{19}{2}$ or 9.5, $\frac{27}{3}$ or 9, $\frac{35}{4}$ or 8.75

There is no constant ratio and the line does not go through the origin. So, there is no direct variation.

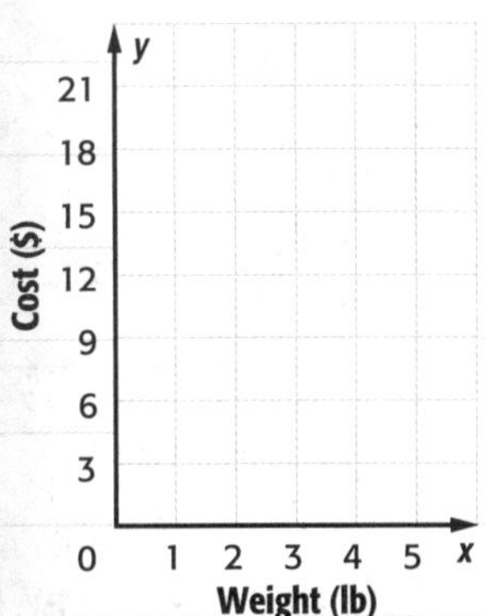

Got It? **Do this problem to find out.**

c. Two pounds of cheese cost $8.40. Show the cost of 1, 2, 3, and 4 pounds of cheese. Is there a direct variation? Explain.

c. ____________

Example

4. Determine whether the linear relationship is a direct variation. If so, state the constant of proportionality.

Time, x	1	2	3	4
Wages ($), y	12	24	36	48

Compare the ratios to check for a common ratio.

$\frac{\text{wages}}{\text{time}} \rightarrow \frac{12}{1} \quad \frac{24}{2} \text{ or } \frac{12}{1} \quad \frac{36}{3} \text{ or } \frac{12}{1} \quad \frac{48}{4} \text{ or } \frac{12}{1}$

Since the ratios are the same, the relationship is a direct variation. The constant of proportionality is $\frac{12}{1}$.

Guided Practice

1. The number of cakes baked varies directly with the number of hours the caterers work. What is the ratio of cakes baked to hours worked? (Examples 1 and 2) ______

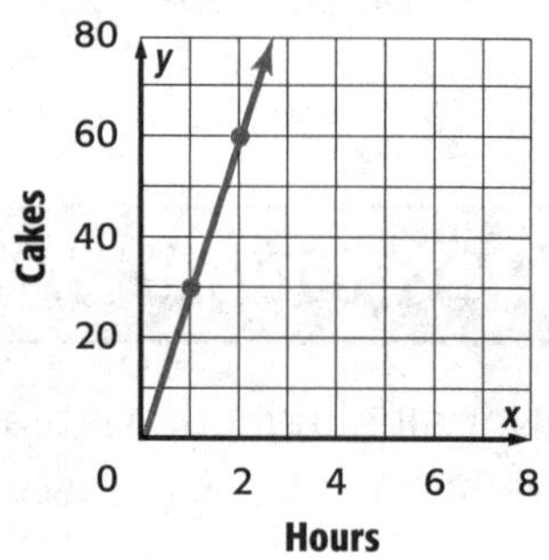

2. An airplane travels 780 miles in 4 hours. Make a table and graph to show the mileage for 2, 8, and 12 hours. Is there a direct variation? Explain.

(Examples 3 and 4) ______

______

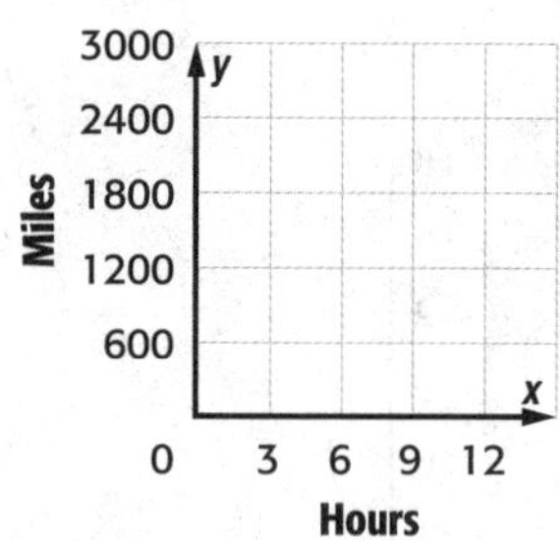

Hours			
Miles			

3. **Building on the Essential Question** How can you determine if a linear relationship is a direct variation from an equation? a table? a graph? ______

______

______

______

Rate Yourself!

How confident are you about direct variation? Check the box that applies.

For more help, go online to access a Personal Tutor.

Independent Practice

Go online for Step-by-Step Solutions

1. Veronica is mulching her front yard. The total weight of mulch varies directly with the number of bags of mulch.

What is the rate of change? (Example 1) ________

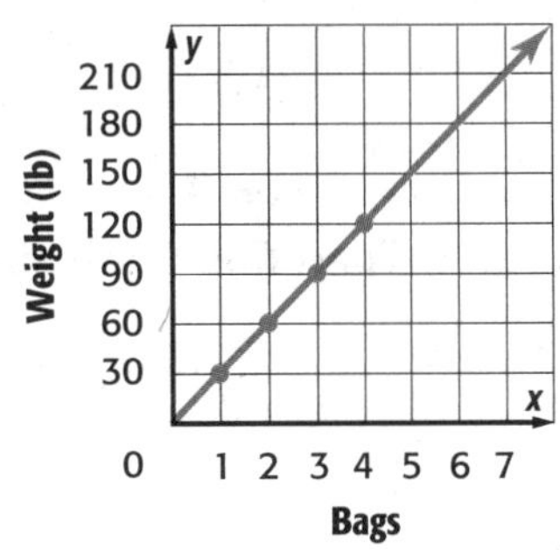

2. The Spanish club held a car wash to raise money. The equation $y = 5x$ represents the amount of money y club members made for washing x cars. Identify the constant of proportionality. Then explain what it represents in this situation. (Example 2) ________

3. A technician charges $25 per hour plus $50 for a house call to repair home computers. Make a table and a graph to show the cost for 1, 2, 3, and 4 hours of home computer repair service. Is there a direct variation? (Example 3)

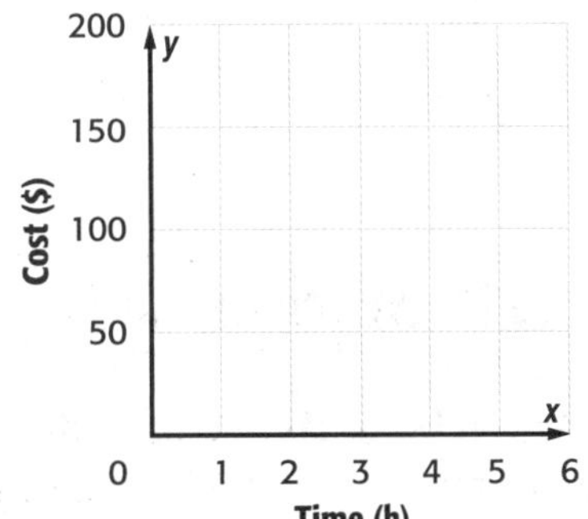

Time (h)				
Charge ($)				

Determine whether each linear relationship is a direct variation. If so, state the constant of proportionality. (Example 4)

4.

Pictures, x	3	4	5	6
Profit, y	24	32	40	48

5.

Minutes, x	185	235	285	335
Cost, y	60	100	140	180

6.

Year, x	5	10	15	20
Height, y	12.5	25	37.5	50

7.

Game, x	2	3	4	5
Points, y	4	5	6	7

8. At a 33-foot depth underwater, the pressure is 29.55 pounds per square inch (psi). At a depth of 66 feet, the pressure reaches 44.4 psi. At what rate is the pressure increasing? ______

Reason Abstractly If y varies directly with x, write an equation for the direct variation. Then find each value.

9. If $y = 14$ when $x = 8$, find y when $x = 12$. ______

10. Find y when $x = 15$ if $y = 6$ when $x = 30$. ______

11. If $y = 6$ when $x = 24$, what is the value of x when $y = 7$? ______

12. Find x when $y = 14$, if $y = 7$ when $x = 8$. ______

H.O.T. Problems Higher Order Thinking

13. **Reason Inductively** Identify two additional values for x and y in a direct variation relationship where $y = 11$ when $x = 18$.

$x =$ ______ $y =$ ______ and $x =$ ______ $y =$ ______

14. **Persevere with Problems** Find y when $x = 14$ if y varies directly with x^2, and $y = 72$ when $x = 6$. ______

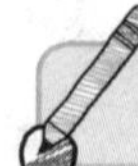

Standardized Test Practice

15. Which of the following relationships represent a direct variation?

Ⓐ

Hours, x	1	2	3	4
Wages (\$), y	10	22	36	50

Ⓑ

Hours, x	1	2	3	4
Wages (\$), y	10	25	30	50

Ⓒ

Hours, x	1	2	3	4
Wages (\$), y	10	20	30	40

Ⓓ

Hours, x	1	2	3	4
Wages (\$), y	6	16	30	48

Extra Practice

16. The money Shelley earns varies directly with the number of dogs she walks. How much does Shelley earn for each dog she walks?

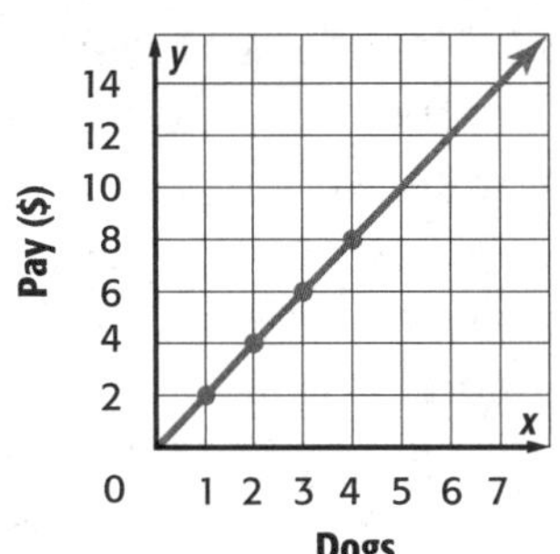

Since the points on the graph lie in a straight line, the rate of change is a constant. The constant ratio is what Shelley earns per dog.

pay ($) → / number of dogs → $\frac{2}{1}, \frac{4}{2}$ *or* $\frac{2}{1}, \frac{6}{3}$ *or* $\frac{2}{1}, \frac{8}{4}$ *or* $\frac{2}{1}$

Shelley earn $2.00 per dog.

17. A cake recipe requires $3\frac{1}{4}$ cups of flour for 13 servings and $4\frac{1}{2}$ cups of flour for 18 servings. How much flour is required to make a cake that serves 28? ______________

Determine whether each linear relationship is a direct variation. If so, state the constant of variation.

18.

Age, x	11	13	15	19
Grade, y	5	7	9	11

____

19.

Price, x	20	25	30	35
Tax, y	4	5	6	7

____

20. CCSS **Multiple Representations** Robert is in charge of the community swimming pool. Each spring he drains it in order to clean it. Then he refills the pool, which holds 120,000 gallons of water. Robert fills the pool at a rate of 10 gallons each minute.

a. **Words** What is the rate at which Robert will fill the pool? Is it constant? ______________

b. **Graph** Graph the relationship on the grid shown.

c. **Algebra** Write an equation for the direct variation.

____

____

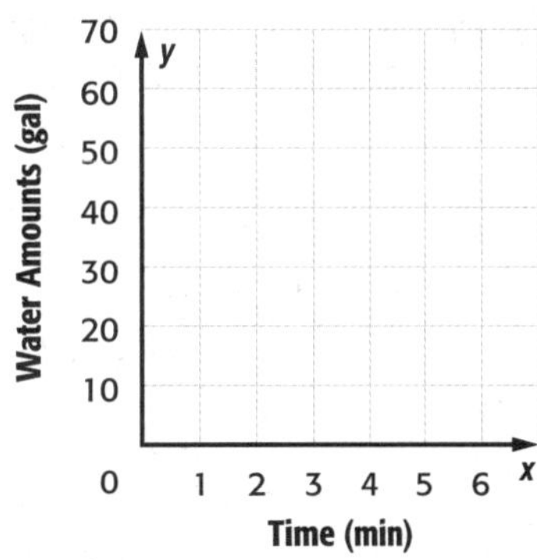

Standardized Test Practice

21. To make lemonade, Andy adds 8 tablespoons of sugar for every 12 ounces of water. If he uses 32 ounces of water, which proportion can he use to find the number of tablespoons of sugar x he should add to make the lemonade?

Ⓐ $\frac{8}{12} = \frac{32}{x}$

Ⓑ $\frac{8}{x} = \frac{32}{12}$

Ⓒ $\frac{8}{12} = \frac{x}{32}$

Ⓓ $\frac{x}{12} = \frac{8}{32}$

22. Anjuli read 22 pages during a 30-minute study hall. At this rate, how many pages would she read in 45 minutes?

Ⓕ 30

Ⓖ 33

Ⓗ 45

Ⓘ 48

23. **Short Response** Determine whether the linear function is a direct variation. If so, state the constant of proportionality. ______

Hours, x	3	5	7	9
Miles, y	108	180	252	324

CCSS Common Core Review

24. The table below shows the number of sheets of paper in various numbers of packages. Graph the data. **6.RP.3b**

Number of Packages	1	2	3	4
Number of Sheets	50	100	150	200

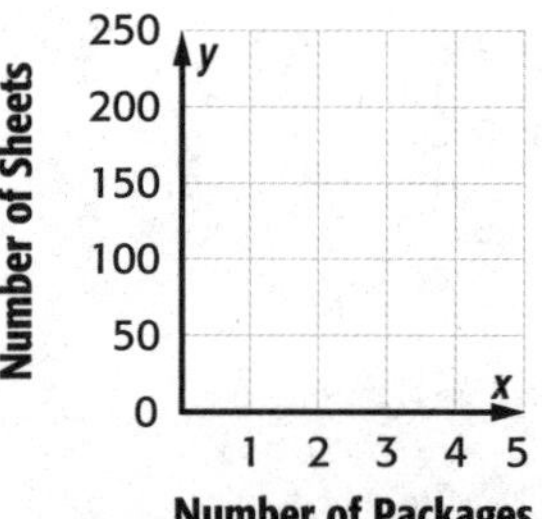

25. The cost of various numbers of tickets to a festival is shown in the table. Graph the data. Then find the slope of the line. Explain what the slope represents. **6.RP.3b**

Number of Tickets	5	10	20	25
Cost ($)	40	80	160	200

______

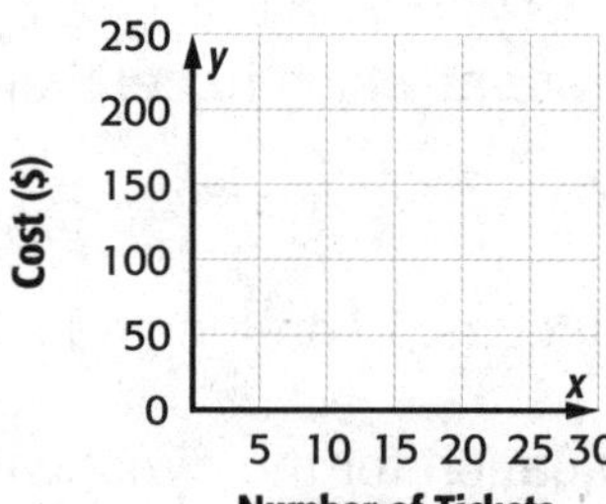

21ST CENTURY CAREER in Engineering

Biomechanical Engineering

Did you know that more than 700 pounds of force are exerted on a 140-pound long-jumper during the landing? Biomechanical engineers understand how forces travel through the shoe to an athlete's foot and how the shoes can help reduce the impact of those forces on the legs. If you are curious about how engineering can be applied to the human body, a career in biomechanical engineering might be a great fit for you.

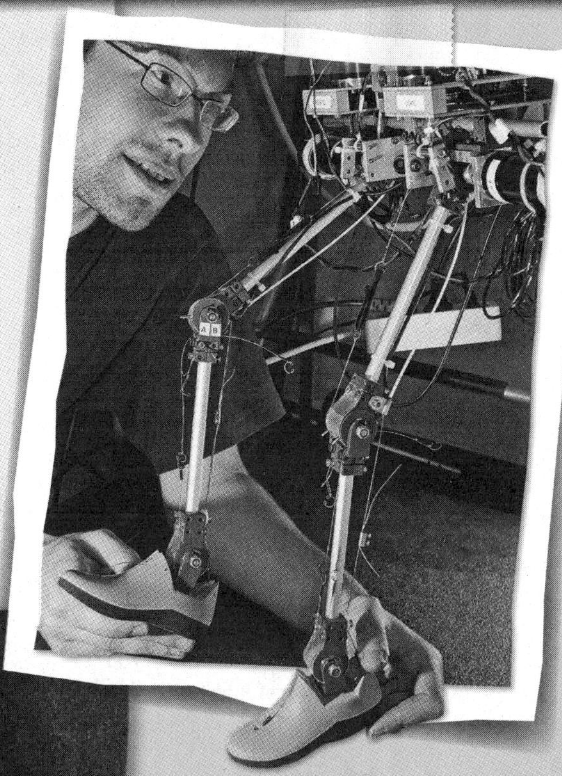

Is This the Career for You?

Are you interested in a career as a biomechanical engineer? Take some of the following courses in high school.

- Biology
- Calculus
- Physics
- Trigonometry

Find out how math relates to a career in Biomechanical Engineering.

Start Off on the Right Foot

Use the information in the graph to solve each problem.

1. Find the constant rate of change for the data shown in the graph below Exercise 2. Interpret its meaning.

2. Is there a proportional relationship between the weight of an athlete and the forces that are generated from running? Explain your reasoning.

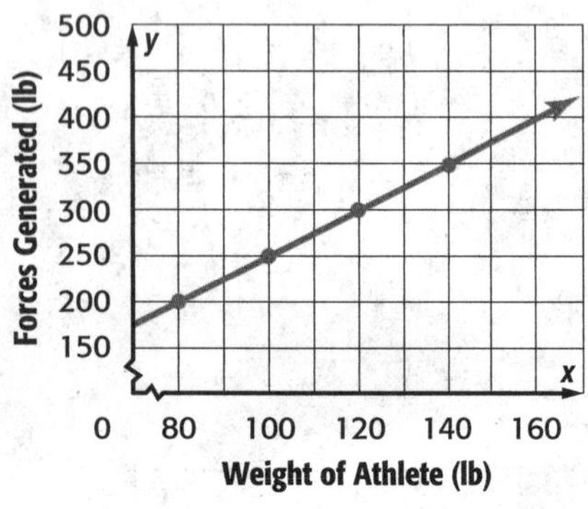

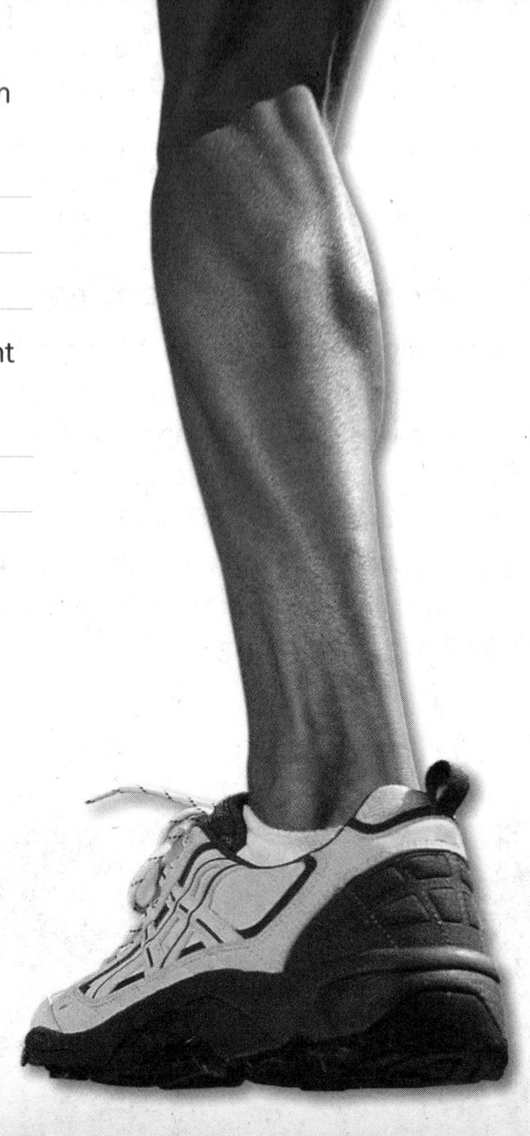

Career Project

It's time to update your career portfolio! Use the Internet or another source to research the fields of biomechanical engineering, biomedical engineering, and mechanical engineering. Write a brief summary comparing and contrasting the fields. Describe how they are all related.

What subject in school is the most important to you? How would you use that subject in this career?

Chapter Review

Vocabulary Check

Complete each sentence using the vocabulary list at the beginning of the chapter. Then circle the word that completes the sentence in the word search.

1. A ____________ is a ratio that compares two quantities with different kinds of units.
2. A rate that has a denominator of 1 unit is called a ____________ rate.
3. A pair of numbers used to locate a point in the coordinate plane is an ____________ pair.
4. (0, 0) represents the ____________.
5. A ____________ fraction has a fraction in the numerator, denominator, or both.
6. A ____________ variation is the relationship between two variable quantities with a constant ratio.
7. The ____________ is the rate of change between any two points on a line.
8. One of the four regions into which a coordinate plane is separated is called a ____________.
9. A ____________ is an equation stating that two ratios or rates are equal.
10. The rate of ____________ describes how one quantity changes in relation to another.
11. ____________ analysis is the process of including units of measurement when you compute.

Key Concept Check

Use Your FOLDABLES

Use your Foldable to help review the chapter.

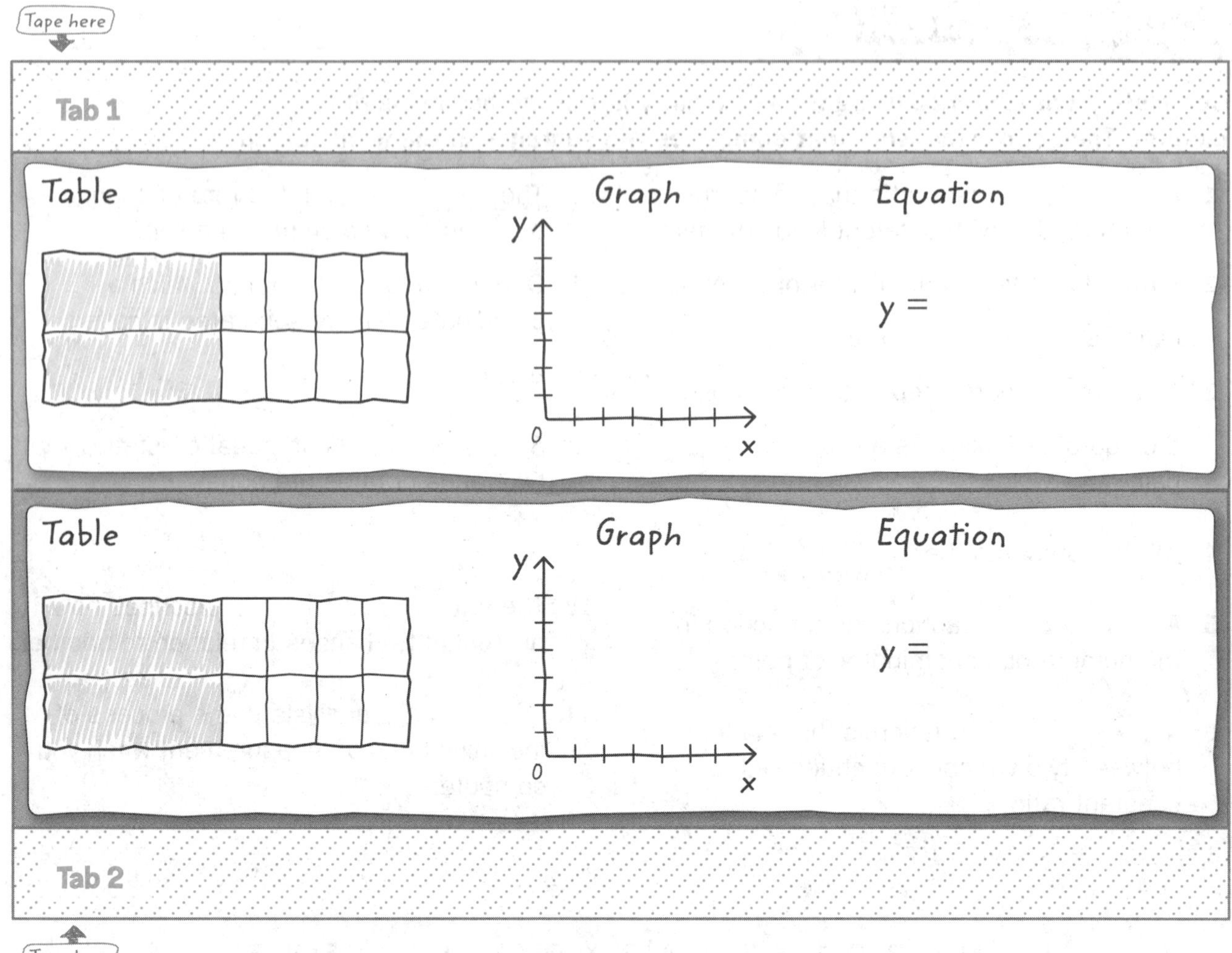

Got it?

Identify the Correct Choice **Write the correct term or number to complete each sentence.**

1. When a rate is simplified so that it has a (numerator, denominator) of 1 unit, it is called a unit rate.

2. If Dinah can skate $\frac{1}{2}$ lap in 15 seconds, she can skate 1 lap in (7.5, 30) seconds.

3. Slope is the ratio of (horizontal change to vertical change, vertical change to horizontal change).

4. When two quantities have a constant ratio, their relationship is called a (direct, linear) variation.

Problem Solving

1. Which bottle of shampoo shown at the right costs less per ounce? (Lesson 1)

Bottle	Price
12 oz	$2.59
16 oz	$3.19

2. An airplane is traveling at an average speed of 245 meters per second. How many kilometers per second is the plane traveling? (Lesson 3)

3. An Internet company charges $30 a month. There is also a $30 installation fee. Is the number of months you can have Internet proportional to the total cost? Explain. (Lesson 4)

4. Damon runs 8 meters in 1 second, 16 meters in 2 seconds, 24 meters in 3 seconds, and 32 meters in 4 seconds. Determine whether Damon's distance is proportional to the number of seconds he runs by graphing on the coordinate grid at the right. Explain your reasoning. (Lesson 5)

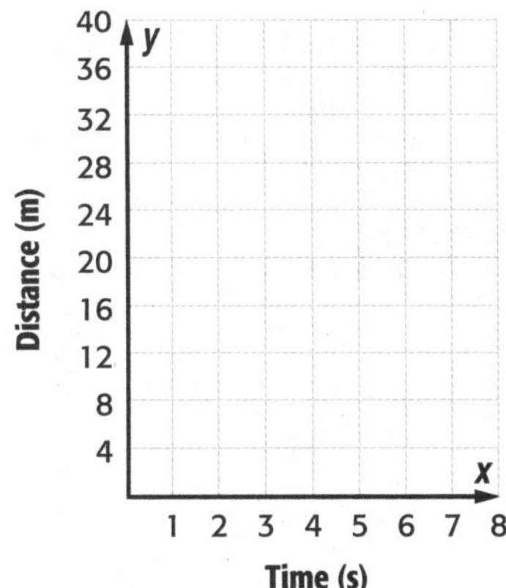

5. Suppose 3 televisions weigh 240.6 pounds. How much do 9 of the same televisions weigh? (Lesson 6)

6. The table and the graph show the amount of rainfall for 2 different days. Which day had a greater rate of change? Explain. (Lesson 8)

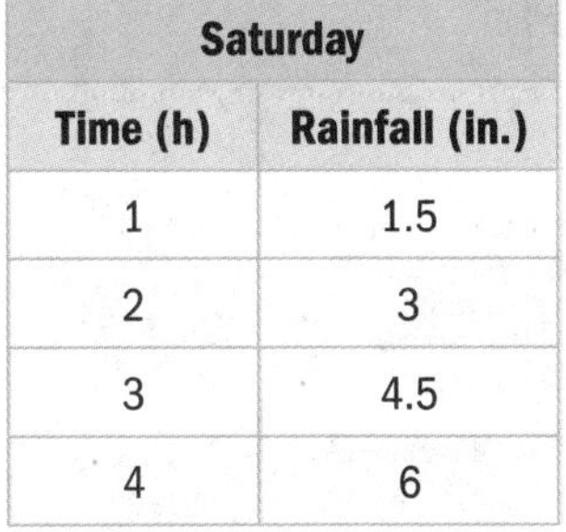

Saturday	
Time (h)	Rainfall (in.)
1	1.5
2	3
3	4.5
4	6

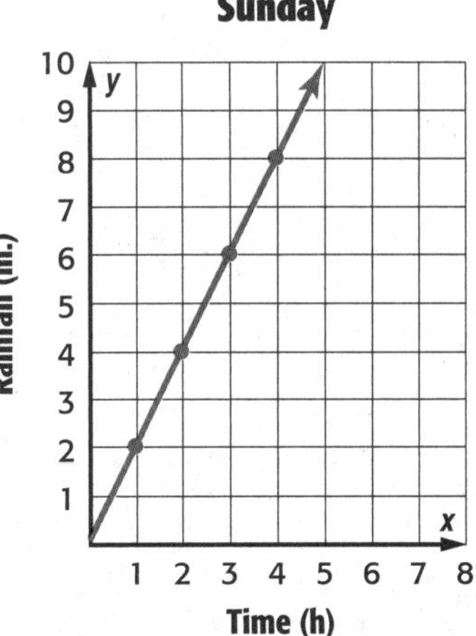

Reflect

Answering the Essential Question

Use what you learned about ratios and proportional reasoning to complete the graphic organizer.

Essential Question

HOW can you show that two objects are proportional...

... with a table?	... with a graph?	... with an equation?

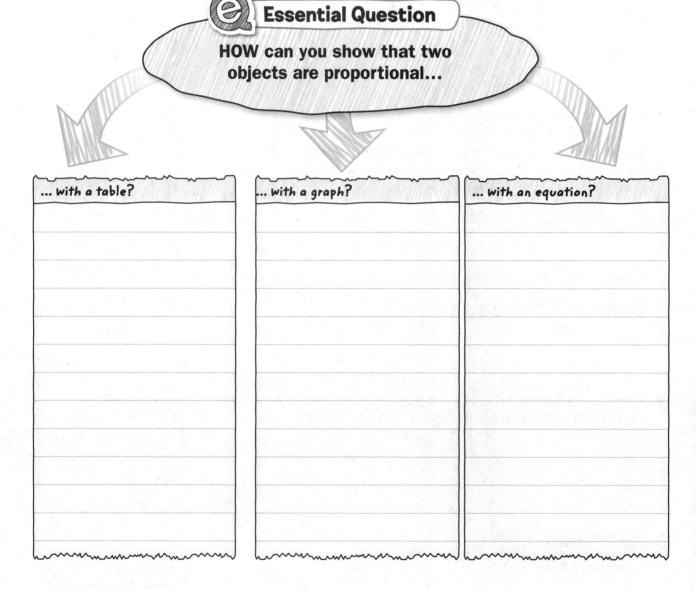

Answer the Essential Question. HOW can you show that two objects are proportional?

Chapter 2 Percents

Essential Question

HOW can percent help you understand situations involving money?

Common Core State Standards

Content Standards
7.RP.2, 7.RP.2c, 7.RP.3, 7.EE.2, 7.EE.3

Mathematical Practices
1, 2, 3, 4, 5, 6

Math in the Real World

Biking The class goal for a biking fundraiser was to make \$300 by the end of the pledge week. Halfway through the week, the students had made \$210. Fill in the graph below to show the percent of the goal achieved.

FOLDABLES® Study Organizer

1 Cut out the Foldable on page FL5 of this book.

2 Place your Foldable on page 180.

3 Use the Foldable throughout this chapter to help you learn about percents.

What Tools Do You Need?

Vocabulary

discount	percent equation	percent of increase	selling price
gratuity	percent error	percent proportion	simple interest
markdown	percent of change	principal	tip
markup	percent of decrease	sales tax	

Study Skill: Studying Math

Draw a Picture Drawing a picture can help you better understand numbers. For example, a *number map* shows how numbers are related to each other.

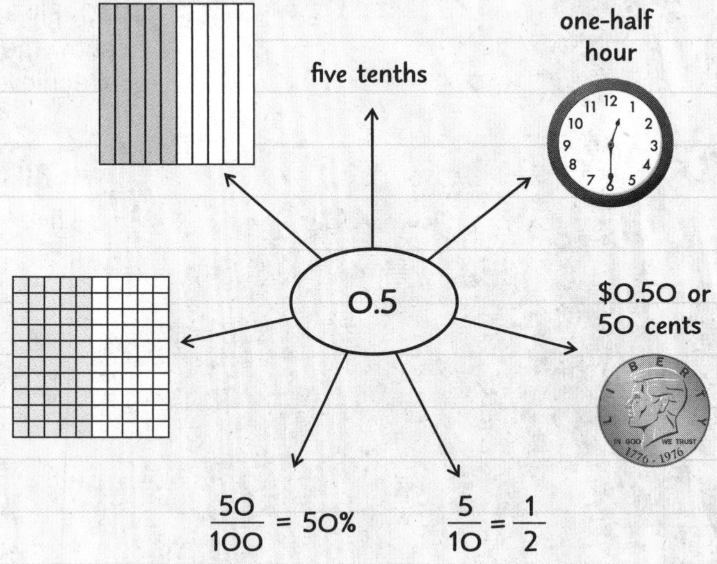

In the space below, make a number map for 0.75.

When Will You Use This?

Your Turn! **You will solve this problem in the chapter.**

Try the Quick Check below.
Or, take the Online Readiness Quiz.

Quick Review

Common Core Review 6.NS.3, 6.RP.3c

Example 1

Evaluate $240 \times 0.03 \times 5$.

$240 \times 0.03 \times 5$

$= 7.2 \times 5$ Multiply 240 by 0.03.

$= 36$ Simplify.

Example 2

Write 0.35 as a percent.

$0.35 = 35\%$ Move the decimal point two places to the right and add the percent symbol.

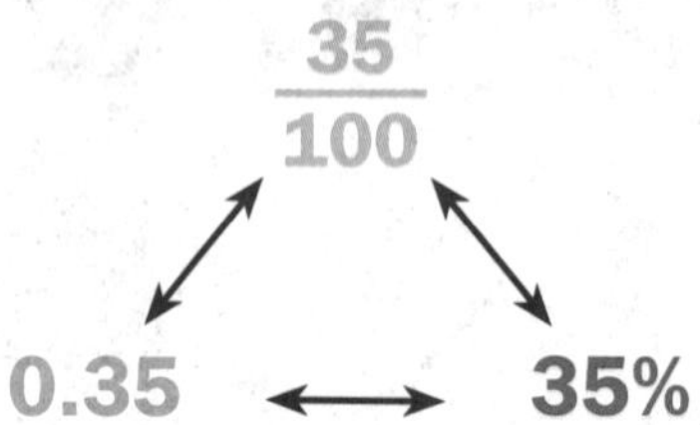

Quick Check

Multiply Decimals **Find each product.**

1. $300 \times 0.02 \times 8 =$ ______

2. $85 \times 0.25 \times 3 =$ ______

3. Suppose Nicole saves $2.50 every day. How much money will she have in 4 weeks? ______

Decimals and Percents **Write each decimal as a percent.**

4. $0.675 =$ ______

5. $0.725 =$ ______

6. $0.95 =$ ______

7. Approximately 0.92 of a watermelon is water. What percent represents this decimal? ______

Which problems did you answer correctly in the Quick Check? Shade those exercise numbers below.

① ② ③ ④ ⑤  ⑦

Inquiry Lab
Percent Diagrams

HOW are percent diagrams used to solve real-world problems?

Content Standards
7.RP.3, 7.EE.3

Mathematical Practices
1, 3, 4

Musical Instruments One fourth of the students in Mrs. Singh's music class chose a guitar as their favorite musical instrument. There are 24 students in Mrs. Singh's music class. How many students chose a guitar as their favorite musical instrument?

What do you know? ______________________________

____

____

What do you need to find? ______________________________

____

Investigation 1

Bar diagrams can be used to represent a part of a whole as a fraction and as a percent.

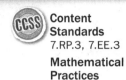

Step 1 The bar diagram represents 100% of the class. Shade the bar diagram to show that $\frac{1}{4}$ or ☐% of the class chose guitar as their favorite instrument.

				100%

|--- ☐ %---|

Step 2 There are ☐ students in Mrs. Singh's music class. Divide the number of students equally into 4 sections. Fill in the number in each section.

|-------------------- 24 students --------------------|

				100%

|--- ☐ %---|

So, ☐ students chose a guitar as their favorite musical instrument.

Investigation 2

Music There are 500 seventh-grade students at Heritage Middle School. Sixty percent of them play a musical instrument. How many seventh-grade students play a musical instrument?

Step 1 Supply the missing information for the second bar.

percent [ ] 100%

students [ ] [ ] total students

Step 2 Divide each bar into ten equal parts. Write 10% in each section of the first bar.

percent **10%** [ ] 100%

students [ ] [ ] [ ] total students

Step 3 Determine what number to write in each section of the second bar. Fill in that number.

percent **10%** [ ] 100%

students [ ] [ ] [ ] total students

Step 4 Shade 60% of the first bar and an equal amount on the second bar.

percent **10%** [ ] 100%

students [ ] [ ] [ ] total students

|----------[ ]----------|

Since [ ] % corresponds to 6 sections, count the number of students in 6 sections. There are [ ] seventh-grade students who play a musical instrument.

Work with a partner. Use bar diagrams to solve each problem.

1. The seventh-grade class at Fort Couch Middle School has a goal of selling 300 tickets to the annual student versus teacher basketball game. The eighth-grade class has a goal of selling 400 tickets.

   a. By the end of the first week, the eighth-grade students sold 30% of their goal. How many tickets has the eighth grade sold? ______

   percent [ ] 100%

   tickets [ ] [ ]

   b. The seventh grade sold 60% of their goal. How many tickets do the students still need to sell? Explain. ______

   ______ [ ] 100%

   ______ [ ] [ ]

2. CCSS **Justify Conclusions** The graph shows the results of a survey asking 500 teens about their allowances. How many teens did *not* receive between \$10 and \$20? Explain.

Weekly Allowance

10% more than \$20

15% less than \$10

75% \$10–\$20

______ [ ] [ ] %

______ [ ] [ ]

Work with a partner to complete the graphic organizer about percent and number bar diagrams. The first one is done for you.

	Percent	Rate per 100	Whole	Part
	30%	$\frac{30}{100}$	150	45
3.	40%	$\frac{40}{100}$	150	
4.	50%	$\frac{50}{100}$	150	

5. Analyze the table above. Do you see any patterns?

CCSS **Model with Mathematics** Write a real-world problem for the bar diagrams shown. Then solve your problem.

6.

10%	10%	10%	10%	10%	10%	10%	10%	10%	10%	100%
25	25	25	25	25	25	25	25	25	25	250

7.

25%	25%	25%	25%	100%
15	15	15	15	60

8. Inquiry HOW are percent diagrams used to solve real-world problems?

Lesson 1

Percent of a Number

What You'll Learn

Scan the rest of the lesson. List two headings you would use to make an outline of the lesson.

- ______________________________
____
- ______________________________
____

Essential Question

HOW can percent help you understand situations involving money?

Common Core State Standards

Content Standards
7.RP.3, 7.EE.3

Mathematical Practices
1, 3, 4

Real-World Link

Pets Some students are collecting money for a local pet shelter. The model shows that they have raised 60% of their $2,000 goal or $1,200.

	Percent	Decimal	Fractions
$2,000	100%	1	$\frac{5}{5}$ or 1
$1,600	80%		
$1,200	60%		
$800	40%		
$400	20%		$\frac{1}{5}$
$0	0%	0	0

FIDO

1. Fill in the decimal and fractional equivalents for each of the percents shown in the model.

2. Use the model to write two multiplication sentences that are equivalent to 60% of 2,000 = 1,200.

____

____

Work Zone

Find the Percent of a Number

To find the percent of a number such as 60% of 2,000, you can use either of the following methods.

- Write the percent as a fraction and then multiply.
- Write the percent as a decimal and then multiply.

Examples

Percent as a Rate
Find a percent of a quantity as a rate per 100.
For example, 5% of a quantity means $\frac{5}{100}$ times the quantity.

1. Find 5% of 300 by writing the percent as a fraction.

Write 5% as $\frac{5}{100}$ or $\frac{1}{20}$. Then find $\frac{1}{20}$ of 300.

$\frac{1}{20}$ of 300 $= \frac{1}{20} \times 300$ — Write a multiplication expression.

$= \frac{1}{\cancel{20}_{1}} \times \frac{\cancel{300}^{15}}{1}$ — Write 300 as $\frac{300}{1}$. Divide out common factors.

$= \frac{1 \times 15}{1 \times 1}$ — Multiply numerators and denominators.

$= \frac{15}{1}$ or 15 — Simplify.

So, 5% of 300 is 15.

2. Find 25% of 180 by writing the percent as a decimal.

Write 25% as 0.25. Then multiply 0.25 and 180.

$$\begin{array}{r} 180 \\ \times\ 0.25 \\ \hline 900 \\ +\ 360 \\ \hline 45.00 \end{array}$$

0.25 ← two decimal places

45.00 ← two decimal places

So, 25% of 180 is 45.

Got It? Do these problems to find out.

Find the percent of each number.

a. 40% of 70

b. 15% of 100

c. 55% of 160

d. 75% of 280

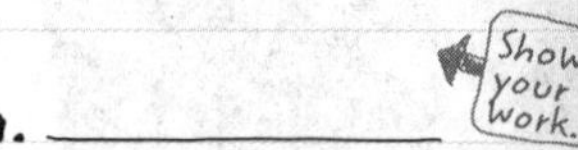

a. ____________

b. ____________

c. ____________

d. ____________

Use Percents Greater Than 100%

Percents that are greater than 100% can be written as improper fractions, mixed numbers, or decimals greater than 1.

$$150\% = \frac{150}{100} = \frac{3}{2} = 1\frac{1}{2} = 1.5$$

Examples

3. Find 120% of 75 by writing the percent as a fraction.

Write 120% as $\frac{120}{100}$ or $\frac{6}{5}$. Then find $\frac{6}{5}$ of 75.

$\frac{6}{5}$ of $75 = \frac{6}{5} \times 75$ — Write a multiplication expression.

$= \frac{6}{\cancel{5}_1} \times \frac{\cancel{75}^{15}}{1}$ — Write 75 as $\frac{75}{1}$. Divide out common factors.

$= \frac{6 \times 15}{1 \times 1}$ — Multiply numerators and denominators.

$= \frac{90}{1}$ or 90 — Simplify.

So, 120% of 75 is 90.

Alternate Method
You can solve Example 3 using a decimal, and you can solve Example 4 using a fraction.

4. Find 150% of 28 by writing the percent as a decimal.

Write 150% as 1.5. Then find 1.5 of 28.

$$\begin{array}{r} 28 \\ \times\ 1.5 \\ \hline 140 \\ +\ 28\ \ \\ \hline 42.0 \end{array}$$

× 1.5 ← one decimal place

42.0 ← one decimal place

So, 150% of 28 is 42.

Got It? Do these problems to find out.

Find each number.

e. 150% of 20

f. 160% of 35

e. ______

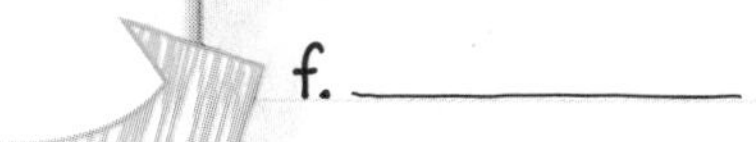

f. ______

Example

5. Refer to the graph. If 275 students took the survey, how many can be expected to have 3 televisions each in their houses?

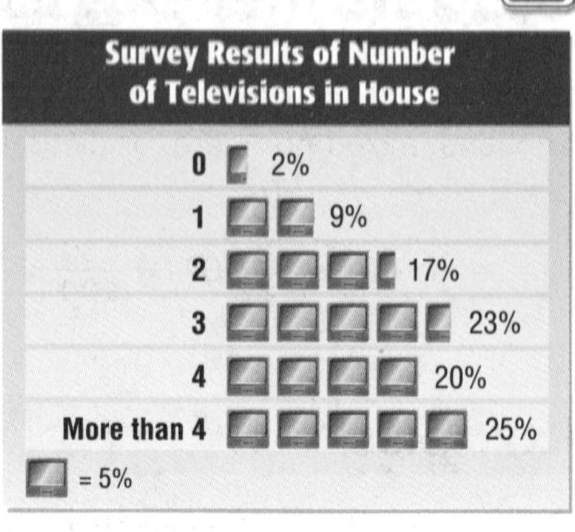

Write the percent as a decimal. Then multiply.

$$23\% \text{ of } 275 = 23\% \times 275$$
$$= 0.23 \times 275$$
$$= 63.25$$

So, about 63 students can be expected to have 3 televisions each.

Commission

Refer to Exercise g. It is common for people who work in the sales industry to earn a commission on the products they sell.

Got It? **Do this problem to find out.**

g. ______

g. Mr. Sudimack earned a 4% commission on the sale of a hot tub that cost $3,755. How much did he earn?

Guided Practice

Find each number. Round to the nearest tenth if necessary. (Examples 1–4)

1. 8% of 50 = ______

2. 95% of 40 = ______

3. 110% of 70 = ______

4. Mackenzie wants to buy a backpack that costs $50. If the tax rate is 6.5%, how much tax will she pay? (Example 5)

______

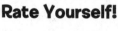

5. **Building on the Essential Question** Give an example of a real-world situation in which you would find the percent of a number. ______

______

______

Rate Yourself!

Are you ready to move on? Shade the section that applies.

For more help, go online to access a Personal Tutor.

Name ______ My Homework ______

Independent Practice

Go online for Step-by-Step Solutions

Find each number. Round to the nearest tenth if necessary. (Examples 1–4)

Show your work.

1. 65% of 186 = ______

2. 45% of $432 = ______

3. 23% of $640 = ______

4. 130% of 20 = ______

5. 175% of 10 = ______

6. 150% of 128 = ______

7. 32% of 4 = ______

8. 5.4% of 65 = ______

9. 23.5% of 128 = ______

10. Suppose there are 20 questions on a multiple-choice test. If 25% of the answers are choice B, how many of the answers are *not* choice B?

(Example 5) ______

11. CCSS **Model with Mathematics** Refer to the graphic novel frame below. Find the dollar amount of the group discount each student would receive at each park.

______

12. In addition to her salary, Ms. Lopez earns a 3% *commission,* or fee paid based on a percent of her sales, on every vacation package that she sells. One day, she sold the three vacation packages shown. Fill in the table for each packages' commission. What was her total commission?

Package	Sale Price	Commission
#1	\$2,375	
#2	\$3,950	
#3	\$1,725	

Copy and Solve **For Exercises 13–21, find each number. Round to the nearest hundredth. Show your work on a separate piece of paper.**

13. $\frac{4}{5}$% of 500

14. $5\frac{1}{2}$% of 60

15. $20\frac{1}{4}$% of 3

16. 1,000% of 99

17. 520% of 100

18. 0.15% of 250

19. 200% of 79

20. 0.3% of 80

21. 0.28% of 50

H.O.T. Problems Higher Order Thinking

22. CCSS **Persevere with Problems** Suppose you add 10% of a number to the number, and then you subtract 10% of the total. Is the result *greater than, less than,* or *equal to* the original number? Explain your reasoning.

23. CCSS **Reason Inductively** When is it easiest to find the percent of a number using a fraction? using a decimal?

Standardized Test Practice

24. Marcos earned \$300 mowing lawns this month. Of his earnings, he plans to spend 18% repairing lawn equipment, put 20% in his savings, and use 35% for camp fees. He will spend the rest. How much will Marcos have left to spend?

Ⓐ \$27.00
Ⓑ \$55.00
Ⓒ \$81.00
Ⓓ \$100.00

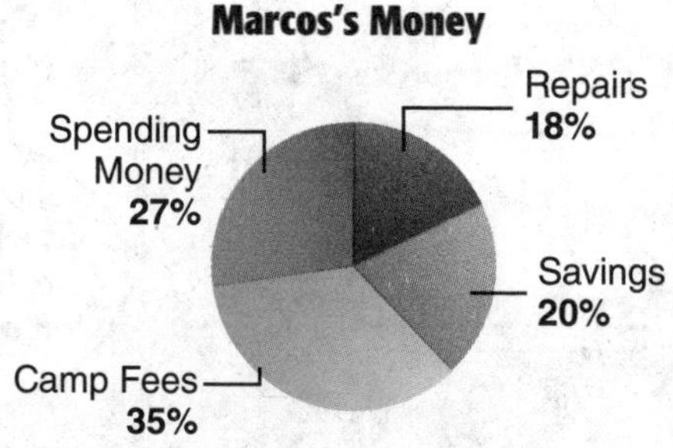

Name ______________________ My Homework

Extra Practice

Find each number. Round to the nearest tenth if necessary.

25. 54% of 85 = 45.9

$0.54 \times 85 = 45.9$

26. 12% of \$230 = \$27.60

$\frac{12}{100} \times 230 = \frac{3}{25} \times 230$

$= \frac{3}{5} \times 46$

$= \frac{138}{5}$ or 27.6

27. 98% of 15 = ______

28. 250% of 25 = ______

29. 108% of \$50 = ______

30. 75.2% of 130 = ______

31. 0.5% of 60 = ______

32. 2.4% of 20 = ______

33. 7.5% of 30 = ______

34. In a recent year, 17.7% of households watched the finals of a popular reality series. There are 110.2 million households in the United States. How many households watched the finals?

______

35. A family pays \$19 each month for Internet access. Next month, the cost will increase by 5% because of an equipment fee. After this increase, what will be the cost for the Internet access?

______

36. CCSS **Persevere with Problems** 250 people were asked to name their favorite fruit.

a. Of those surveyed, how many people prefer peaches?

______

b. Which type of fruit did more than 100 people prefer?

______

Favorite Fruit	
Berries	44%
Peaches	32%
Cherries	24%

Standardized Test Practice

37. **Short Response** Tanner has 200 baseball cards. Of those, 42% are in mint condition. How many of the cards are in mint condition? ____________

____

38. The table shows the results of a survey of 200 movie rental customers.

Favorite Type of Movie	Percent of Customers
Comedy	15
Mystery	10
Horror	46
Science Fiction	29

How many customers prefer horror movies?

Ⓐ 20
Ⓑ 30
Ⓒ 46
Ⓓ 92

39. The graph shows the Ramirez family budget. Their budget is based on a monthly income of $3,000.

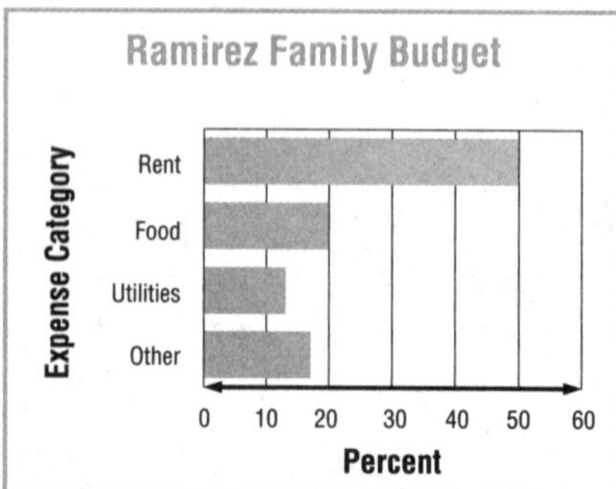

Which of the following is *true*?

Ⓕ The family budgeted $1,000 for rent.
Ⓖ The family budgeted $600 for food.
Ⓗ The family budgeted $100 more for utilities than for other expenses.
Ⓘ The family budgeted $900 more for food than for rent.

Common Core Review

Multiply. 6.NS.3

40. 1.7 × 54 = ____________

41. 1.5 × 3.65 = ____________

42. 49.6 × 2.7 = ____________

43. Trent spent 50 minutes at the neighbor's house. He spent $\frac{2}{5}$ of the time swimming. How many minutes did Trent spend swimming? 5.NF.4 ____________

50 min

____

44. There are 240 seventh-graders at Yorktown Middle School. Two-thirds of the students participate in after-school activities. How many students participate in after-school activities? 5.NF.4

____

Lesson 2

Percent and Estimation

What You'll Learn

Scan the rest of the lesson. List two headings you would use to make an outline of the lesson.

•

•

Essential Question

HOW can percent help you understand situations involving money?

Common Core State Standards

Content Standards
7.RP.3, 7.EE.3

Mathematical Practices
1, 3, 4, 5

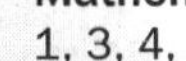

Real-World Link

Music Suppose 200 people are surveyed to find out how they learned to play an instrument. The results are shown in the table below.

Type of Teaching	Actual Percent	Estimated Percent	Fraction
Private Lessons	42%	40%	$\frac{2}{5}$
Lessons at School	32%		
Self-Taught	26%		

1. Estimate each percent. Choose an estimate that can be represented by a fraction that is easy to use. Then, write each estimated percent as a fraction in simplest form.

2. About how many people took lessons at school?

3. Sarah estimates the percent of people who taught themselves to play an instrument as 25%, and then she found $\frac{1}{4}$ of 200. Would her answer be less than or greater than the actual number of people who were self taught? Explain.

Work Zone

Estimate the Percent of a Number

Sometimes an exact answer is not needed when using percents. One way to estimate the percent of a number is to use a fraction.

Another method for estimating the percent of a number is first to find 10% of the number and then multiply.

$$70\% = 7 \cdot 10\%$$

So, 70% equals 7 times 10% of a number.

Examples

1. Jodi has paid 62% of the $500 she owes for her loan. Estimate 62% of 500.

$62\% \text{ of } 500 \approx 60\% \text{ of } 500$ — $62\% \approx 60\%$

$\approx \frac{3}{5} \cdot 500$ — $60\% = \frac{6}{10}$ or $\frac{3}{5}$

≈ 300 — Multiply.

So, 62% of 500 is about 300.

What are two ways to estimate 22% of 130? Explain below.

2. Marita and four of her friends ordered a pizza that cost $14.72. She is responsible for 20% of the bill. About how much money will she need to pay?

Step 1 Find 10% of $15.00.

$10\% \text{ of } \$15.00 = 0.1 \cdot \$15.00$

$= \$1.50$ — To multiply by 10%, move the decimal point one place to the left.

Step 2 Multiply.

20% of $15.00 is 2 times 10% of $15.00.

$2 \cdot \$1.50 = \$3.00$

So, Marita should pay about $3.00.

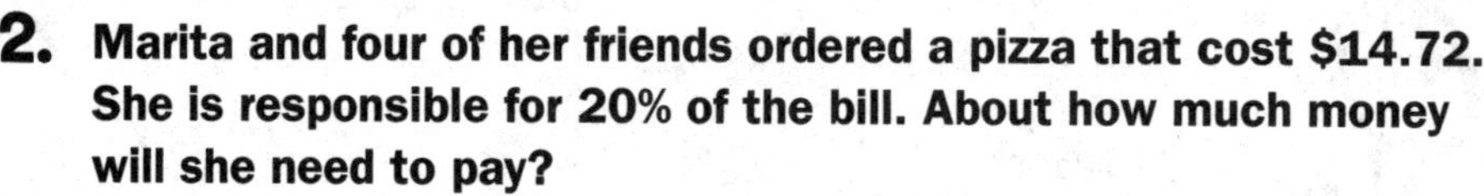

Got It? Do these problems to find out.

a. Estimate 42% of 120.

b. Dante plans to put 80% of his paycheck into a savings account and spend the other 20%. His paycheck this week was $295. About how much money will he put into his savings account?

a. ________

b. ________

Percents Greater Than 100 or Less Than 1

You can also estimate percents of numbers when the percent is greater than 100 or less than 1.

Check for Reasonableness
When the percent is greater than 100, the estimate will always be greater than the number.

Example

3. Estimate 122% of 50.

122% is about 120%.

$120\% \text{ of } 50 = 100\% \text{ of } 50 + 20\% \text{ of } 50$ — 120% = 100% + 20%

$= (1 \cdot 50) + \left(\frac{1}{5} \cdot 50\right)$ — 100% = 1 and $20\% = \frac{1}{5}$

$= 50 + 10 \text{ or } 60$ — Simplify.

So, 122% of 50 is about 60.

Got It? Do these problems to find out.

c. 174% of 200 **d.** 298% of 45 **e.** 347% of 80

c. ____________

d. ____________

e. ____________

Real World Example

4. There are 789 seventh grade students at Washington Middle School. About $\frac{1}{4}$% of the seventh grade students have traveled overseas. What is the approximate number of seventh grade students that have traveled overseas? Explain.

$\frac{1}{4}$% is one fourth of 1%. 789 is about 800.

$1\% \text{ of } 800 = 0.01 \cdot 800$ — Write 1% as 0.01.

$= 8$ — To multiply by 1%, move the decimal point two places to the left.

One fourth of 8 is $\frac{1}{4} \cdot 8$ or 2.

So, about 2 seventh grade students have traveled overseas.

Got It? Do this problem to find out.

f. A county receives $\frac{3}{4}$% of a state sales tax. About how much money would the county receive from the sale of a computer that costs $1,020?

f. ____________

Example

5. Last year, 639 students attended a summer camp. Of those who attended this year, 0.5% also attended summer camp last year. About how many students attended the summer camp two years in a row?

0.5% is half of 1%.

$1\% \text{ of } 639 = 0.01 \cdot 639$

≈ 6

So, 0.5% of 639 is about $\frac{1}{2}$ of 6 or 3.

About 3 students attended summer camp 2 years in a row.

Guided Practice

Estimate. (Examples 1–4)

1. 52% of 10 ≈ ______

2. 79% of 489 ≈ ______

3. 151% of 70 ≈ ______

4. $\frac{1}{2}$% of 82 ≈ ______

5. Of the 78 teenagers at a youth camp, 63% have birthdays in the spring. About how many teenagers have birthdays in the spring? (Example 2)

______

6. About 0.8% of the land in Maine is federally owned. If Maine has 19,847,680 acres, about how many acres are federally owned? (Example 5) ______

7. **Building on the Essential Question** How can you estimate the percent of a number? ______

Rate Yourself!

How confident are you about estimating percents? Shade the ring on the target.

For more help, go online to access a Personal Tutor.

Name ______________________ My Homework ______________________

Independent Practice

Go online for Step-by-Step Solutions 

Estimate. (Examples 1–4)

1. 47% of 70 ≈ ______

2. 39% of 120 ≈ ______

3 21% of 90 ≈ ______

4. 65% of 152 ≈ ______

5. 72% of 238 ≈ ______

6. 132% of 54 ≈ ______

7. 224% of 320 ≈ ______

8. $\frac{3}{4}$% of 168 ≈ ______

9. 0.4% of 510 ≈ ______

10. **Financial Literacy** Carlie spent $42 at the salon. Her mother loaned her the money. Carlie will pay her mother 15% of $42 each week until the loan is repaid. About how much will Carlie pay each week? (Example 2)

______

11 The United States has 12,383 miles of coastline. If 0.8% of the coastline is located in Georgia, about how many miles of coastline are in Georgia? (Example 5)

______

12. CCSS **Persevere with Problems** Use the graph shown.

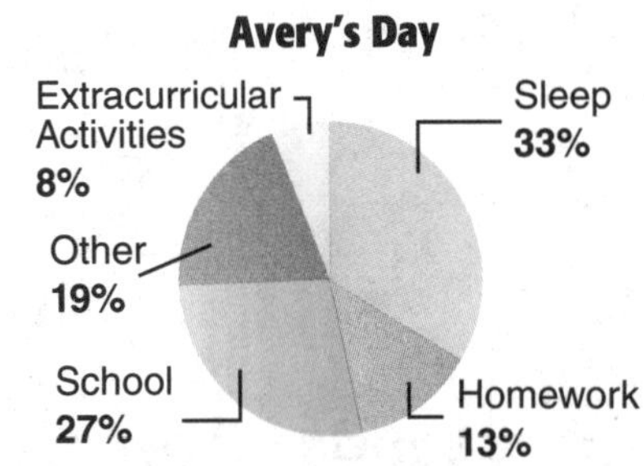

a. About how many more hours does Avery spend sleeping than doing the activities in the "other" category? Justify your answer.

______

b. What is the approximate number of minutes Avery spends each day on extracurricular activities?

______

Estimate.

13. 67% of 8.7 ≈ ______

14. 54% of 76.8 ≈ ______

15. 10.5% of 238 ≈ ______

16. The average white rhinoceros gives birth to a single calf that weighs about 3.8% as much as its mother. If the mother rhinoceros weighs 3.75 tons, about how many pounds does its calf weigh? ______

17. The students at Monroe Junior High sponsored a canned food drive. The seventh-grade class collected 129% of its canned food goal.
   a. About how many canned foods did the seventh-graders collect if their goal was 200 cans? ______
   b. About how many canned foods did the seventh-graders collect if their goal was 595 cans? ______

H.O.T. Problems Higher Order Thinking

18. CCSS **Persevere with Problems** Explain how you could find $\frac{3}{8}$% of $800.

19. CCSS **Use MathTools** Is an estimate for the percent of a number *always*, *sometimes*, or *never* greater than the actual percent of the number? Give an example or a counterexample to support your answer.

Standardized Test Practice

20. Mallory is buying bedroom furniture for $1,789.43. The dresser is 39.7% of the total cost. Which is the best estimate for the cost of the dresser?

Ⓐ $540　Ⓑ $630　Ⓒ $720　Ⓓ $810

Name ______ My Homework ______

Extra Practice

Estimate.

21. 76% of 180 ≈ 135

$\frac{3}{4} \cdot 180 = 135$ or

Homework Help

$0.1 \cdot 180 = 18$

$7.5 \cdot 18 = 135$

22. 57% of 29 ≈ 18

$\frac{3}{5} \cdot 30 = 18$ or

$0.1 \cdot 30 = 3$

$6 \cdot 3 = 18$

23. 92% of 104 ≈ ______

24. $\frac{1}{2}$% of 412 ≈ ______

25. 0.9% of 74 ≈ ______

26. 32% of 89.9 ≈ ______

27. You use 43 muscles to frown. When you smile, you use 32% of these same muscles. About how many muscles do you use when you smile?

______

28. CCSS **Justify Conclusions** The coastline of the Atlantic Coast is 2,069 miles long. Approximately $\frac{6}{10}$% of the coastline lies in New Hampshire. About how many miles of the coastline lie in New Hampshire? Explain how you estimated.

______

______

29. The table shows the number of passes attempted and the percent completed by the top quarterbacks in the NFL for a recent season.

NFL Quarterbacks

Player	Passes Attempted	Percent Completed
T. Brady	578	69
P. Manning	515	65
T. Romo	520	64
D. Garrard	325	64

a. Estimate the number of passes that Tom Brady completed.

______

b. Is your estimate greater or less than the actual number of passes he completed? Explain. ______

______

c. Without calculating, determine whether Tony Romo or David Garrard completed more passes. Justify your reasoning.

______

______

Standardized Test Practice

30. The graph shows the results of a survey of 510 students.

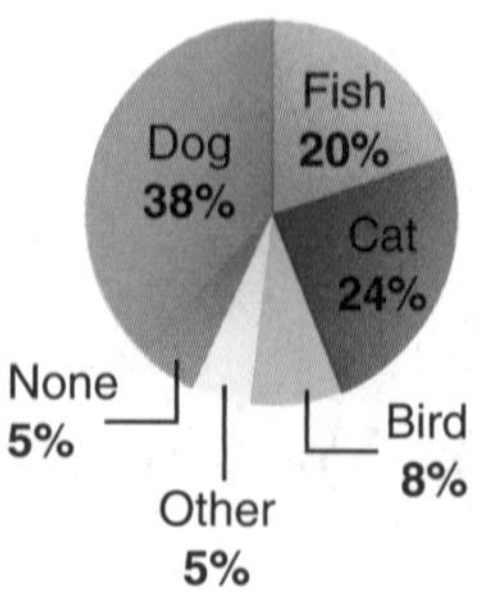

Which is the best estimate for the percent of students who prefer cats?

Ⓐ 75
Ⓑ 125
Ⓒ 225
Ⓓ 450

31. Abbey asked 50 students to vote for the school issue that was most important to them. The results are shown below.

Issue	Votes
Library use	10%
Time to change classes	12%
Use of electronics	18%
Lunch room rules	20%
Dress code	40%

About how many students chose "Time to change classes" as the most important issue?

Ⓕ 3
Ⓖ 5
Ⓗ 9
Ⓘ 12

Common Core Review

Solve each equation. Show your work. 6.EE.7

32. $5n = 120$

33. $1{,}200 = 4a$

34. $6x = 39$

35. Marquita created the design at the right. She created the design from 8 equal-size rectangles. Write a fraction in simplest form that represents the yellow portion of the design. 5.NF.1

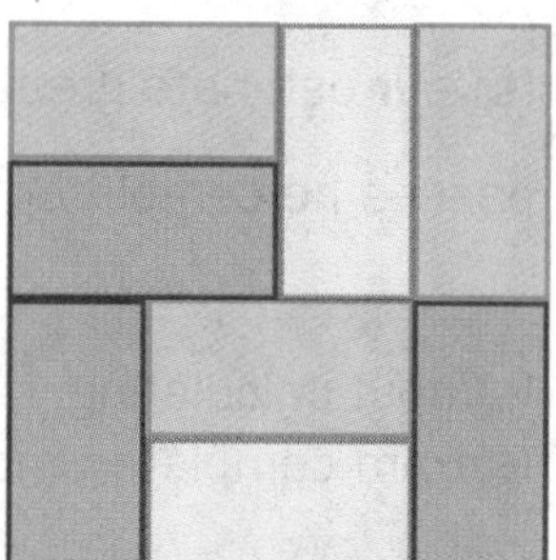

36. Write three fractions equivalent to $\frac{3}{5}$. 5.NF.1

Inquiry Lab

Find Percents

HOW is percent used to solve real-world problems?

Content Standards
7.RP.3, 7.EE.3

Mathematical Practices
1, 3, 4

Drama The eighth grade had 300 tickets to sell to the school play and the seventh grade had 250 tickets to sell. One hour before the show, the eighth grade had sold 225 tickets and the seventh grade had sold 200 tickets. Complete the investigation below to find which grade sold the greater percent of tickets.

Investigation

Step 1 The bar diagrams below show 100% for each grade. Label the total tickets to be sold above each bar. Divide each bar into 10 equal sections. So, each section will represent 10%.

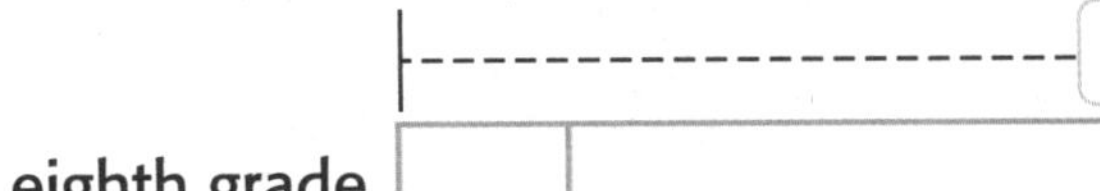

☐ tickets

eighth grade ☐ 100%

☐ tickets

seventh grade ☐ 100%

Step 2 Find the number that belongs in each section for both of the bars. Then write that number in the sections.

Eighth grade:

300 ÷ 10 = ☐

Seventh grade:

250 ÷ 10 = ☐

Step 3 Find the number of sections to shade for each bar. Then shade the sections.

Eighth grade:

225 ÷ 30 = ☐

Seventh grade:

200 ÷ 25 = ☐

The eighth grade sold ☐ % of their tickets. The seventh grade sold ☐ % of their tickets.

The ________ grade sold the greater percent of their tickets.

Collaborate

Work with a partner. Show your work using bar diagrams.

1. CCSS **Model with Mathematics** Vanlue Middle School has 600 students and Memorial Middle School has 450 students. Vanlue has 270 girls and Memorial has 225 girls. Which school has the greater percent of girls? Explain. ______

______

______

______

Vanlue: [ ] students — 100%

Memorial: [ ] students — 100%

Analyze

Work with a partner to answer the following question.

2. CCSS **Model with Mathematics** Seventy-five students were in the audience for a 3-D screening of a movie. Fifty students were in the audience for a 2-D screening of the same movie. If the students aren't the only audience members, describe a situation in which the percent of students who went to the 2-D screening is greater than the percent of students who went to the 3-D screening. ______

______

______

______

Reflect

3. Inquiry HOW is percent used to solve real-world problems? ______

______

______

______

Lesson 3

The Percent Proposition

What You'll Learn

Scan the lesson. Predict two things you will learn about the percent proportion.

-
-

Essential Question

HOW can percent help you understand situations involving money?

Vocabulary

percent proportion

Common Core State Standards

Content Standards
7.RP.3

Mathematical Practices
1, 3, 4

Real-World Link

Monster Trucks The tires on a monster truck weigh approximately 2 tons. The entire truck weighs about 6 tons.

1. Write the ratio of tire weight to total weight as a fraction.

$\frac{\text{Part}}{\text{Whole}} = \underline{\qquad\qquad} = \frac{\square}{\square}$

2. Represent the fraction above by shading in the model.

3. Write the fraction as a decimal to the nearest hundredth.

4. About what percent of the monster truck's weight is the tires?

Work Zone

Key Concept: Use the Percent Proportion

Type	Example	Proportion
Find the Percent	What percent of 5 is 4?	$\frac{4}{5} = \frac{n}{100}$
Find the Part	What number is 80% of 5?	$\frac{p}{5} = \frac{80}{100}$
Find the Whole	4 is 80% of what number?	$\frac{4}{w} = \frac{80}{100}$

In a **percent proportion**, one ratio or fraction compares part of a quantity to the whole quantity. The other ratio is the equivalent percent written as a fraction with a denominator of 100.

4 out of 5 is 80%

$$\frac{\text{part}}{\text{whole}} \rightarrow \frac{4}{5} = \frac{80}{100} \Big\} \text{percent}$$

Example

The Percent Proportion

The word *of* is usually followed by the whole.

1. What percent of $15 is $9?

Words What percent of $15 is $9?

Variable Let n represent the percent.

Proportion $\frac{\text{part}}{\text{whole}} \rightarrow \frac{9}{15} = \frac{n}{100} \Big\}$ percent

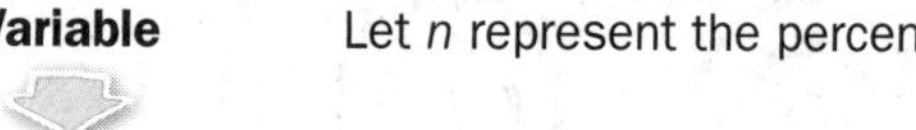

$\frac{9}{15} = \frac{n}{100}$ Write the proportion.

$9 \cdot 100 = 15 \cdot n$ Find the cross products.

$900 = 15n$ Simplify.

$\frac{900}{15} = \frac{15n}{15}$ Divide each side by 15.

$60 = n$

So, $9 is 60% of $15.

a. ______________

b. ______________

Got It? Do these problems to find out.

a. What percent of 25 is 20?

b. $12.75 is what percent of $50?

Example

2. What number is 40% of 120?

Words	What number is 40% of 120?
Variable	Let p represent the part.
Proportion	part → whole → $\frac{p}{120} = \frac{40}{100}$ } percent

$\frac{p}{120} = \frac{40}{100}$ Write the proportion.

$p \cdot 100 = 120 \cdot 40$ Find the cross products.

$100p = 4{,}800$ Simplify.

$\frac{100p}{100} = \frac{4{,}800}{100}$ Divide each side by 100.

$p = 48$ So, 48 is 40% of 120.

Got It? **Do these problems to find out.**

c. What number is 5% of 60? **d.** 12% of 85 is what number?

Show your work.

c. ______

d. ______

Example

3. 18 is 25% of what number?

Words	18 is 25% of what number?
Variable	Let w represent the whole.
Proportion	part → whole → $\frac{18}{w} = \frac{25}{100}$ } percent

$\frac{18}{w} = \frac{25}{100}$ Write the proportion.

$18 \cdot 100 = w \cdot 25$ Find the cross products.

$1{,}800 = 25w$ Simplify.

$\frac{1{,}800}{25} = \frac{25w}{25}$ Divide each side by 25.

$72 = w$ So, 18 is 25% of 72.

STOP and Reflect

In the proportion $\frac{3}{20} = \frac{15}{100}$, identify the part, whole, and percent.

part = ______

whole = ______

percent = ______

Got It? **Do these problems to find out.**

e. 40% of what number is 26? **f.** 84 is 75% of what number?

e. ______

f. ______

Example

4. The average adult male Western Lowland gorilla eats about 33.5 pounds of fruit each day. How much food does the average adult male gorilla eat each day?

Western Lowland Gorilla's Diet	
Food	**Percent**
Fruit	67%
Seeds, leaves, stems, and pith	17%
Insects/ insect larvae	16%

You know that 33.5 pounds is the part. You need to find the whole.

$\frac{33.5}{w} = \frac{67}{100}$ Write the proportion.

$33.5 \cdot 100 = w \cdot 67$ Find the cross products.

$3{,}350 = 67w$ Simplify.

$\frac{3{,}350}{67} = \frac{67w}{67}$ Divide each side by 67.

$50 = w$

The average adult male gorilla eats 50 pounds of food each day.

Guided Practice

Find each number. Round to the nearest tenth if necessary. (Examples 1–3)

1. What percent of \$90 is \$9? ______

2. What number is 2% of 35? ______

3. 62 is 90.5% of what number? ______

4. Brand A cereal contains 10 cups of cereal. How many more cups of cereal are in Brand B cereal? (Example 4)

5. **Building on the Essential Question** How can you use the percent proportion to solve real-world problems?

Rate Yourself!

How confident are you about using the percent proportion? Shade the ring on the target.

For more help, go online to access a Personal Tutor.

FOLDABLES *Time to update your Foldable!*

Independent Practice

Go online for Step-by-Step Solutions

Find each number. Round to the nearest tenth if necessary. (Examples 1–3)

1. What percent of 60 is 15? __________

2. What number is 15% of 60? __________

3. 9 is 12% of which number? __________

4. 12% of 72 is what number? __________

5. What percent of 50 is 18? __________

6. 12 is 90% of what number? __________

7. A pair of sneakers is on sale as shown. This is 75% of the original price. What was the original price of the shoes? (Example 4) __________

8. Of the 60 books on a bookshelf, 24 are nonfiction. What percent of the books are nonfiction? (Example 4) __________

Find each number. Round to the nearest hundredth if necessary.

9. 40 is 50% of what number? __________

10. 12.5% of what number is 24? __________

11. What percent of 300 is 0.6? __________

12. What number is 0.5% of 8? __________

Find each number. Round to the nearest hundredth if necessary.

13. **STEM** Use the table shown.

Planet	Radius (km)
Mercury	2,440
Mars	3,397
Jupiter	71,492

a. Mercury's radius is what percent of Jupiter's radius?

b. If the radius of Mars is about 13.7% of Neptune's radius, what is the radius of Neptune?

c. Earth's radius is about 261.4% of Mercury's radius. What is the radius of Earth?

H.O.T. Problems Higher Order Thinking

14. **CCSS Reason Inductively** Seventy percent of the 100 students in a middle school cafeteria bought their lunch. Some of the students that bought their lunch leave the cafeteria to attend an assembly. Now only 60% of the remaining students bought their lunch. How many students are remaining in the cafeteria? Explain.

15. **CCSS Persevere with Problems** Without calculating, arrange the following from greatest to least value. Justify your reasoning.

20% of 100, 20% of 500, 5% of 100

Standardized Test Practice

16. One hundred ninety-two students were surveyed about their favorite kind of TV programs. The results are shown in the table. Which kind of program did 25% of the students report as their favorite?

Ⓐ Music
Ⓑ Reality
Ⓒ Comedy
Ⓓ Sports

Favorite TV Programs	
Kind	**Number**
Music	48
Reality	44
Comedy	41
Sports	36
Drama	23

Name ________________ My Homework ________________

Extra Practice

Find each number. Round to the nearest tenth if necessary.

17. What number is 25% of 180? 45

$\frac{n}{180} = \frac{25}{100}$

$\frac{n}{180} = \frac{1}{4}$

$4n = 180$

$n = 45$

18. \$3 is what percent of \$40? 7.5%

$\frac{3}{40} = \frac{p}{100}$

$40p = 300$

$p = 7.5\%$

19. 9 is 45% of what number? ________

20. 75 is 20% of what number? ________

21. What percent of 60 is 12? ________

22. What number is 5% of 46? ________

23. CCSS **Justify Conclusions** Roman has 2 red pencils in his backpack. If this is 25% of the total number of pencils, how many pencils are in his backpack? Explain.

____

24. CCSS **Justify Conclusions** Eileen and Michelle scored 48% of their team's points. If their team had a total of 50 points, how many points did they score? Explain.

____

Find each number. Round to the nearest hundredth if necessary.

25. What percent of 25 is 30? ________

26. What number is 8.2% of 50? ________

Standardized Test Practice

27. Of the 273 students in a school, 95 volunteered to work the book sale. About what percent of the students volunteered?

Ⓐ 35%

Ⓑ 65%

Ⓒ 70%

Ⓓ 75%

28. Brian ate 3 granola bars. This is 25% of the total number of granola bars in a box. How many granola bars are in a box?

Ⓕ 10

Ⓖ 11

Ⓗ 12

Ⓘ 13

29. **Short Response** The types of flowers shown in the table make up an arrangement. What percent of the flowers in the arrangement are roses?

Flower Arrangement	
Lilies	4
Roses	15
Snapdragons	6

Common Core Review

Multiply. 5.NF.4

30. $\frac{1}{2} \times \frac{2}{3} =$ ______

31. $\frac{3}{5} \times \frac{1}{4} =$ ______

32. $\frac{2}{7} \times \frac{1}{6} =$ ______

Divide. 6.NS.1

33. $\frac{2}{5} \div \frac{3}{4} =$ ______

34. $\frac{1}{3} \div \frac{5}{6} =$ ______

35. $\frac{1}{5} \div \frac{5}{7} =$ ______

36. A store had a sale as shown at the right. A pair of shoes cost $80. What is $\frac{3}{4}$ of $80? 5.NF.4

Lesson 4

The Percent Equation

What You'll Learn

Scan the lesson. Predict two things you will learn about using the percent equation.

- ______________________________
- ______________________________

Essential Question

HOW can percent help you understand situations involving money?

Vocabulary

percent equation

Common Core State Standards

Content Standards
7.RP.2, 7.RP.2c, 7.RP.3, 7.EE.3

Mathematical Practices
1, 2, 3, 4

Vocabulary Start-Up

You have used a percent proportion to find the missing part (p), percent (n), or whole (w). You can also use a **percent equation**. The percent equation is part = percent • whole.

Label the diagram that shows the relationship between the percent proportion and the percent equation with the terms *part, whole,* and *percent.* Use each term once.

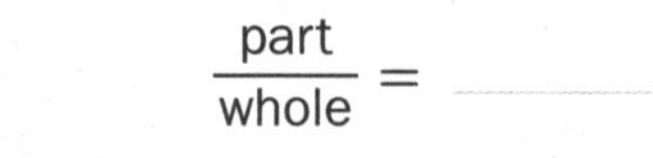

$\frac{\text{part}}{\text{whole}} =$ ________ Write the percent proportion.

$\frac{\text{part}}{\text{whole}} \cdot \text{whole} = \text{percent} \cdot$ ________ Multiply each side by the whole.

________ $= \text{percent} \cdot \text{whole}$ Divide out common factors to obtain the percent equation.

Real-World Link

A survey found that 16% of all seventh-graders at Lincoln Middle school think that tarantulas are the scariest creatures. There are 150 seventh-graders at the school. How would you write a percent equation to find how many seventh-graders said that tarantulas are the scariest creatures?

$\square = 0.16 \cdot \square$

Key Concept: Use the Percent Equation

Type	Example	Equation
Find the Percent	3 is what percent of 6?	$3 = n \cdot 6$
Find the Part	What number is 50% of 6?	$p = 0.5 \cdot 6$
Find the Whole	3 is 50% of what number?	$3 = 0.5 \cdot w$

Work Zone

You can use the percent equation to solve problems that involve percent.

3 is 50% of 6

$$\underbrace{3}_{\text{part}} = \underbrace{0.5}_{\text{percent}} \times \underbrace{6}_{\text{whole}}$$

Note that the percent is written as a decimal.

Example

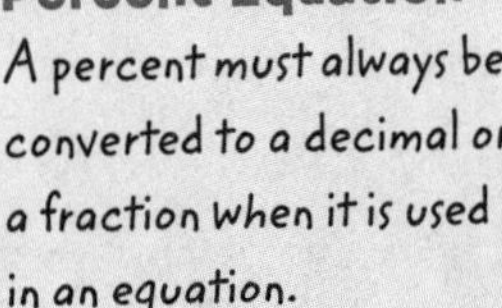

Percent Equation
A percent must always be converted to a decimal or a fraction when it is used in an equation.

1. **What number is 12% of 150?**

Do you need to find the percent, part, or whole? ______

Estimate $0.10 \cdot 150 = 15$

part = percent • whole

$p = 0.12 \cdot 150$ Write the percent equation. 12% = 0.12

$p = 18$ Multiply.

So, 18 is 12% of 150.

18 is close to the estimate of 15. So, the answer is reasonable. You can also check your answer using the percent proportion.

Check $\frac{18}{150} \stackrel{?}{=} \frac{12}{100}$

$18 \cdot 100 \stackrel{?}{=} 150 \cdot 12$

$1{,}800 = 1{,}800$ ✓

Got It? Do these problems to find out.

Write an equation for each problem. Then solve.

a. What is 6% of 200?

b. Find 72% of 50.

c. What is 14% of 150?

d. Find 50% of 70.

a. ______

b. ______

c. ______

d. ______

Example

2. 21 is what percent of 40?

Do you need to find the percent, part, or whole? ________

Estimate $\frac{21}{40} \approx \frac{1}{2}$ or 50%

part = percent · whole

$21 = n \cdot 40$ Write the percent equation.

$\frac{21}{40} = \frac{40n}{40}$ Divide each side by 40.

$0.525 = n$ Simplify.

So, 21 is 52.5% of 40.

Check 52.5% ≈ 50% ✓

Percent
Remember to write the decimal as a percent in your final answer.

Got It? Do these problems to find out.

Write an equation for each problem. Then solve. Round to the nearest tenth if necessary.

e. What percent of 40 is 9?

f. 27 is what percent of 150?

e. ________

f. ________

Example

3. 13 is 26% of what number?

Do you need to find the percent, part, or whole? ________

Estimate $\frac{1}{4}$ of $48 = 12$

part = percent · whole

$13 = 0.26 \cdot w$ Write the percent equation. 26% = 0.26

$\frac{13}{0.26} = \frac{0.26w}{0.26}$ Divide each side by 0.26.

$50 = w$ Simplify.

So, 13 is 26% of 50.

Check 50 ≈ 48. ✓

Got It? Do these problems to find out.

Write an equation for each problem. Then solve. Round to the nearest tenth if necessary.

g. 39 is 84% of what number?

h. 26% of what number is 45?

g. ________

h. ________

Example

4. A survey found that 25% of people aged 18–24 gave up their home phone and only use a cell phone. If 3,264 people only use a cell phone, how many people were surveyed?

Words	3,264 people is 25% of what number of people?
Variable	Let w represent the number of people.
Equation	$3{,}264 = 0.25 \cdot w$

$3{,}264 = 0.25 \cdot w$ — Write the percent equation. 25% = 0.25

$\frac{3{,}264}{0.25} = \frac{0.25w}{0.25}$ — Divide each side by 0.25. Use a calculator.

$13{,}056 = w$ — Simplify.

About 13,056 people were surveyed.

Guided Practice

Write an equation for each problem. Then solve. Round to the nearest tenth if necessary. (Examples 1–3)

1. What number is 88% of 300? 264.

2. 24 is what percent of 120? 20%

3. 3 is 12% of what number? 0.25

4. A local bakery sold 60 loaves of bread in one day. If 65% of these were sold in the afternoon, how many loaves were sold in the afternoon? (Example 4) 48

5. **Building on the Essential Question** When might it be easier to use the percent equation rather than the percent proportion? ______

Rate Yourself!

Are you ready to move on? Shade the section that applies.

YES ? NO

For more help, go online to access a Personal Tutor.

FOLDABLES Time to update your Foldable!

Independent Practice

Go online for Step-by-Step Solutions

Write an equation for each problem. Then solve. Round to the nearest tenth if necessary. (Examples 1–3)

1. 75 is what percent of 150? ________

2. 84 is 60% of what number? ________

3. What number is 65% of 98? ________

4. Find 39% of 65. ________

5. Find 24% of 25. ________

6. What number is 53% of 470? ________

7. Ruben bought 6 new books for his collection. This increased his collection by 12%. How many books did he have before his purchases? (Example 4)

____

8. A store sold 550 video games during the month of December. If this made up 12.5% of its yearly video game sales, about how many video games did the store sell all year? (Example 4)

____

9. CCSS **Persevere with Problems** About 142 million people in the United States watch online videos (Some people are in more than one category). Use the graph that shows what type of videos they watch.

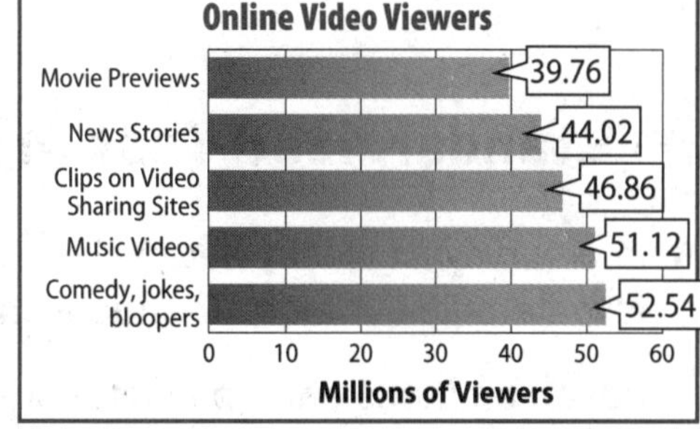

   a. About what percent of viewers watch comedy, jokes, and bloopers? ________

   b. About what percent watch news stories?

   ____

Write an equation for each problem. Then solve. Round to the nearest tenth if necessary.

10. Find 135% of 64. ______

11. What number is 0.4% of 82.1? ______

12. 450 is 75.2% of what number? ______

13. What percent of 200 is 230? ______

H.O.T. Problems Higher Order Thinking

14. CCSS **Model with Mathematics** Write a percent problem for which the percent is greater than 100 and the part is known. Use the percent equation to solve your problem to find the whole.

______

15. CCSS **Persevere with Problems** If you need to find the percent of a number, explain how you can predict whether the part will be *less than*, *greater than*, or *equal* to the number.

______

Standardized Test Practice

16. In a survey, students were asked to choose their favorite take-out food. The table shows the results.

Based on these data, predict how many out of 1,800 students would choose sandwiches.

Ⓐ 504
Ⓑ 576
Ⓒ 680
Ⓓ 720

Favorite Take-Out Food

Type of Food	Percent
Pizza	40
Sandwiches	32
Chicken	28

Name ______________________ My Homework ______________________

Extra Practice

Write an equation for each problem. Then solve. Round to the nearest tenth if necessary.

17. 9 is what percent of 45? 20%

$9 = n \times 45$

$\frac{9}{45} = \frac{45n}{45}$

$0.2 \text{ or } 20\% = n$

18. What percent of 96 is 26? 27.1%

$26 = n \times 96$

$\frac{26}{96} = \frac{96n}{96}$

$0.271 \text{ or } 27.1\% = n$

19. What percent of 392 is 98? ________

20. 30 is what percent of 64? ________

21. 33% of what number is 1.45? 4.344

22. 84 is 75% of what number? 112

23. 17 is 40% of what number? 4.25

24. 80% of what number is 64? 80

25. The length of Giselle's arm is 27 inches. The length of her lower arm is 17 inches. About what percent of Giselle's arm is her lower arm?

____

26. Approximately 0.02% of North Atlantic lobsters are born bright blue in color. Out of 5,000 North Atlantic lobsters, how many would you expect to be blue in color?

2500

27. Financial Literacy Suppose you earn $6 per hour at your part-time job. What will your new hourly rate be after a 2.5% raise? Explain.

____

____

Standardized Test Practice

28. If 60% of a number is 18, what is 90% of the number?

Ⓐ 3
Ⓑ 16
Ⓒ 27
Ⓓ 30

29. Which equation represents the statement below?

80% of what number is 64?

Ⓕ $64 = 0.8 \times w$
Ⓖ $80 = 64 \times w$
Ⓗ $64 = 80 \times w$
Ⓘ $64 = 8.0 \times w$

30. Taryn's grandmother took her family out to dinner. If the dinner was $74 and Taryn's dinner was 20% of the bill, how much was Taryn's dinner?

Ⓐ $6.80
Ⓑ $7.20
Ⓒ $9.50
Ⓓ $14.80

Common Core Review

Fill in each ◯ with <, >, or = to make a true statement. 5.NBT.3b

31. 5.56 ◯ $5\frac{5}{7}$

32. 4.027 ◯ 4.0092

33. 88% ◯ 0.9

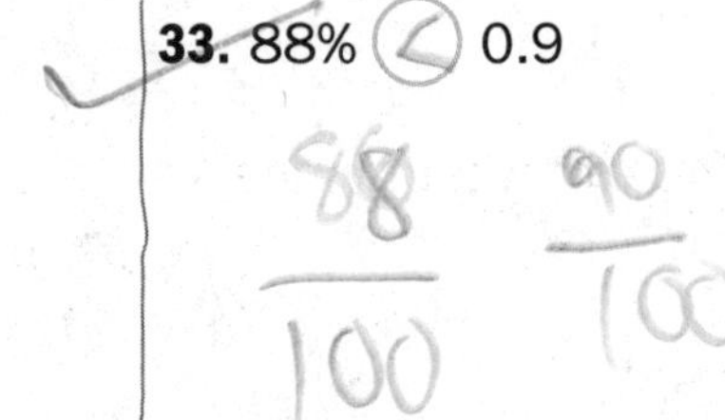

34. Last week, a jacket cost $39.50. This week, the price of the jacket decreased by $4.50. What is the price of the jacket after the decrease? How much would it cost to purchase 2 jackets this week? 5.NBT.5

Use the graph to solve.

35. What number represents 100% of the fall student athletes?

Explain. 6.SP.5a

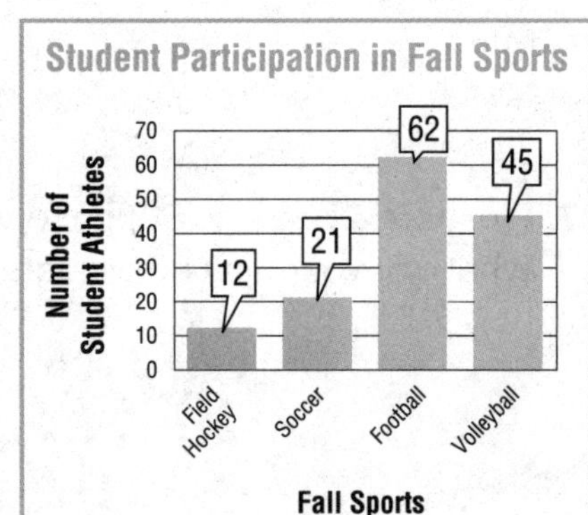

Problem-Solving Investigation

Determine Reasonable Answers

Content Standards
7.RP.3, 7.EE.3
Mathematical Practices
1, 3, 4

Case #1 Vacations

Wesley's family spent $1,400 on a trip to the Grand Canyon. They spent 30% of the total on a sightseeing helicopter flight. Wesley estimates that his family spent about $450 on the flight.

Determine whether Wesley's estimate is reasonable.

Understand *What are the facts?*

- Wesley's family spent $1,400 on vacation.
- Thirty percent of the total was spent on a helicopter flight.
- Wesley estimates that 30% is about $450.

Plan *What is your strategy to solve this problem?*

Make a bar diagram that represents 100%.

Solve *How can you apply the strategy?*

Fill in each section of the diagram with 10% of $1,400.

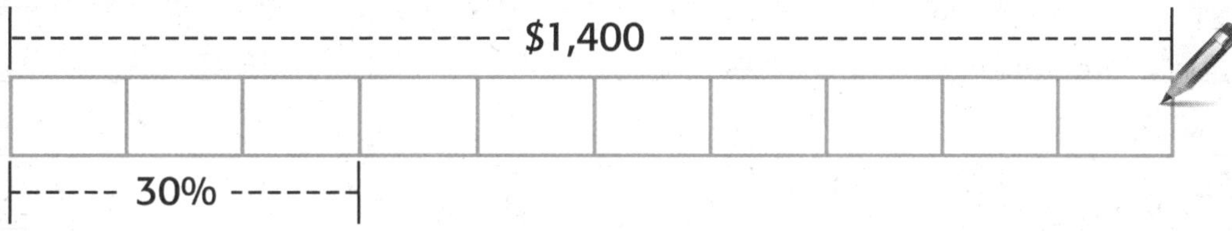

Add three sections to get a total of ______.
So, the helicopter flight cost $420.

Check *Does the answer make sense?*

Wesley estimated the cost of the helicopter flight to be $450.
Since $450 is close to $420, his estimate was reasonable.

Analyze the Strategy

Make a Conjecture How can you use $\frac{1}{3}$ to determine if Wesley's estimate is reasonable? Explain.

Case #2 Don't forget the tip!

At a local Italian restaurant, Brett's bill was $17.50. He decides to leave a 20% tip for the server.

Is $4 a reasonable tip?

Understand

Read the problem. What are you being asked to find?

I need to ______________________________.

Is there any information that you do *not* need to know?

I do not need to know ______________________________.

Plan

Choose a problem-solving strategy.

I will use the ______________________________ strategy.

Solve

Use your problem-solving strategy to solve the problem.

Use a bar diagram to represent 100%. Divide it into ☐ parts. Each part represents ☐% or $☐.

|------------- $17.50 -------------|

100%

|-- tip --|

Two parts, or ☐%, equal $☐ + $☐ = $☐.

Is $4 a reasonable tip? ______________________________

Check

Use information from the problem to check your answer.

____

____

Collaborate Work with a small group to solve the following cases. Show your work on a separate piece of paper.

Case #3 Travel

A travel agency surveyed 140 families about their favorite vacation spots.

Is 60, 70, or 80 families a reasonable estimate for the number of families that did not choose Hawaii?

Case #4 Exercise

A survey showed that 61% of middle school students do some kind of physical activity every day.

Suppose there are 828 middle school students in your school. Would the number of students who exercise be about 300, 400, or 500?

Case #5 School

Of 423 students, 57.6% live within 5 miles of the school.

What is a reasonable estimate for the number of students living within 5 miles of the school?

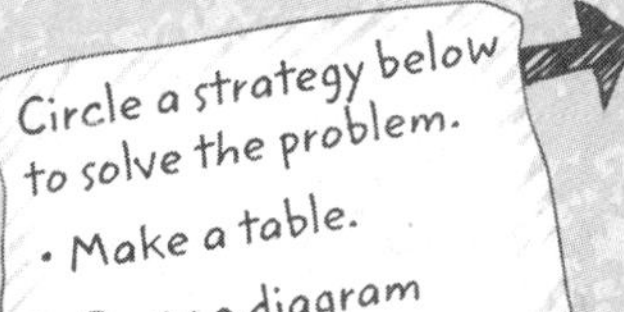

Circle a strategy below to solve the problem.
• Make a table.
• Draw a diagram
• Act it out.

Case #6 Bowling

In bowling, you get a spare when you knock down the ten pins in two throws.

How many possible ways are there to get a spare?

Mid-Chapter Check

Vocabulary Check

1. Fill in the blank in the sentence below with the correct term. (Lesson 4)

   The ______________________ states that the part equals the percent multiplied by the whole.

Skills Check and Problem Solving

Find each number. Round to the nearest tenth if necessary. (Lessons 1 and 3)

Show your work

2. What percent of 84 is 12? ________

3. 15 is 25% of what number? ________

Estimate. (Lesson 2)

4. 20% of 392 ________

5. 78% of 112 ________

Write an equation for each problem. Then solve. Round to the nearest tenth if necessary. (Lesson 4)

6. CCSS **Use Math Tools** A computer costs $849.75 and the hard drive is 61.3% of the total cost. What is a reasonable estimate for the cost of the hard drive? (Lesson 2)

   ____

7. What number is 35% of 72? ________

8. 16.1 is what percent of 70? ________

9. **Standardized Test Practice** Ayana has 220 coins in her piggy bank. Of those, 45% are pennies. How many coins are *not* pennies? (Lesson 1)

   Ⓐ 121 Ⓒ 99
   Ⓑ 116 Ⓓ 85

Inquiry Lab

Percent of Change

HOW can you use a bar diagram to show a percent of change?

Content Standards
7.RP.3, 7.EE.3

Mathematical Practices
1, 3, 4

Admission The admission price for the state fair has increased by 50% in the last five years. The admission price was \$6 five years ago. What is the current admission price? Do the Investigation below to find out.

Investigation

Use a bar diagram to solve.

Step 1 The bar diagram represents 100%.

price 5 years ago = \$6	100%

Since $50\% = \frac{1}{2}$, divide the bar diagram in half. Fill in the missing information.

price 5 years ago = ____

\$3		100%

Step 2 The admission price increased by 50%. Complete the bar diagram that represents 150% of the price 5 years ago.

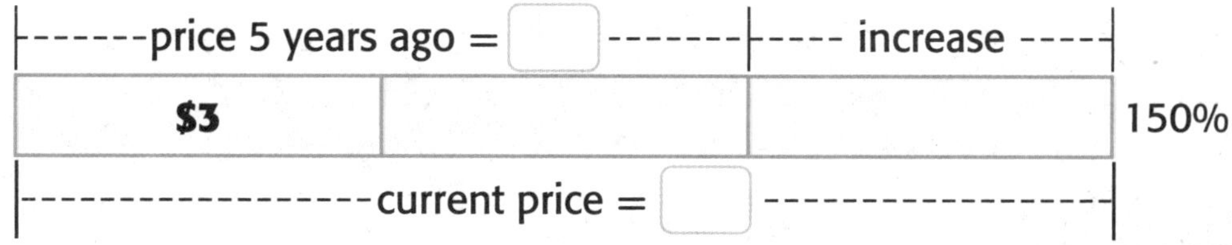

So, the current admission price is ____ + ____ or ____.

Work with a partner to solve the following problems.

1. The height of a tree was 8 feet. After a year, the tree's height increased by 25%. Use a bar diagram to find the new height of the tree? ____

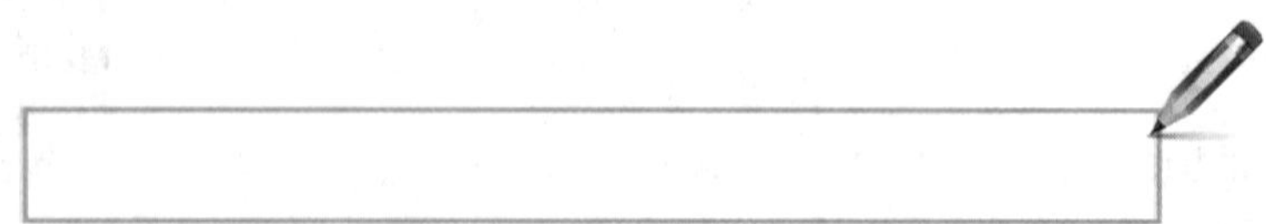

2. **Model with Mathematics** The model below describes the following scenario: Ryan put $160 in a bank account. After 2 months, the total in his account decreased by 25%. Fill in the amount in Ryan's account after 2 months.

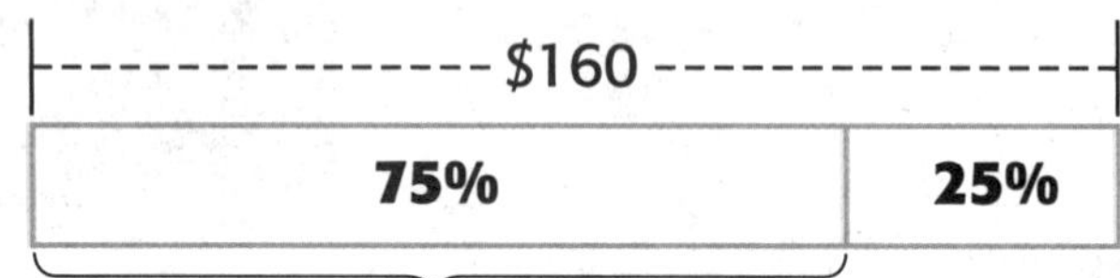

The amount in Ryan's account after 2 months is ____.

Work with a partner to answer the following questions.

3. Find 125% of 8 two different ways.

4. **Reason Inductively** Refer to Exercises 1 and 3. How can you find the new amount of a quantity that has increased over a period of time?

5. Inquiry HOW can you use a bar diagram to show a percent of change?

Lesson 5

Percent of Change

What You'll Learn

Scan the lesson. Predict two things you will learn about percent of change.

-

-

Essential Question

HOW can percent help you understand situations involving money?

Vocabulary

percent of change
percent of increase
percent of decrease
percent error

Common Core State Standards

Content Standards
7.RP.3, 7.EE.3

Mathematical Practices
1, 3, 4, 5, 6

Real-World Link

Watch

Speed Racer The Indy 500 is one of the world's great motor races. The table shows the average speed of the winning race cars for various years.

Year	Speed (mph)
1922	94
1955	128
2010	162

1. Write the ratio $\frac{\text{speed increase from 1922 to 1955}}{\text{speed in 1922}}$.

   Then write the ratio as a percent.

   Round to the nearest whole percent.

   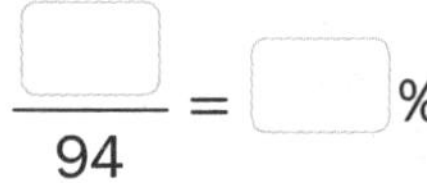

   $\frac{\square}{94} = \square\%$

2. Write the ratio $\frac{\text{speed increase from 1955 to 2010}}{\text{speed in 1955}}$.

   Then write the ratio as a percent.

   Round to the nearest whole percent.

   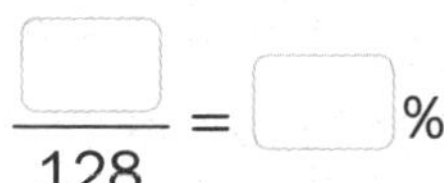

   $\frac{\square}{128} = \square\%$

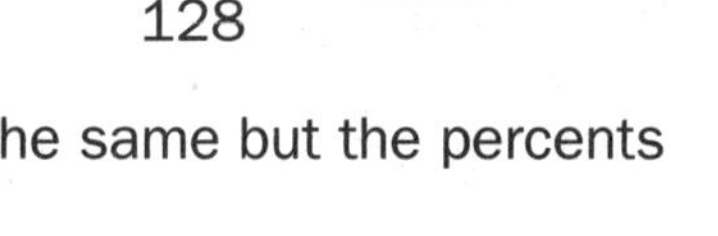

3. Why are the amounts of increase the same but the percents different?

Key Concept

Percent of Change

Work Zone

Words A **percent of change** is a ratio that compares the change in quantity to the original amount.

Equation $\text{percent of change} = \dfrac{\text{amount of change}}{\text{original amount}}$

When you compare the amount of change to the original amount in a ratio, you are finding the percent of change. The percent of change is based on the original amount.

If the original quantity is increased, then it is called a **percent of increase**. If the original quantity is decreased, then it is called a **percent of decrease**.

$$\text{percent of increase} = \frac{\text{amount of increase}}{\text{original amount}}$$

$$\text{percent of decrease} = \frac{\text{amount of decrease}}{\text{original amount}}$$

Examples

1. Find the percent of change in the cost of gasoline from 1980 to 2010. Round to the nearest whole percent if necessary.

Since the 2010 price is greater than the 1980 price, this is a percent of increase.

Step 1 Find the amount of increase.
$2.95 − $1.30 = $1.65

Step 2 Find the percent of increase.

$$\begin{aligned} \text{percent of increase} &= \frac{\text{amount of increase}}{\text{original amount}} \\ &= \frac{\$1.65}{\$1.30} && \text{Substitution} \\ &\approx 1.27 && \text{Simplify.} \\ &\approx 127\% && \text{Write 1.27 as a percent.} \end{aligned}$$

The cost of gasoline increased by about 127% from 1980 to 2010.

Percents
In the percent of change formula, the decimal repesenting the percent of change must be written as a percent.

2. Yusuf bought a DVD recorder for $280. Now, it is on sale for $220. Find the percent of change in the price. Round to the nearest whole percent if necessary.

Since the new price is less than the original price, this is a percent of decrease.

Step 1 Find the amount of decrease.
$\$280 - \$220 = \$60$

Step 2 Find the percent of decrease.

$$\text{percent of decrease} = \frac{\text{amount of decrease}}{\text{original amount}}$$

$$= \frac{\$60}{\$280} \quad \text{Substitution}$$

$$\approx 0.21 \quad \text{Simplify.}$$

$$\approx 21\% \quad \text{Write 0.21 as a percent.}$$

The price of the DVD recorder decreased by about 21%.

Percent of Change

Always use the original amount as the whole when finding percent of change.

Got It? Do these problems to find out.

a. Find the percent of change from 10 yards to 13 yards.

b. The price of a radio was $20. It is on sale for $15. What is the percent of change in the price of a radio?

a. ____________

b. ____________

Percent Error

Key Concept

Words The **percent error** is a ratio that compares the inaccuracy of an estimate, or amount of error, to the actual amount.

Equation $\text{percent error} = \frac{\text{amount of error}}{\text{actual amount}}$

Finding the percent error is similar to finding the percent of change. Instead of finding the amount of increase or decrease, you will find the amount an estimate is greater or less than the actual amount.

Suppose you guess there are 300 gum balls in a jar, but there are actually 400.

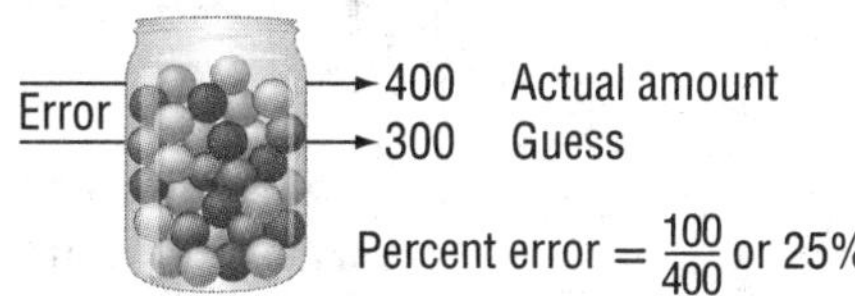

Example

3. Ahmed wants to practice free-throws. He estimates the distance from the free-throw line to the hoop and marks it with chalk. Ahmed's estimate was 13.5 feet. The actual distance should be 15 feet. Find the percent error.

Step 1 Find the amount of error.
15 feet − 13.5 feet = 1.5 feet

Step 2 Find the percent error.

$$\text{percent error} = \frac{\text{amount of error}}{\text{actual amount}}$$

$$= \frac{1.5}{15} \quad \text{Substitution.}$$

$$= 0.1 \text{ or } 10\% \quad \text{Simplify.}$$

So, the percent error is 10%.

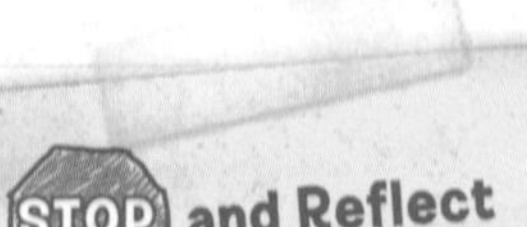

STOP and Reflect

How is finding a percent error like finding a percent of change?

Got It? Do this problem to find out.

c. Find the percent error if the estimate is $230 and the actual amount is $245. Round to the nearest whole percent.

c. ____________

Guided Practice

Find each percent of change. Round to the nearest whole percent if necessary. State whether the percent of change is an *increase* or a *decrease*. (Examples 1 and 2)

1. 30 inches to 24 inches ____________

2. $126 to $150 ____________

3. Jessie estimates the weight of her cat to be 10 pounds. The actual weight of the cat is 13.75 pounds. Find the percent error. (Example 3) ____________

4. **Building on the Essential Question** Explain how two amounts of change can be the same but the percents of change can be different.

____

____

____

Rate Yourself!

How confident are you about percent of change? Check the box that applies.

For more help, go online to access a Personal Tutor.

Name ______________________ My Homework ______________________

Independent Practice

Go online for Step-by-Step Solutions

Find each percent of change. Round to the nearest whole percent if necessary. State whether the percent of change is an *increase* or a *decrease*. (Examples 1 and 2)

1. 15 yards to 18 yards

2. 100 acres to 140 acres

3. $15.60 to $11.70

4. 125 centimeters to 87.5 centimeters

5. 1.6 hours to 0.95 hour

6. 132 days to 125.4 days

Be Precise Find the percent error. (Example 3)

7. Each week, Mr. Jones goes to the grocery store. Mr. Jones estimates that he will spend $120 when he goes to the grocery store this week. He actually spends $94.

8. Marcus estimates that 230 people will attend choir concert. There was an actual total of 300 people who attended the choir concert.

For each situation, find each percent of change. Round to the nearest whole percent if necessary. State whether the percent of change is an *increase* or a *decrease*. (Examples 1 and 2)

9. Three months ago, Santos could walk 2 miles in 40 minutes. Today he can walk 2 miles in 25 minutes.

10. Last school year the enrollment of Genoa Middle School was 465 students. This year the enrollment is 525.

11. Refer to the rectangle at the right. Suppose the side lengths are doubled.

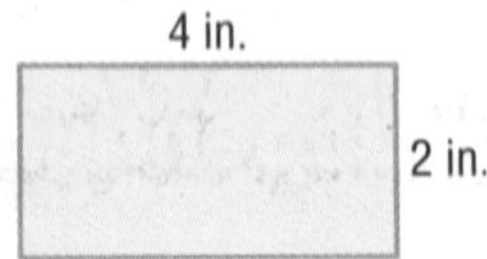

a. Find the percent of change in the perimeter. ____________

b. Find the percent of change in the area. ____________

12. CCSS **Use Math Tools** Find examples of data reflecting change over a period of time in a newspaper or magazine, on television, or on the Internet. Determine the percent of change. Explain whether the data show a percent of increase or decrease.

13. Use the graph shown to find the percent of change in CD sales from 2011 to 2012. ____________

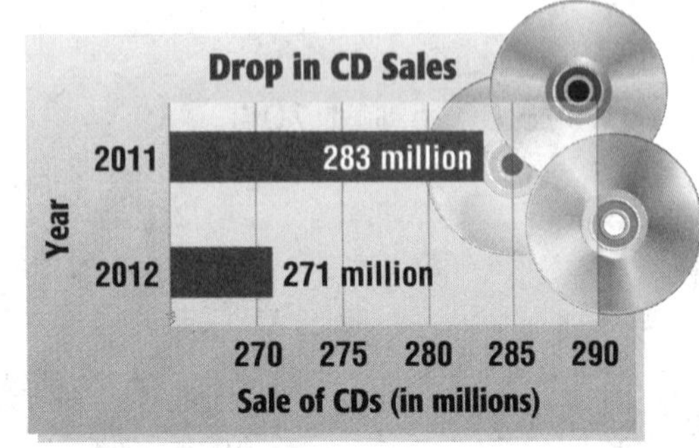

H.O.T. Problems Higher Order Thinking

14. CCSS **Persevere with Problems** The costs of two sound systems were decreased by $10. The original costs of the systems were $90 and $60. Without calculating, which had a greater percent of decrease? Explain.

15. CCSS **Find the Error** Dario is finding the percent of change from $52 to $125. Find his mistake and correct it.

Standardized Test Practice

16. Which of the following represents the least percent of change?

Ⓐ A coat that was originally priced at $90 is now $72.

Ⓑ A puppy who weighed 6 ounces at birth now weighs 96 ounces.

Ⓒ A child grew from 54 inches to 60 inches in 1 year.

Ⓓ A savings account increased from $500 to $550 in 6 months.

Name ______________________ My Homework ______________________

Extra Practice

Find each percent of change. Round to the nearest whole percent. State whether the percent of the change is an *increase* or *decrease*.

17. \$12 to \$6

50%; decrease

$12 - 6 = 6$

$\frac{6}{12} = 0.5$ or 50%

Homework Help

18. 48 notebooks to 14 notebooks

19. \$240 to \$320

20. 624 feet to 702 feet

21. The table shows the number of youth 7 years and older who played soccer from 2004 to 2012.

a. Find the percent of change from 2008 to 2012. Round to the nearest tenth of a percent. Is it an increase or decrease?

b. Find the percent of change from 2006 to 2008. Round to the nearest tenth of a percent. Is it an increase or decrease?

Playing Soccer

Year	Number (millions)
2004	12.9
2006	13.7
2008	13.3
2010	14.0
2012	13.8

22. Shoe sales for a certain company were \$25.9 billion. Sales are expected to increase by about 20% in the next year. Find the projected amount of shoe sales next year. ______

23. CCSS **Be Precise** Eva estimates that 475 songs will fit on her MP3 player. The actual amount of songs that fit is 380. Find the percent error. ______

24. The table shows Catalina's babysitting hours. She charges \$6.50 per hour. Write a sentence that compares the percent of change in the amount of money earned from April to May to the percent of change in the amount of money earned from May to June. Round to the nearest percent if needed.

Month	Hours Worked
April	30
May	35
June	45

Standardized Test Practice

25. **Short Response** A music video Web site received 5,000 comments on a new song they released. After the artist performed the song on television, the number of comments increased by 30% the next day. How many new comments were on the Web site at the end of the next day?

______

26. Students in a reading program gradually increased the amount of time they read. The first week, they read 20 minutes per day. Each week thereafter, they increased their reading time by 50% until they read an hour per day. In what week of the program did the students begin reading an hour per day?

Ⓐ Week 2
Ⓑ Week 3
Ⓒ Week 4
Ⓓ Week 5

27. **Short Response** Find the percent of change in the perimeter of the square below if its side length is tripled. ______

3 cm

3 cm

Common Core Review

Find each sum. 6.NS.3

28. 1.5 + 2.25 = ______

29. 32.5 + 13.43 = ______

30. \$66.99 + \$8.15 = ______

31. The distances around Earth at the equator and through the North and South Poles are shown at the right. How many miles would you travel if you circled Earth along both routes? 6.NS.3

______

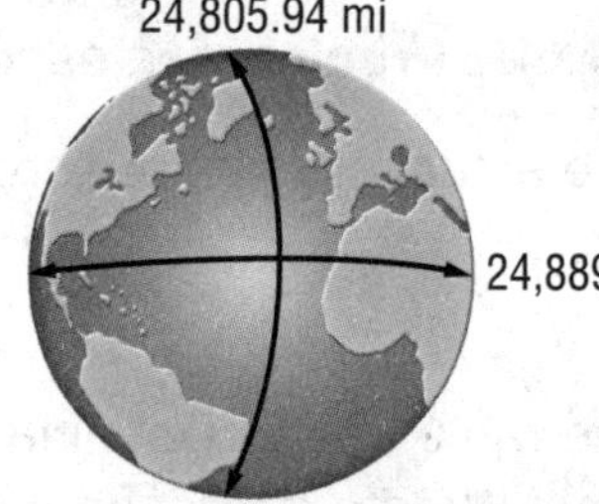

32. The table shows the prices of various grocery items. What is the cost of 3 cans of chicken broth and 2 cans of vegetable soup? 6.NS.3

______

Item	Price (\$)
chicken broth	0.99
chili	2.49
vegetable soup	1.49

Lesson 6

Sales Tax, Tips, and Markup

What You'll Learn

Scan the rest of the lesson. List two headings you would use to make an outline of the lesson.

- ______________________________
- ______________________________

Essential Question

HOW can percent help you understand situations involving money.

Vocabulary

sales tax
tip
gratuity
markup
selling price

Common Core State Standards

Content Standards
7.RP.3, 7.EE.2, 7.EE.3

Mathematical Practices
1, 3, 4

Real-World Link

Kayaks Alonso plans to buy a new kayak that costs \$2,100. But when he buys the kayak, it actually costs more because he lives in a county where there is a 7% sales tax.

You can find the amount of tax on an item by multiplying the price by the tax percentage.

1. Circle the amount below that shows the amount of tax Alonso will pay for the kayak.

   \$350 \$235 \$147

2. Use the amount of tax from Exercise 1 to fill in the receipt at the right. Then find the total cost Alonso will pay for the kayak.

3. Multiply 1.07 and \$2,100. How does the result compare to your answer in Exercise 2?

4. On Alonso's kayaking trip, hiring a guide costs \$50. Alonso wants to give the guide a 10% tip. Explain how to find the amount of the tip.

Jimmie's Kayaks

Kayak	______
Sales Tax +	______
Total	______

Work Zone

Sales Tax and Total Cost

Sales tax is an additional amount of money charged on items that people buy. The total cost of an item is the regular price plus the sales tax.

Example

Watch Tutor

1. Drew wants to buy exercise equipment that costs \$140 and the sales tax is 5.75%. What is the total cost of the equipment?

Method 1 **Add sales tax to the regular price.**

First, find the sales tax.

Let t represent the sales tax.

part = percent × whole	Write the percent equation.
$t = 0.0575 \times 140$	5.75% = 0.0575
$t = 8.05$	Multiply.

Next, add the sales tax to the regular price.
\$8.05 + \$140 = \$148.05

Method 2 **Add the percent of tax to 100%.**

100% + 5.75% = 105.75%	Add the percent of tax to 100%.

Let t represent the total.

part = percent × whole	Write the percent equation.
$t = 1.0575 \times 140$	105.75% = 1.0575
$t = \$148.05$	Multiply.

The total cost of the exercise equipment is \$148.05.

Got It? **Do this problem to find out.**

a. What is the total cost of a sweatshirt if the regular price is \$42 and the sales tax is $5\frac{1}{2}$%?

a. __________

Tips and Markups

A **tip** or **gratuity** is a small amount of money in return for a service. The total price is the regular price of the service plus the tip.

A store sells items for more than it pays for those items. The amount of increase is called the **markup**. The **selling price** is the amount the customer pays for an item.

Examples

2. A customer wants to tip 15% on a restaurant bill that is $35. What will be the total bill with tip?

Method 1 **Add the tip to the regular price.**

First, find the tip. Let t represent the tip.

part = percent × whole

$t = 0.15 \times 35$ 15% = 0.15

$t = 5.25$ Multiply.

Next, add the tip to the bill.

$5.25 + $35 = $40.25   Add.

Method 2 **Add the percent of tip to 100%.**

100% + 15% = 115%   Add the percent of tip to 100%.

The total cost is 115% of the bill. Let t represent the total.

part = percent × whole

$t = 1.15 \times 35$ 115% = 1.15

$t = 40.25$ Multiply.

Using either method, the total cost of the bill with tip is $40.25.

Mental Math

10% of a number can be found by moving the decimal one place to the left. 10% of $20 is $2. So, 20% of $20 is $4.

3. A haircut costs $20. Sales tax is 4.75%. Is $25 sufficient to cover the haircut with tax and a 15% tip?

Sales tax is 4.75% and the tip is 15%, so together they will be 19.75%.

Let t represent the tax and tip.

part = percent × whole

$t = 0.1975 \times 20$ 0.15 + 0.0475 = 0.1975

$t = 3.95$ Multiply.

$20 + $3.95 = $23.95   Add.

Since $23.95 < $25, $25 is sufficient to cover the total cost.

Got It? **Do these problems to find out.**

b. Scott wants to tip his taxicab driver 20%. If his commute costs $15, what is the total cost?

c. Find the total cost of a spa treatment of $42 including 6% tax and 20% tip.

b. ____________

c. ____________

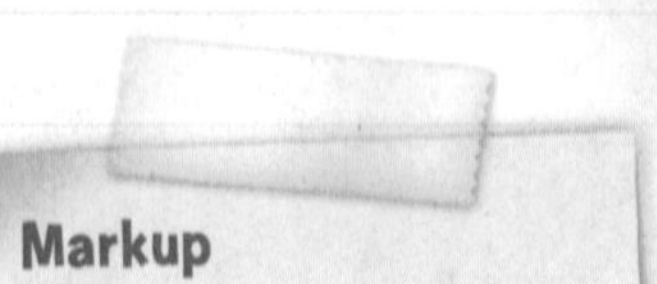

Example

4. A store pays $56 for a GPS navigation system. The markup is 25%. Find the selling price.

First, find the markup. Let m represent the markup.

part = percent × whole	Write the percent equation.
$m = 0.25 \times 56$	25% = 0.25
$m = 14$	Mulitply.

Next, add the markup to the amount the store pays.

$14 + $56 = $70	Add.

The selling price of the GPS navigation system is $70.

Markup

In Example 4, you could find the selling price by finding 125% of the amount the store pays.

Got It? **Do this problem to find out.**

d. A store pays $150 for a portable basketball backboard and the markup is 40%. What is the selling price?

d. ________

Guided Practice

Find the total cost to the nearest cent. (Examples 1 and 2)

1. $2.95 notebook; 5% tax ________

2. $28 lunch; 15% tip ________

3. Jaimi went to have a manicure that cost $30. She wanted to tip the technician 20% and tax is 5.75%. How much did she spend total for the manicure? (Example 3)

____

4. Find the selling price of a $62.25 karaoke machine with a 60.5% markup. (Example 4) ________

5. **Building on the Essential Question** Describe two methods for finding the total price of a bill that includes a 20% tip. Which method do you prefer? ________

____

____

Rate Yourself!

How well do you understand finding sales tax, tips, and markups? Circle the image that applies.

Clear

Somewhat Clear

Not So Clear

For more help, go online to access a Personal Tutor.

Independent Practice

Go online for Step-by-Step Solutions

Find the total cost to the nearest cent. (Examples 1 and 2)

1. \$58 bill; 20% tip ______

2. \$43 dinner; 18% gratuity ______

3 \$1,500 computer; 7% tax ______

4. \$46 shoes; 2.9% tax ______

5 **Financial Literacy** A restaurant bill comes to \$28.35. Find the total cost if the tax is 6.25% and a 20% tip is left on the amount before tax. (Example 3) ______

6. Toru takes his dog to be groomed. The fee to groom the dog is \$75 plus 6.75% tax. Is \$80 enough to pay for the service? Explain. (Example 3) ______

7. Find the selling price of a \$270 bicycle with a 24% markup. (Example 4) ______

8. Find the selling price of a \$450 painting with a 45% markup. (Example 4) ______

9. What is the sales tax on the chair shown if the tax rate is 5.75%? ______

10. A store pays $10 for a bracelet, and the markup is 115%. A customer will also pay a $5\frac{1}{2}$% sales tax. What will be the total cost of the bracelet to the nearest cent? ______

H.O.T. Problems Higher Order Thinking

11. CCSS **Persevere with Problems** The Leather Depot buys a coat from a supplier for $90 wholesale and marks up the price by 40%. If the retail price is $134.82, what is the sales tax? ______

12. CCSS **Model with Mathematics** Give an example of the regular price of an item and the total cost including sales tax if the tax rate is 5.75%.

______

______

13. CCSS **Which One Doesn't Belong?** In each pair, the first value is the regular price of an item and the second value is the price with gratuity. Identify the pair that does not belong with the other three. Explain your reasoning to a classmate.

$30, $34.50	$54, $64.80	$16, $18.40	$90, $103.50

______

______

Standardized Test Practice

14. Prices for several cell phones are listed in the table below. The table shows the regular price *p* and the price with tax *t*.

Phone	Regular Price (*p*)	Price with Tax (*t*)
Flip phone	$80	$86.40
Slide phone	$110	$118.80
Video phone	$120	$129.60

Which formula can be used to calculate the price with tax?

Ⓐ $t = p \times 0.8$

Ⓑ $t = p - 0.8$

Ⓒ $t = p \times 0.08$

Ⓓ $t = p \times 1.08$

Name ______ My Homework ______

Extra Practice

Find the total cost to the nearest cent.

15. \$99 CD player; 5% tax \$103.95

$0.05 \times 99 = 4.95$

\$99.00
+ 4.95
\$103.95

16. \$13 haircut; 15% tip \$14.95

$0.15 \times 13 = 1.95$

\$13.00
+ 1.95
\$14.95

17. \$7.50 meal; 6.5% tax ______

18. \$39 pizza order; 15% tip ______

19. \$89.75 scooter; $7\frac{1}{4}$% tax ______

20. \$8.50 yoga mat; 75% markup ______

21. CCSS **Reason Inductively** Diana and Sujit clean homes for a summer job. They charge \$70 for the job plus 5% for supplies. A homeowner gave them a 15% tip. Did they receive more than \$82 for their job? Explain.

______

______

22. CCSS **Find the Error** Jamar is finding the selling price of a pair of \$40 skates with a 30% markup. Find his mistake and correct it.

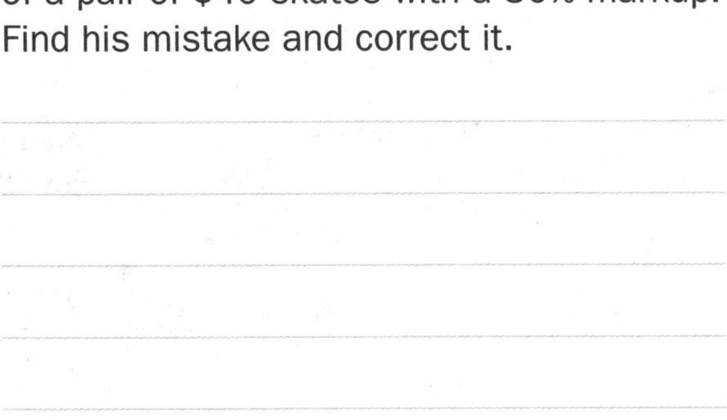

Standardized Test Practice

23. Ms. Taylor bought a water tube to pull behind her boat. The tube cost $87.00 and 9% sales tax was added at the register. Ms. Taylor gave the cashier five $20 bills. How much change should she have received?

Ⓐ $4.83
Ⓑ $5.17
Ⓒ $94.83
Ⓓ $117.00

24. A trampoline costs $220 and the sales tax is 6.25%. What is the total cost of the trampoline?

Ⓕ $13.75
Ⓖ $15.75
Ⓗ $233.75
Ⓘ $240

25. **Short Response** The same pair of boots are at different stores. The cost and sales tax on the boots at each store are shown in the table. In which store would you pay less for the boots? Explain.

Store	Price	Tax
A	$54.90	6%
B	$53.25	7%

Common Core Review

Solve. 6.NS.3

26. 45 − 4.5 = ______

27. 89 − 31.15 = ______

28. $102 − $25.75 = ______

29. Renata paid $35.99 for a dress. The dress was on sale for $14.01 off its regular price. What was the regular price of the dress? 6.NS.3

30. Mr. Durant bought Console B for $20.99 off the advertised price. Find the total amount Mr. Durant paid. 6.NS.3

Game Console	Advertised Price ($)
A	128.99
B	138.99
C	148.99

Lesson 7
Discount

What You'll Learn

Scan the lesson. List two real-world scenarios in which you would use discounts.

- ______________________________
- ______________________________

Essential Question

HOW can percent help you understand situations involving money?

Vocabulary

discount
markdown

Common Core State Standards

Content Standards
7.RP.3, 7.EE.3

Mathematical Practices
1, 3, 4, 5

Real-World Link

Water Parks A pass at a water park is $58 dollars at the beginning of the season. The cost of the pass decreases each month.

1. Each month 10% is taken off the price of a season pass. Find the discounted price for August by completing the fill-ins below.

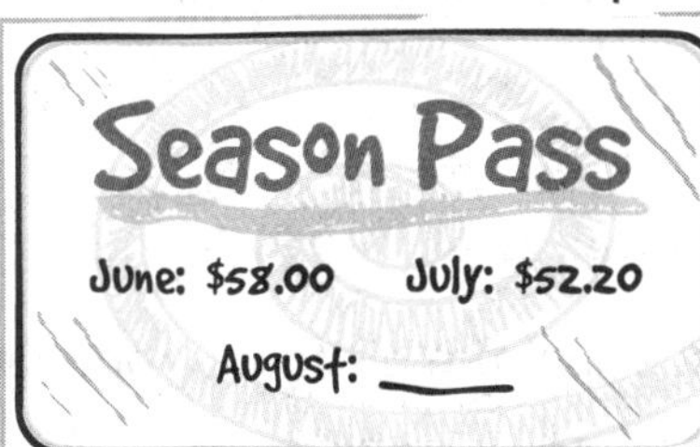

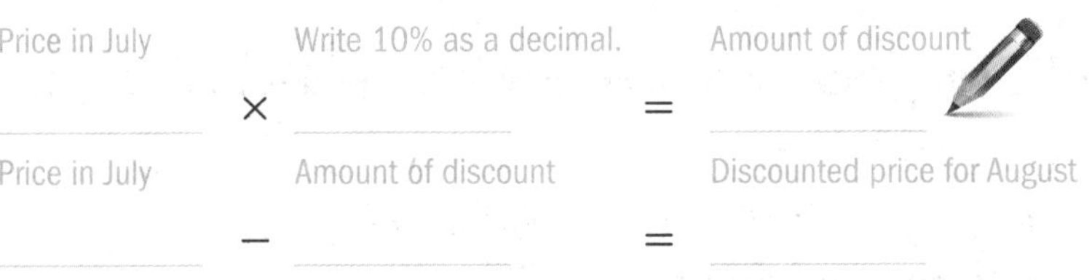

Price in July ____ × Write 10% as a decimal. ____ = Amount of discount ____

Price in July ____ − Amount of discount ____ = Discounted price for August ____

2. Multiply 0.9 and $52.20. How does the result compare to your answer in Exercise 1?

____

3. Write the definition of *discount* in your own words.

____

Work Zone

Find Sale Price and Original Price

Discount or **markdown** is the amount by which the regular price of an item is reduced. The sale price is the regular price minus the discount.

Example

1. A DVD normally costs \$22. This week it is on sale for 25% off the original price. What is the sale price of the DVD?

Original Price \$22

Sale Price | 25%

Method 1 **Subtract the discount from the regular price.**

First, find the amount of the discount.

Let d represent the discount.

part = percent × whole		Write the percent equation.
$d = 0.25 \times 22$		25% = 0.25.
$d = 5.50$		Multiply.

Next, subtract the discount from the regular price.

\$22 – \$5.50 = \$16.50

Method 2 **Subtract the percent of discount from 100%.**

100% – 25% = 75% — Subtract the discount from 100%.

The sale price is 75% of the regular price.

Let s represent the sale price.

part = percent × whole		Write the percent equation.
$s = 0.75 \times 22$		75% = 0.75
$s = 16.50$		Multiply.

The sale price of the DVD is \$16.50.

Got It? **Do this problem to find out.**

a. A shirt is regularly priced at \$42. It is on sale for 15% off of the regular price. What is the sale price of the shirt?

a. ______

Example

2. A boogie board that has a regular price of \$69 is on sale at a 35% discount. What is the sale price with 7% tax?

Step 1 Find the amount of the discount.

Let d represent the discount.

part = percent × whole	Write the percent equation.
$d = 0.35 \times 69$	35% = 0.35
$d = 24.15$	Multiply.

Step 2 Subtract the discount from the regular price.

\$69 − \$24.15 = \$44.85

Step 3 The percent of tax is applied after the discount is taken.

7% of \$44.85 = 0.07 • 44.85	Write 7% as a decimal.
= 3.14	The tax is \$3.14.
\$44.85 + \$3.14 = \$47.99	Add the tax to the sale price.

The sale price of the boogie board including tax is \$47.99.

Got It? **Do this problem to find out.**

b. A CD that has a regular price of \$15.50 is on sale at a 25% discount. What is the sale price with 6.5% tax?

b. ____________

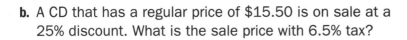

Example

3. A cell phone is on sale for 30% off. If the sale price is \$239.89, what is the original price?

The sale price is 100% − 30% or 70% of the original price.

Let p represent the original price.

part = percent × whole	
$239.89 = 0.7 \times p$	
$\frac{239.89}{0.7} = \frac{0.7p}{0.7}$	Divide each side by 0.7.
$342.70 = p$	Simplify.

The original price is \$342.70.

Percent Equation

Remember that in the percent equation, the percent must be written as a decimal. Since the sale price is 70% of the original price, use 0.7 to represent 70% in the percent equation.

Got It? **Do this problem to find out.**

c. Find the original price if the sale price of the cell phone is \$205.50.

c. ____________

Example

4. Clothes Are Us and Ratcliffe's are having sales. At Clothes Are Us, a pair of sneakers is on sale for 40% off the regular price of $50. At Ratcliffe's, the same brand of sneakers is discounted by 30% off of the regular price of $40. Which store has the better sale price? Explain.

Find the sale price of the sneakers at each store.

Clothes Are Us	**Ratcliffe's**
60% of \$50 $= 0.6 \times \$50$	70% of \$40 $= 0.7 \times \$40$
$= \$30$	$= \$28$
The sale price is \$30.	The sale price is \$28.

Since \$28 < \$30, the sale price at Ratcliffe's is the better buy.

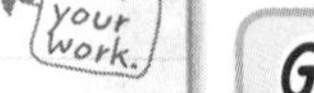

Show your work.

Got It? **Do this problem to find out.**

d. If the sale at Clothes Are Us was 50% off, which store would have the better buy? Explain.

d. ____________

Guided Practice

1. Mary and Roberto bought identical backpacks at different stores. Mary's backpack originally cost \$65 and was discounted 25%. Roberto's backpack originally cost \$75 and was on sale for 30% off of the original price. Which backpack was the better buy? Explain. (Examples 1, 2, and 4)

Show your work.

2. A pair of in-line skates is on sale for \$90. If this price represents a 9% discount from the original price, what is the original price to the nearest cent? (Example 3)

3. **Building on the Essential Question** Describe two methods for finding the sale price of an item that is discounted 30%.

Rate Yourself!

Are you ready to move on? Shade the section that applies.

I have a few questions.

I'm ready to move on.

I have a lot of questions.

For more help, go online to access a Personal Tutor.

Independent Practice

Go online for Step-by-Step Solutions

Find the sale price to the nearest cent. (Examples 1 and 2)

1. \$64 jacket; 20% discount ______

2. \$1,200 TV; 10% discount ______

3 \$7.50 admission; 20% off; 5.75% tax ______

4. \$4.30 makeup; 40% discount; 6% tax ______

5 A bottle of hand lotion is on sale for \$2.25. If this price represents a 50% discount from the original price, what is the original price to the nearest cent? (Example 3)

______

6. A tennis racket at Sport City costs \$180 and is discounted 15%. The same model racket costs \$200 at Tennis World and is on sale for 20% off. Which store is offering the better deal? Explain. (Example 4)

______

______

7. CCSS **Model with Mathematics** Refer to the graphic novel frame below.

a. Find the price that a student would pay including the group discount for each amusement park. ______

______

b. Which is the best deal? ______

8. The Wares want to buy a new computer. The regular price is $1,049. The store is offering a 20% discount and a sales tax of 5.25% is added after the discount. What is the total cost? ______

Find the original price to the nearest cent.

9. calendar: discount, 75%; sale price, $2.25 ______

10. telescope: discount, 30%; sale price, $126 ______

11. **Use Math Tools** Compare and contrast tax and discount.

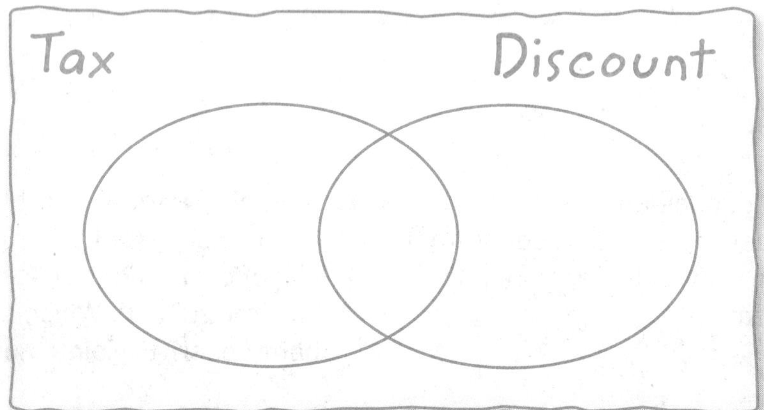

H.O.T. Problems Higher Order Thinking

12. **Model with Mathematics** Give an example of the sale price of an item and the total cost including sales tax if the tax rate is 5.75% and the item is 25% off. ______

13. **Persevere with Problems** A store is having a sale in which all items are discounted 20%. Including tax, Colin paid $21 for a picture. If the sales tax rate is 5%, what was the original price of the picture? ______

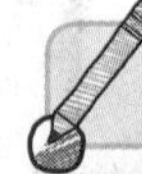

Standardized Test Practice

14. A computer software store is having a sale. The table shows the regular price r and the sale price s of various items. Which formula can be used to calculate the sale price?

Item	Regular Price (r)	Sale Price (s)
A	$5.00	$4.00
B	$8.00	$6.40
C	$10.00	$8.00
D	$15.00	$12.00

Ⓐ $s = r \times 0.2$
Ⓑ $s = r - 0.2$
Ⓒ $s = r \times 0.8$
Ⓓ $s = r - 0.8$

Extra Practice

Find the sale price to the nearest cent.

15. \$119.50 skateboard; 20% off; 7% tax \$102.29

$0.20 \times \$119.50 = \$23.90$
$\$119.50 - \$23.90 = \$95.60$
$0.07 \times \$95.60 = \$6.69$
$\$95.60 + \$6.69 = \$102.29$

16. \$40 sweater; 33% discount ______

17. \$199 MP3 player; 15% discount ______

18. \$12.25 pen set; 60% discount ______

19. Mrs. Robinson bought a novel at a bookstore on sale for 20% off its regular price of \$29.99. Mr. Chang bought the same novel at a different bookstore for 10% off its regular price of \$25. Which person received the better discount? Explain. ______

20. CCSS **Multiple Representations** An online store is having a sale on digital cameras. The table shows the regular price and the sale price for the cameras.

Camera Model	Regular Price	Sale Price	Discount
A	\$97.99	\$83.30	
B	\$102.50	\$82.00	
C	\$75.99	\$65.35	
D	\$150.50	\$135.45	

a. Words Write a rule that can be used to find the percent of decrease for any of the cameras.

b. Table Complete the table for the discount.

c. Numbers Which model has the greatest percent discount?

Standardized Test Practice

21. A chair that costs $210 was reduced by 40% for a one-day sale. After the sale, the sale price was increased by 40%. What is the price of the chair?

Ⓐ $176.40
Ⓑ $185.30
Ⓒ $205.50
Ⓓ $210.00

22. Carmen paid $10.50 for a T-shirt at the mall. It was on sale for 30% off. What was the original price before the discount?

Ⓕ $3.15
Ⓖ $7.35
Ⓗ $15.00
Ⓘ $35.00

CCSS Common Core Review

Find the percent of change. Round to the nearest whole percent if necessary. State whether the percent of change is an *increase* or *decrease*.
7.RP.3

23. 35 birds to 45 birds

24. 60 inches to 38 inches

25. $2.75 to $1.80

26. Complete the table to express each number of months in years. Write in simplest form. The first one is done for you. 6.RP.3a

Number of Months	1	2	3	4	6
Time in Years	$\frac{1}{12}$				

27. Carlos, Karen, and Beng saved money for an overseas trip. Carlos saved for $1\frac{1}{2}$ years. Karen saved for $1\frac{1}{3}$ years. Beng saved for $1\frac{1}{6}$ years. For how many months did each person save? 5.MD.1

Lesson 8

Financial Literacy: Simple Interest

What You'll Learn

Scan the lesson. Predict two things you will learn about financial literacy.

- ______________________________
- ______________________________

Essential Question

HOW can percent help you understand situations involving money?

Vocabulary

principal
simple interest

Common Core State Standards

Content Standards
7.RP.3, 7.EE.3

Mathematical Practices
1, 3, 4

Vocabulary Start-Up

Principal is the amount of money deposited or borrowed. **Simple interest** is the amount paid or earned for the use of money.

The simple interest formula is shown below. Fill in the diagram using the correct words from the word bank.

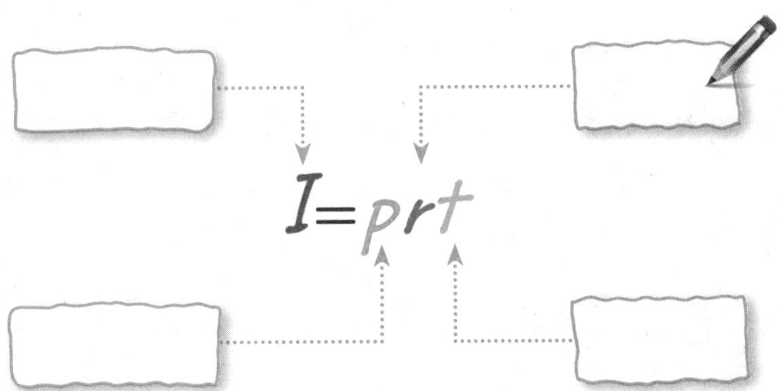

Word Bank
interest
principal
rate
time

Real-World Link

Mrs. Ramirez is investing $400 in a savings account at a simple interest rate of 2% to purchase a laptop computer. She plans on investing the money for 18 months.

Based on this real-world situation, fill in the blanks with the correct numbers. Write the rate as a decimal. Time is expressed in years.

principal = ____ rate = ____ time = ____ years

Key Concept Simple Interest Formula

Words Simple interest I is the product of the principal p, the annual interest rate r, and the time t, expressed in years.

Symbols $I = prt$

Work Zone

If you have a savings account, the bank pays you interest for the use of your money. Use the formula $I = prt$ to find the amount of interest that will be earned.

Examples

Arnold puts $580 into a savings account. The account pays 3% simple interest. How much interest will he earn in each amount of time?

1. 5 years

$I = prt$	Formula for simple interest
$I = 580 \cdot 0.03 \cdot 5$	Replace p with \$580, r with 0.03, and t with 5.
$I = 87$	Simplify.

So, Arnold will earn $87 in interest in 5 years.

2. 6 months

6 months $= \frac{6}{12}$ or 0.5 year	Write the time as years.
$I = prt$	Formula for simple interest
$I = 580 \cdot 0.03 \cdot 0.5$	$p = \$580, r = 0.03, t = 0.5$
$I = 8.7$	Simplify.

So, Arnold will earn $8.70 in interest in 6 months.

Got It? Do these problems to find out.

a. Jenny puts $1,560 into a savings account. The account pays 2.5% simple interest. How much interest will she earn in 3 years?

b. Marcos invests $760 into a savings account. The account pays 4% simple interest. How much interest will he earn after 5 years?

a. ____________

b. ____________

Interest on Loans and Credit Cards

If you borrow money from a bank, you pay the bank interest for the use of their money. You also pay interest to a credit card company if you have an unpaid balance. Use the formula $I = prt$ to find the amount of interest owed.

Examples

3. Rondell's parents borrow $6,300 from the bank for a new car. The interest rate is 6% per year. How much simple interest will they pay if they take 2 years to repay the loan?

$I = prt$ Formula for simple interest

$I = 6{,}300 \cdot 0.06 \cdot 2$ Replace p with $6,300, r with 0.06, and t with 2.

$I = 756$ Simplify.

Rondell's parents will pay $756 in interest in 2 years.

4. Derrick's dad bought new tires for $900 using a credit card. His card has an interest rate of 19%. If he has no other charges on his card and does not make a payment, how much money will he owe after one month?

$I = prt$ Formula for simple interest

$I = 900 \cdot 0.19 \cdot \frac{1}{12}$ Replace p with $900, r with 0.19, and t with $\frac{1}{12}$.

$I = 14.25$ Simplify.

The interest owed after one month is $14.25.

So, the total amount owed would be $900 + $14.25 or $914.25.

and Reflect

Explain in the space below how you would find the simple interest on a $500 loan at a 6% interest rate for 18 months.

Got It? Do these problems to find out.

c. Mrs. Hanover borrows $1,400 at a rate of 5.5% per year. How much simple interest will she pay if it takes 8 months to repay the loan?

d. An office manager charged $425 worth of office supplies on a credit card. The credit card has an interest rate of 9.9%. How much money will he owe at the end of one month if he makes no other charges on the card and does not make a payment?

c. ____________

d. ____________

Example

5. Luis is taking out a car loan for $5,000. He plans on paying off the car loan in 2 years. At the end of 2 years, Luis will have paid $300 in interest. What is the simple interest rate on the car loan?

$I = prt$	Formula for simple interest
$300 = 5{,}000 \cdot r \cdot 2$	Replace I with 300, p with 5,000, and t with 2.
$300 = 10{,}000r$	Simplify.
$\frac{300}{10{,}000} = \frac{10{,}000r}{10{,}000}$	Divide each side by 10,000.
$0.03 = r$	

The simple interest rate is 0.03 or 3%.

Got It? **Do this problem to find out.**

e. Maggie is taking out a student loan for $2,600. She plans on paying off the loan in 3 years. At the end of 3 years, Maggie will have paid $390 in interest. What is the simple interest rate on the student loan?

e. ________

Guided Practice

1. The Masters family financed a computer that cost $1,200. If the interest rate is 19%, how much will the family owe for the computer after one month if no payments are made? (Examples 1–4) ________

2. Samantha received a loan from the bank for $4,500. She plans on paying off the loan in 4 years. At the end of 4 years, Samantha will have paid $900 in interest. What is the simple interest rate on the bank loan? (Example 5)

____

3. **Building on the Essential Question** How can you use a formula to find simple interest?

____

Rate Yourself!

How confident are you about using the simple interest formula? Check the box square that applies.

For more help, go online to access a Personal Tutor.

Independent Practice

Go online for Step-by-Step Solutions eHelp 

Find the simple interest earned to the nearest cent for each principal, interest rate, and time. (Examples 1 and 2)

1. $640, 3%, 2 years ______

2. $1,500, 4.25%, 4 years ______

3. $580, 2%, 6 months ______

4. $1,200, 3.9%, 8 months ______

Find the simple interest paid to the nearest cent for each loan amount, interest rate, and time. (Example 3)

5. $4,500, 9%, 3.5 years ______

6. $290, 12.5%, 6 months ______

7. Leon charged $75 at an interest rate of 12.5%. How much will Leon have to pay after one month if he makes no payments? (Example 4)

______

8. Jamerra received a $3,000 car loan. She plans on paying off the loan in 2 years. At the end of 2 years, Jamerra will have paid $450 in interest. What is the simple interest rate on the car loan? (Example 5)

______

9. CCSS **Justify Conclusions** Pablo has $4,200 to invest for college.

   a. If Pablo invests $4,200 for 3 years and earns $630, what is the simple interest rate? ______

   b. Pablo's goal is to have $5,000 after 4 years. Is this possible if he invests with a rate of return of 6%? Explain. ______

______

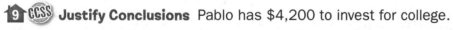

10. **Financial Literacy** The table shows interest owed for a home improvement loan based on how long it takes to pay off the loan.

a. What is the simple interest owed on $900 for 9 months? ______

b. Find the simple interest owed on $2,500 for 18 months. ______

c. Find the simple interest owed on $5,600 for 6 months. ______

Time	Rate
6 months	2.4%
9 months	2.9%
12 months	3.0%
18 months	3.1%

H.O.T. Problems Higher Order Thinking

11. CCSS **Justify Conclusions** Suppose you earn 3% on a $1,200 deposit for 5 years. Explain how the simple interest is affected if the rate is increased by 1%. What happens if the time is increased by 1 year?

______

______

______

12. CCSS **Persevere with Problems** Dustin bought a $2,000 computer with a credit card. The minimum payment each month is $35. Each month 1% of the unpaid balance is added to the amount he owes.

a. If Dustin pays only $35 the first month, what will he owe the second month? ______

b. If Dustin makes the minimum payment, what will he owe the third month? ______

Standardized Test Practice

13. Jada invests $590 in a money market account. Her account pays 7.2% simple interest. If she does not add or withdraw any money, how much interest will Jada's account earn after 4 years of simple interest?

Ⓐ $75.80
Ⓑ $158.67
Ⓒ $169.92
Ⓓ $220.67

Extra Practice

Find the simple interest earned to the nearest cent for each principal, interest rate, and time.

14. \$1,050, 4.6%, 2 years \$96.60

$I = prt$

$I = \$1{,}050 \cdot 0.046 \cdot 2$

$I = 96.60$

15. \$500, 3.75%, 4 months ______

16. \$250, 2.85%, 3 years ______

17. \$3,000, 5.5%, 9 months ______

Find the simple interest paid to the nearest cent for each loan amount, interest rate, and time.

18. \$1,000, 7%, 2 years ______

19. \$725, 6.25%, 1 year ______

20. \$2,700, 8.2%, 3 months ______

21. \$175.80, 12%, 8 months ______

22. Jake received a student loan for \$12,000. He plans on paying off the loan in 5 years. At the end of 5 years, Jake will have paid \$3,600 in interest. What is the simple interest rate on the student loan?

______

Standardized Test Practice

23. Mei-Ling invested $2,000 in a simple interest account for 3 years. At the end of 3 years, she had earned $150 in interest. What was the simple interest rate of the account?

Ⓐ 0.025%
Ⓑ 0.25%
Ⓒ 2.5%
Ⓓ 25%

24. Mr. Sprockett borrows $3,500 from his bank to buy a used car. The loan has a 7.4% annual simple interest rate. If it takes Mr. Sprockett two years to pay back the loan, what is the total amount he will be paying?

Ⓕ $3,012
Ⓖ $3,598
Ⓗ $4,018
Ⓘ $4,550

Common Core Review

Label the number line below from 0 to 10. Then graph each number. 6.NS.6

25. 2.5

26. $8\frac{1}{4}$

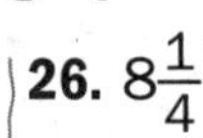

27. 5.9

28. $\frac{1}{1}$

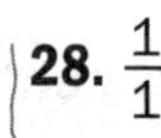

29. Johnna walks 5.4 blocks to school. Belinda walks 5.6 blocks to school. Assume the blocks are the same length. Who walks a longer distance to school? Justify your reasoning. 6.NS.7

30. Use the Commutative and Associative Properties of Addition to mentally find (11 + 64) + 9. Then justify each step. 6.EE.3

Inquiry Lab

Spreadsheet: Compound Interest

HOW is compound interest different from simple interest?

Content Standards
7.RP.3

Mathematical Practices
1, 3, 5

College Jin Li's parents deposit $2,000 in a college savings account. The account pays an interest rate of 4% compounded annually. Complete the Investigation to find how much money will be in the account after 9 years.

Investigation

Compound interest is interest earned on the original principal and on interest earned in the past. At the end of each time period, the interest earned is added to the principal, which becomes the new principal for the next time period.

A computer spreadsheet is a useful tool for quickly performing calculations involving compound interest. To perform a calculation in a spreadsheet cell, first enter the equals sign. For example, enter =A4+B4 to find the sum of Cells A4 and B4.

Create a spreadsheet like the one shown.

Compound Interest

◇	A	B	C	D
1	Rate	0.04		
2				
3	Principal	Interest	New Principal	Time (YR)
4	$2000.00	$80.00	$2080.00	1
5	$2080.00	$83.20	$2163.20	2
6	$2163.20	$86.53	$2249.73	3
7	$2249.73	$89.99	$2339.72	4
8	$2339.72	$93.59	$2433.31	5
9	$2433.31	$97.33	$2530.64	6
10	$2530.64	$101.23	$2631.86	7
11	$2631.86	$105.27	$2737.14	8
12				

Sheet 1 | Sheet 2 | Sheet 3

The interest rate is entered as a decimal.

The spreadsheet evaluates the formula A4×B1.

The interest is added to the principal every year. The spreadsheet evaluates the formula A4+B4.

What formula would the spreadsheet use to find the new principal at the end of Year 9? ______________________

So, the account will have a balance of ______________ after 9 years.

Work with a partner. Create spreadsheets for the situations below. Then answer the questions.

1. Lakeesha deposits $1,500 into a Young Savers account. The account receives 4% interest compounded annually. What is the balance in Lakeesha's account after 2 years? after 3 years?

   2 Years: ____________ 3 Years: ____________

2. Michael deposits $2,650 into an account. The interest rate on the account is 6% compounded annually. What is the balance in Michael's account after 2 years? after 3 years?

   2 Years: ____________ 3 Years: ____________

Work with a partner to answer the following question.

3. CCSS **Reason Inductively** Suppose you deposit $1,000 into a bank account paying 4.75% interest compounded annually. At the same time, a friend deposits $1,000 in a separate account that pays 5% simple interest. You and your friend withdraw your money from the accounts after 6 years. Predict which account made more money. Explain.

4. Inquiry HOW is compound interest different from simple interest?

21ST CENTURY CAREER in Video Game Design

Video Game Designer

Are you passionate about computer gaming? You might want to explore a career in video game design. A video game designer is responsible for a game's concept, layout, character development, and game-play. Game designers use math and logic to compute how different parts of a game will work.

College & Career READINESS

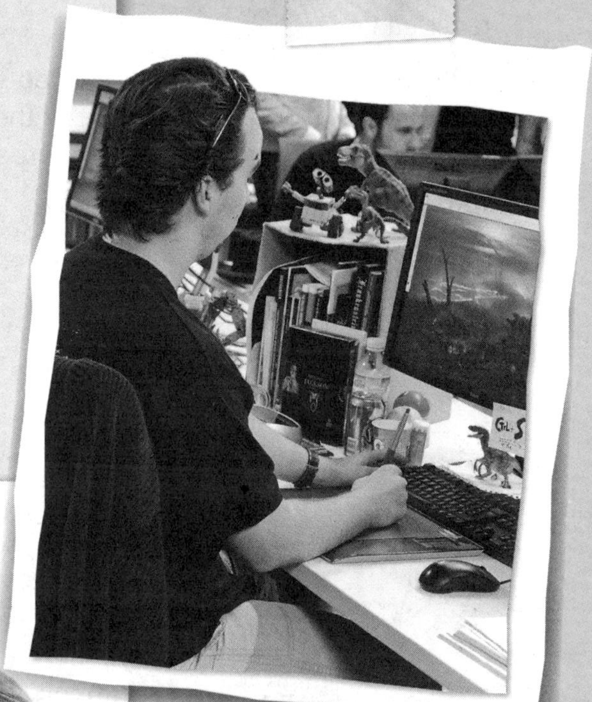

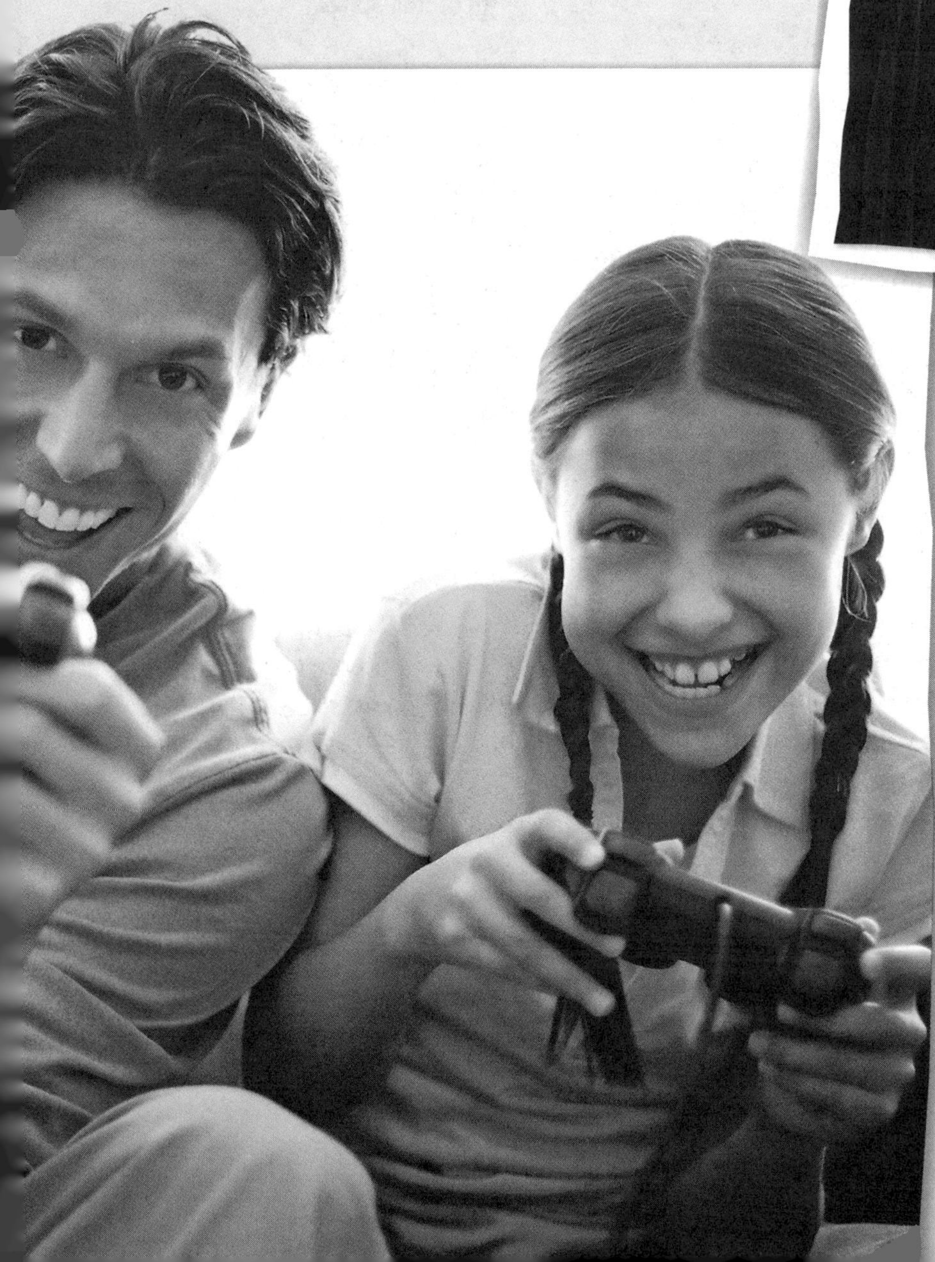

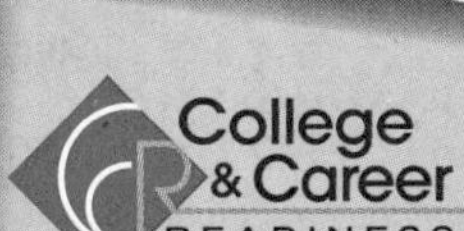

Is This the Career for You?

Are you interested in a career as a video game designer? Take some of the following courses in high school.

- 3-D Digital Animation
- Introduction to Computer Literacy
- Introduction to Game Development

Find out how math relates to a career in Video Game Design.

All Fun and Games

Use the information in the circle graph and the table to solve the problems below.

1. How many of the top 20 video games sold were sports games? ________
2. Out of the top 20 video games sold, how many more music games were there than racer games? ________
3. In Week 1, the total sales for a video game were $2,374,136. What percent of the total sales was from the United States? Round to the nearest whole percent. ________
4. Find the percent of change in sales of the video game from Week 1 to Week 3 in Japan. Round to the nearest whole percent. ________
5. Which country had a greater percent decrease in sales from Week 1 to Week 2: Japan or the United States? Explain.

____

____

____

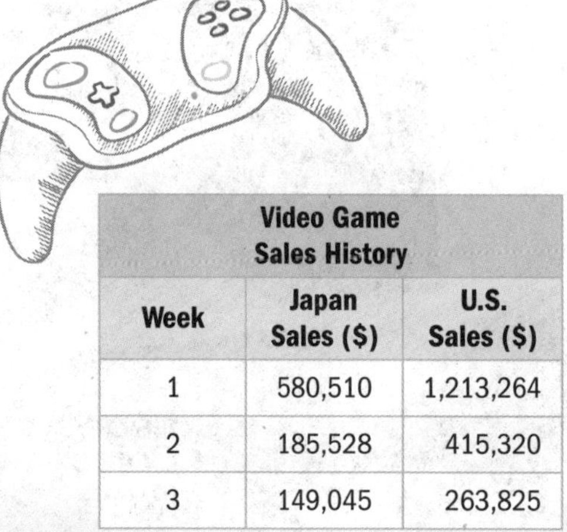

Video Game Sales History		
Week	Japan Sales ($)	U.S. Sales ($)
1	580,510	1,213,264
2	185,528	415,320
3	149,045	263,825

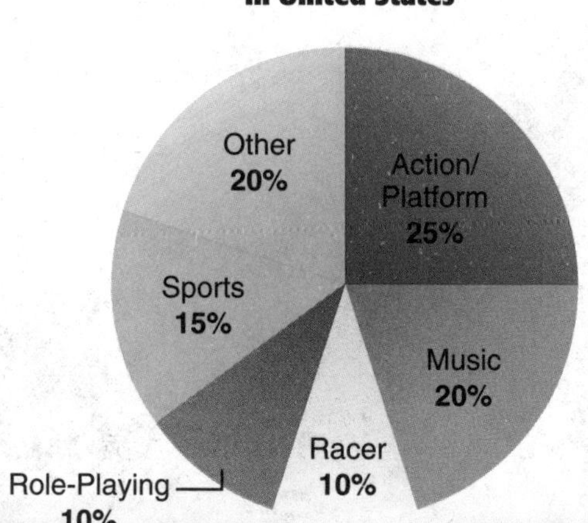

Career Project

It's time to update your career portfolio! Choose one of your favorite video games. Make a list of what you think are the best features of the game. Then describe any changes that you, as a video game designer, would make to the game.

____

____

____

____

____

List the strengths you have that would help you succeed in this career.

- ________
- ________
- ________
- ________
- ________

Chapter Review 

Vocabulary Check 

Complete the crossword puzzle using the vocabulary list at the beginning of the chapter.

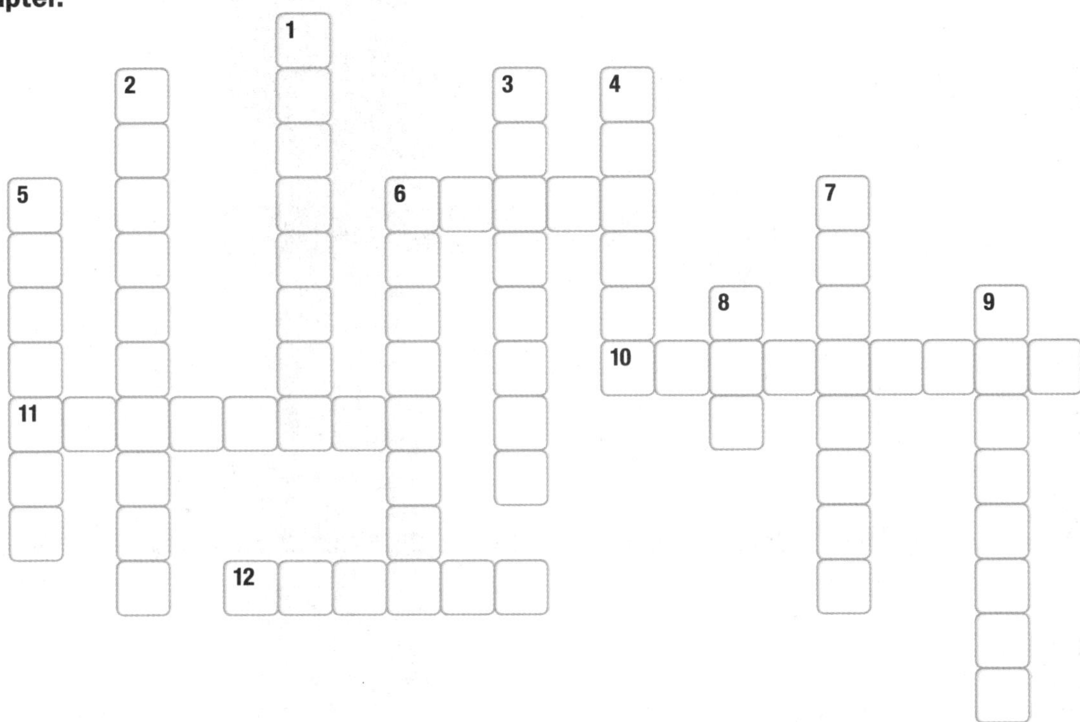

Down

1. type of percent when the final amount is greater than the original amount
2. statement that two ratios are equal
3. amount that the regular price is reduced
4. difference between what a store pays for an item and what a customer pays
5. price that a customer pays for an item
6. mathematical sentence stating that two expressions are equal
7. another term for the term in 3 down
8. gratuity
9. additional amount of money charged to items that people buy

Across

6. type of percent that compares the inaccuracy of an estimate to the actual amount
10. amount of money deposited or borrowed
11. amount paid or earned for the use of money
12. type of percent that compares the final and original amounts

Key Concept Check

Use Your FOLDABLES

Use your Foldable to help review the chapter.

Tape here

Percents

Examples

Examples

Got it?

Match each equation with its solution.

1. $0.15w = 45$

2. $15 = 20n$

3. $0.3(60) = p$

4. $600n = 750$

5. $0.15(80) = x$

6. $20 = 0.8w$

a. 125%

b. 12

c. 300

d. 18

e. 75%

f. 25

Problem Solving

1. CCSS **Justify Conclusions** The table shows the results of a survey in which 175 students were asked what type of food they wanted for a class party. How many students chose Italian food? Explain. (Lesson 1)

Type of Food	Percent
Subs	32%
Tex-Mex	56%
Italian	12%

2. A soccer team lost 30% of its games. Suppose the team won 14 games. How many games did the team play? (Lesson 3) ______

3. Tyree bought a collectible comic book for $49.62 last year. This year, he sold it for $52.10. Find the percent of change of the price of the comic book. Round to the nearest percent. (Lesson 5) ______

4. CCSS **Justify Conclusions** A restaurant bill comes to $42.75. Suppose the sales tax is 6% and a 15% tip is left on the amount after the tax is added. How much in all did the customer pay? Explain. (Lesson 6)

5. A new radio is priced at $30. An electronics store has an end-of-the-year sale. All of the items in the store are discounted by 40%. What is the sale price of the radio? (Lesson 7) ______

6. **Financial Literacy** Aleta deposited $450 into a savings account earning 3.75% simple interest. How much interest will she earn in 6 years? Explain. (Lesson 8)

Reflect

Answering the Essential Question

Use what you learned about percent to complete the graphic organizer. For each situation, circle an arrow to show if the final amount would be greater or less than the original amount. Then write a real-world percent problem and an equation that models it.

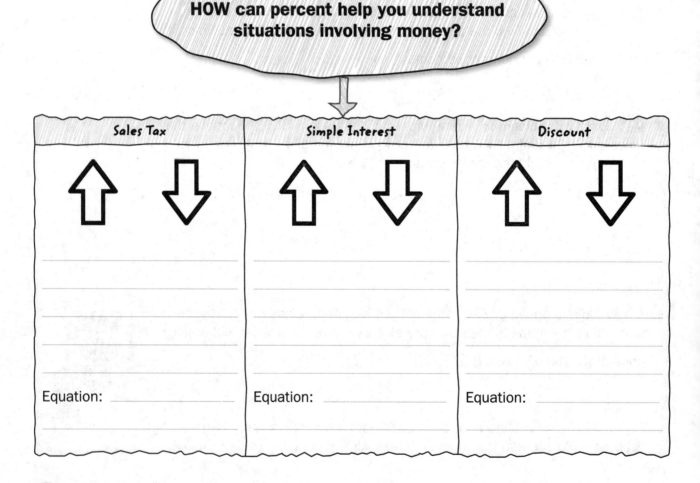

Essential Question

HOW can percent help you understand situations involving money?

Sales Tax	Simple Interest	Discount
⇧ ⇩	⇧ ⇩	⇧ ⇩
Equation:	Equation:	Equation:

Answer the Essential Question. HOW can percent help you understand situations involving money?

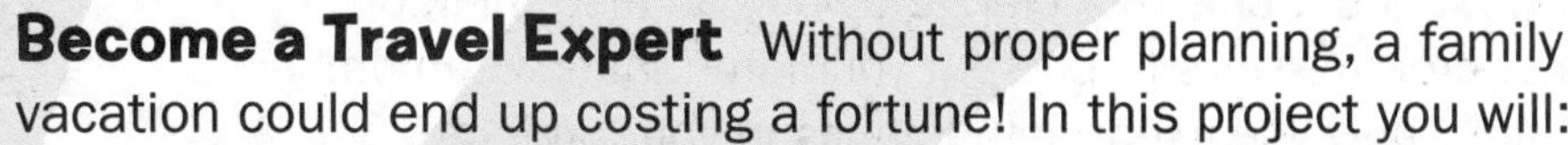

Watch

Become a Travel Expert Without proper planning, a family vacation could end up costing a fortune! In this project you will:

- **Collaborate** with your classmates as you research the cost of a family vacation.
- **Share** the results of your research in a creative way.
- **Reflect** on how you use mathematics to describe change and model real-world situations.

By the end of this project, you will be ready to plan a family vacation without breaking the bank.

Collaborate

Go Online Work with your group to research and complete each activity. You will use your results in the Share section on the following page.

1. Research the cost for a family of four to fly round trip to a destination of your choosing. Record the cost of a flight that is nonstop and one that has at least one extra stop. Make sure to include the cost of the tax.

2. Research two different rental cars that would be available at a local company. Compare the miles per gallon (mpg) that each car averages on the highway. How much gas would you use for each car if you were going to be traveling 450 miles on your trip?

3. If you are traveling out of the country you will need to know the current exchange rates. Record the exchange rate for three different countries. How much is \$100 worth in those countries?

4. Choose a vacation spot that is a city in the United States. Find a popular restaurant for tourists in your city and look up their menu online. Calculate the cost for a dinner that feeds four people. Don't forget the tip.

5. Different states have different sales tax rates. Choose three different states. Research the sales tax rate for each of those states. Then, determine the total cost of buying jeans that cost \$50 plus the sales tax.

Share

With your group, decide on a way to share what you have learned about the cost of a family vacation. Some suggestions are listed below, but you can also think of other creative ways to present your information. Remember to show how you used mathematics in your project!

- Use your creative writing skills to write journal entries or blogs. Your writing should describe how you were able to save money while traveling on your vacations.
- Act as a travel agent to put together one domestic and one international travel package for a family of four. Create a digital brochure to explain each package.

Check out the note on the right to connect this project with other subjects.

with Language Arts

Financial Literacy *Imagine that you are the director of tourism for your state. Write a script for a commercial that is trying to encourage tourists to visit. Your script should include:*

- *unique activities found in your state*
- *ways of traveling in your state*

Reflect

6. **Answer the Essential Question** How can you use mathematics to describe change and model real-world situations?

   a. How did you use what you learned about ratios and proportional reasoning to describe change and model the real-world situations in this project?

   b. How did you use what you learned about percents to describe change and model the real-world situations in this project?

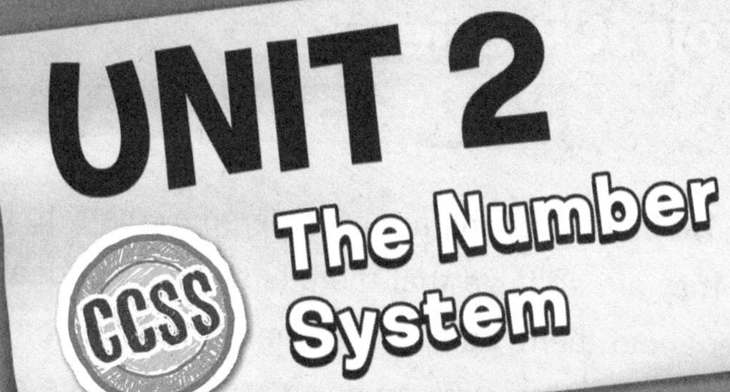

Essential Question

HOW can mathematical ideas be represented?

Chapter 3 Integers

Negative integers can be used in everyday contexts that involve values below zero. In this chapter, you will add, subtract, multiply, and divide integers.

Chapter 4 Rational Numbers

Every quotient of integers (with non-zero divisor) is a rational number. In this chapter, you will solve multi-step real-life problems by performing operations on rational numbers.

Unit Project Preview

Explore the Ocean Depths Oceans are considered by many to be the last frontier on Earth. They are so huge that there is still a great deal that we have yet to discover about them. Think about the tallest thing on Earth. Whatever it is, it doesn't even come close to being as deep as the ocean. However, as new technologies advance, exploration continues deeper and deeper into the ocean allowing us to find many new sea creatures.

At the end of Chapter 4, you'll complete a project about the oceans. But for now, jot down four sea creatures in the table and list at what depth of the ocean you think they live.

Ocean Creatures	
Creature	**Depth (m)**

Chapter 3
Integers

Essential Question

WHAT happens when you add, subtract, multiply, and divide integers?

Common Core State Standards

Content Standards
7.NS.1, 7.NS.1a, 7.NS.1b, 7.NS.1c, 7.NS.1d, 7.NS.2, 7.NS.2a, 7.NS.2b, 7.NS.2c, 7.NS.3, 7.EE.3

Mathematical Practices
1, 2, 3, 4, 5, 6, 7, 8

Math in the Real World

Penguins can stay under water up to 20 minutes at a time, sometimes diving to a depth of −275 feet. The number 20 is a positive integer; −275 is a negative integer.

On the graph below, graph a point at the maximum depth a penguin can dive.

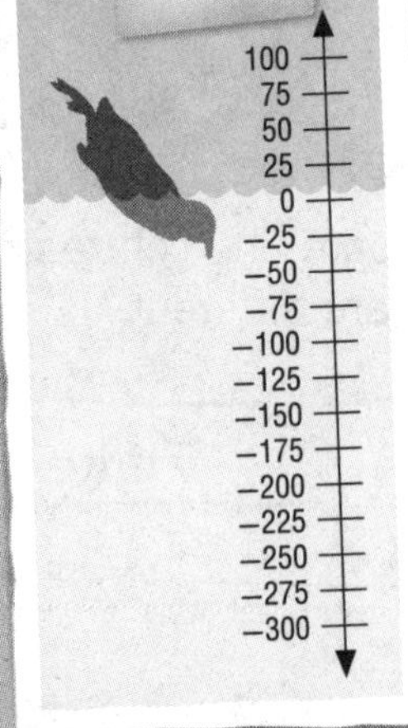

FOLDABLES® Study Organizer

Cut out the Foldable on page FL7 of this book.

Place your Foldable on page 254.

Use the Foldable throughout this chapter to help you learn about integers.

What Tools Do You Need?

Vocabulary

absolute value	integer	positive integer
additive inverse	negative integer	zero pair
graph	opposites	

Study Skill: Writing Math

Compare and Contrast When you *compare*, you notice how things are alike. When you *contrast*, you notice how they are different. Here are two cell phone plans.

Compare and contrast the monthly plans. Make a list of how they are alike and how they are different.

Alike/Compare	Different/Contrast

When Will You Use This?

Your Turn! **You will solve this problem in the chapter.**

Try the Quick Check below.
Or, take the Online Readiness Quiz.

Quick Review

Common Core Review 5.OA.1, 6.NS.6

Example 1

Evaluate $48 \div (6 + 2)5$.

Follow the order of operations.

$48 \div (6 + 2)5$

$= 48 \div 8 \cdot 5$ Add 6 and 2.

$= 6 \cdot 5$ Divide 48 by 8.

$= 30$ Multiply.

Example 2

Graph and label $M(6, 3)$ on the coordinate plane.

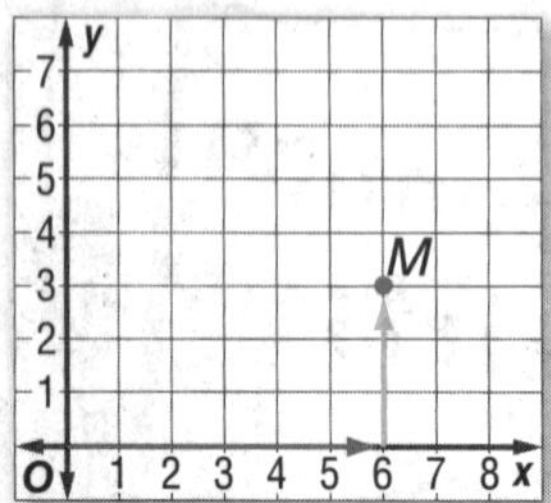

Start from the origin. Point M is located 6 units to the right and 3 units up.

Draw a dot and label the point.

Quick Check

Order of Operations **Evaluate each expression.**

1. $54 \div (6 + 3) =$ ______

2. $(10 + 50) \div 5 =$ ______

3. $18 + 2(4 - 1) =$ ______

Coordinate Graphing **Graph and label each point on the coordinate grid. The first one is done for you.**

4. $A(1, 1)$ **5.** $B(2, 8)$

6. $C(8, 1)$ **7.** $D(3, 4)$

8. $E(1, 5)$ **9.** $G(7, 6)$

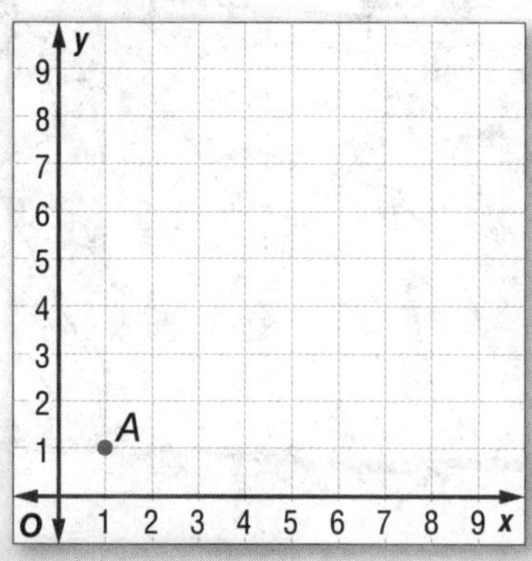

How Did You Do?

Which problems did you answer correctly in the Quick Check? Shade those exercise numbers below.

1 2 3 4 5 6 7 8 9

Lesson 1

Integers and Absolute Value

What You'll Learn

Scan the lesson. List two headings you would use to make an outline of the lesson.

- ____________________
- ____________________

Essential Question

WHAT happens when you add, subtract, multiply, and divide integers?

Vocabulary

integer
negative integer
positive integer
graph
absolute value

Common Core State Standards

Content Standards
Preparation for 7.NS.3

Mathematical Practices
1, 3, 4, 5

Vocabulary Start-Up

Numbers like 5 and −8 are called integers. An **integer** is any number from the set {..., −4, −3, −2, −1, 0, 1, 2, 3, 4, ...}, where ... means *continues without end*.

Complete the graphic organizer.

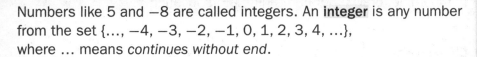

Describe It	Picture It
	−3 −2 −1 0 1 2 3
List Some Examples	**List Some NonExamples**

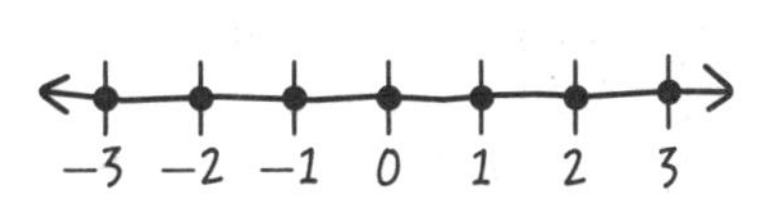
integer

Real-World Link

- The bottom of a snowboarding halfpipe is 5 meters below the top. Circle the integer you would you use to represent this position?

  5 or −5

- Describe another situation that uses negative integers. ____________________

Awesome halfpipe!

Work Zone

Identify and Graph Integers

Negative integers are integers less than zero. They are written with a − sign.

Positive integers are integers greater than zero. They can be written with a + sign.

Zero is neither negative nor positive.

Integers can be graphed on a number line. To **graph** an integer on the number line, draw a dot on the line at its location.

Examples

Write an integer for each situation.

1. **an average temperature of 5 degrees below normal**

Because it represents *below* normal, the integer is −5.

2. **an average rainfall of 5 inches above normal**

Because it represents *above* normal, the integer is +5 or 5.

Got It? **Do these problems to find out.**

Show your work.

Write an integer for each situation.

a. 6 degrees above normal **b.** 2 inches below normal

a. ____________

b. ____________

Example

3. **Graph the set of integers {4, −6, 0} on a number line.**

Draw a number line. Then draw a dot at the location of each integer.

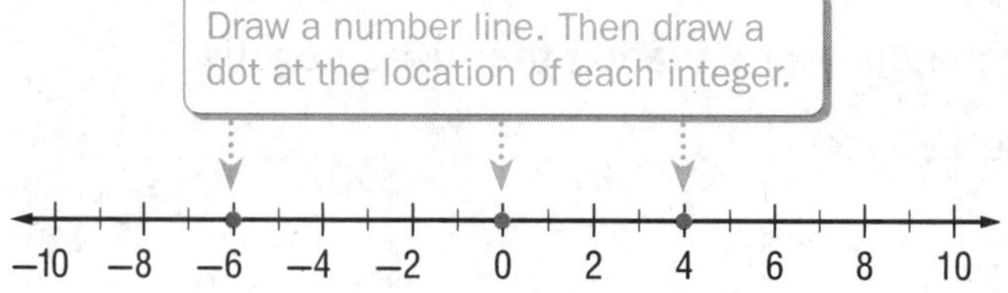

Got It? Do these problems to find out.

Graph each set of integers on a number line.

c. $\{-2, 8, -7\}$

d. $\{-4, 10, -3, 7\}$

c. ____________

d. ____________

Absolute Value

Key Concept

Words The absolute value of a number is the distance between the number and zero on a number line.

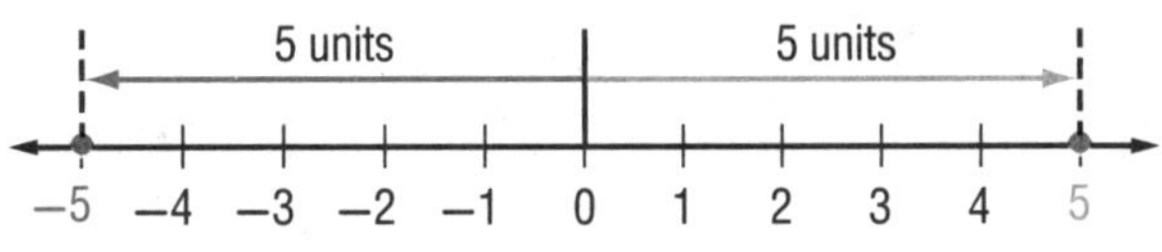

Examples $|-5| = 5$ $|5| = 5$

On the number line in the Key Concept box, notice that –5 and 5 are each 5 units from 0, even though they are on opposite sides of 0. Numbers that are the same distance from zero on a number line have the same **absolute value**.

Examples

Evaluate each expression.

4. $|-4|$

The graph of –4 is 4 units from 0.

So, $|-4| = 4$.

5. $|-5| - |2|$

$|-5| - |2| = 5 - 2$ $\quad |-5| = 5, |2| = 2$

So, $|-5| - |2| = 3$.

Order of Operations

The absolute value bars are considered to be a grouping symbol. When evaluating $|-5| - |2|$, evaluate the absolute values before subtracting.

Got It? Do these problems to find out.

e. $|8|$

f. $2 + |-3|$

g. $|-6| - 5$

e. ____________

f. ____________

g. ____________

Example

6. Nick climbs 30 feet up a rock wall and then climbs 22 feet down to a landing area. The number of feet Nick climbs can be represented using the expression $|30| + |-22|$. How many feet does Nick climb?

$|30| + |-22| = 30 + |-22|$ The absolute value of 30 is 30.

$= 30 + 22$ or 52 The absolute value of −22 is 22. Simplify.

So, Nick climbs 52 feet.

Guided Practice

Write an integer for each situation. (Examples 1 and 2)

1. a deposit of $16 ______

2. a loss of 11 yards ______

3. 6°F below zero ______

Evaluate each expression. (Examples 4–6)

4. $|-9| =$ ______

5. $|18| - |-10| =$ ______

6. $|-11| - |-6| =$ ______

7. Graph the set of integers {11, −5, −8} on a number line. (Example 3)

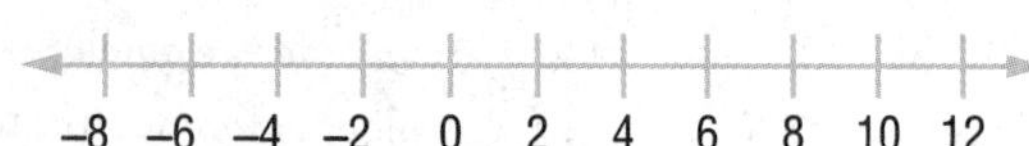

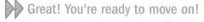

8. **Building on the Essential Question** Why is the absolute value of a number positive? Explain your reasoning. ______

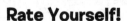

Rate Yourself!

☐ I understand integers and absolute value.

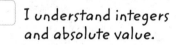

Great! You're ready to move on!

☐ I still have some questions about integers and absolute value.

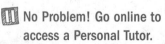

No Problem! Go online to access a Personal Tutor.

Independent Practice

Go online for Step-by-Step Solutions

Write an integer for each situation. (Examples 1and 2)

1. a profit of $9 ______

2. a bank withdrawal of $50 ______

3. 53°C below zero ______

4. 7 inches more than normal ______

Graph each set of integers on a number line. (Example 3)

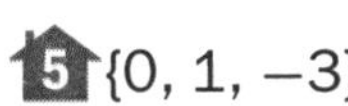

5. {0, 1, −3}

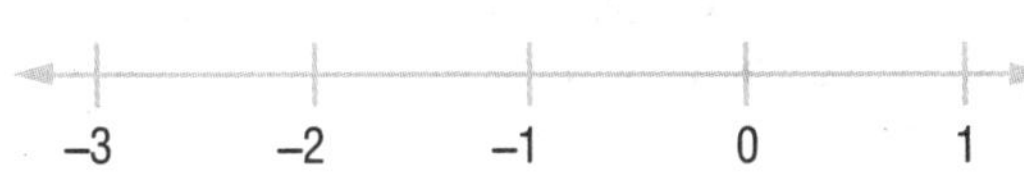

6. {−5, −1, 10, −9}

Evaluate each expression. (Examples 4 and 5)

7. $|10| =$ ______

8. $|-7| - 5 =$ ______

9. $1 + |7| =$ ______

10. The number of yards a football team moves on the field can be represented using the expression $|8| + |-4|$. How many yards does the football team move? (Example 6)

______

11. In golf, scores are often written in relationship to *par*, the average score for a round at a certain course. Write an integer to represent a score that is 7 under par. (Examples 1and 2)

______

12. A scuba diver descended 10 feet, 8 feet, and 11 feet. The total number of feet can be represented using the expression $|-10| + |-8| + |-11|$. What is the total number of feet the scuba diver descended?

13. **CCSS Use Math Tools** Mr. Chavez spent \$199.99 for a new smart phone, \$39.99 on a carrying case, and \$59.99 on accessories. The expression $|-199.99| + |-39.99| + |-59.99|$ represents the total amount that Mr. Chavez spent. How much did Mr. Chavez spend altogether? Check your answer using estimation.

H.O.T. Problems Higher Order Thinking

14. **CCSS Reason Inductively** If $|x| = 3$, what is the value of x?

15. **CCSS Persevere with Problems** Two numbers A and B are graphed on a number line. Is it *always*, *sometimes*, or *never* true that $A - |B| \leq A + B$ and $A > |B|$. Explain.

16. **CCSS Which One Doesn't Belong?** Identify the expression that is not equal to the other three. Explain your reasoning.

$15 - |-5|$	$|-4| + 6$	$-|7 + 3|$	$|-10|$

Standardized Test Practice

17. Which integer represents the temperature shown on the thermometer?

Ⓐ −11°F
Ⓑ −10°F
Ⓒ 10°F
Ⓓ 11°F

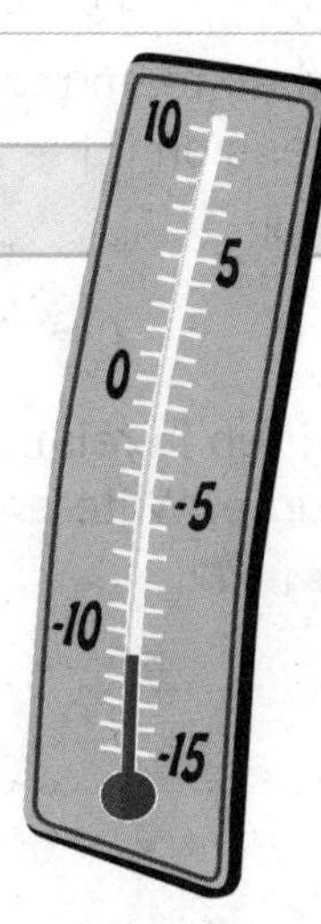

Name ____________________ My Homework ____________________

Extra Practice

Write an integer for each situation.

18. 2 feet below flood level −2

Because it represents below flood level, the integer is −2.

19. an elevator goes up 12 floors ______

Model with Mathematics **Graph each set of integers on a number line.**

20. {3, −7, 6}

−10 −8 −6 −4 −2 0 2 4 6 8 10

21. {−2, −4, −6, −8}

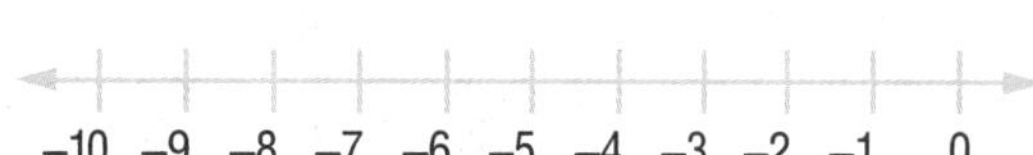

−10 −9 −8 −7 −6 −5 −4 −3 −2 −1 0

Evaluate each expression.

22. $|-12| =$ ______

23. $7 + |4| =$ ______

24. $|-9| + |-5| =$ ______

25. $|-10| \div 2 \times |5| =$ ______

26. $12 - |-8| + 7 =$ ______

27. $|27| \div 3 - |-4| =$ ______

28. Jasmine's pet guinea pig gained 8 ounces in one month. Write an integer to describe the amount of weight her pet gained.

______

Standardized Test Practice

29. Which point has a coordinate with the greatest absolute value?

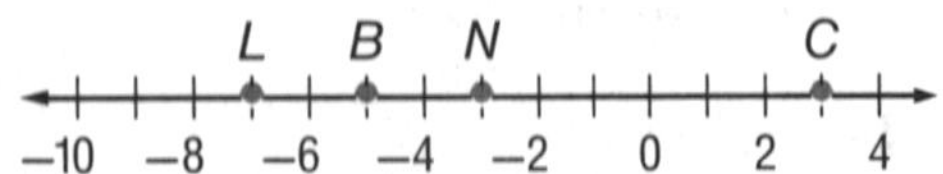

Ⓐ Point *B*

Ⓑ Point *C*

Ⓒ Point *L*

Ⓓ Point *N*

30. Which of the following statements about these real-world situations is *not* true?

Ⓕ A \$100 check deposited in a bank can be represented by +100.

Ⓖ A loss of 15 yards in a football game can be represented by −15.

Ⓗ A temperature of 20 below zero can be represented by −20.

Ⓘ A submarine diving 300 feet under water can be represented by +300.

31. Short Response Rachel recorded the low temperatures for one week in the table. On which day was the low temperature the farthest from 0°F? ______

Low Temperatures							
Day	Sunday	Monday	Tuesday	Wednesday	Thursday	Friday	Saturday
Temperature (°F)	2	−6	4	−8	2	0	−1

Common Core Review

Write the ordered pair corresponding to each point graphed at the right. Then state the quadrant or axis location of each point. 6.NS.6

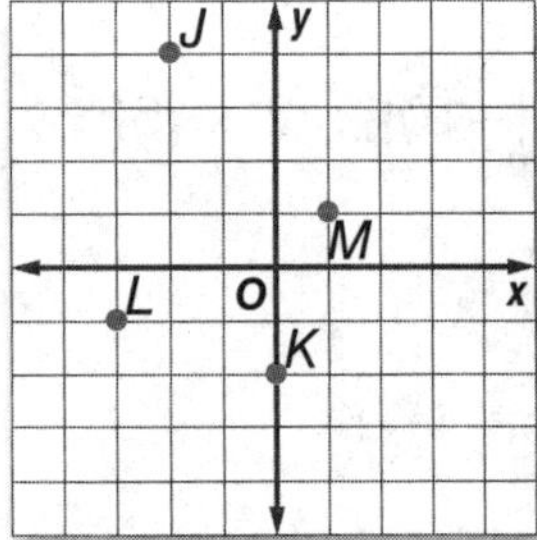

32. *J* ______

33. *K* ______

34. *L* ______

35. *M* ______

Graph and label each point on the coordinate plane. 6.NS.6c

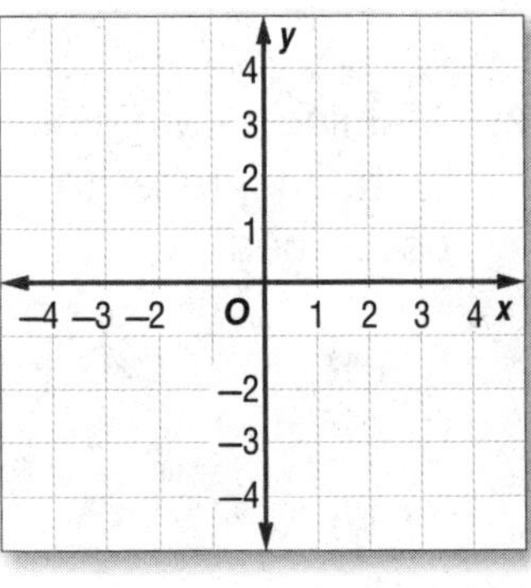

36. *A*(2, 4)

37. *B*(−3, 1)

38. *C*(2, 0)

39. *D*(−3, −3)

Inquiry Lab

Add Integers

WHEN is the sum of two integers a negative number?

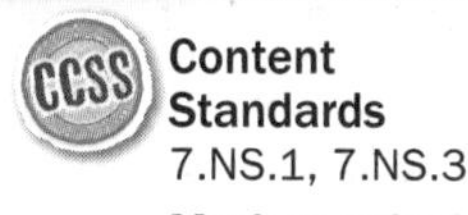

Content Standards
7.NS.1, 7.NS.3

Mathematical Practices
1, 3, 7

Football In football, forward progress is represented by a positive integer. Losing yardage is represented by a negative integer. On the first play, a team lost 5 yards. On the second play, the team lost 2 yards. What was the team's total yardage on the two plays? Find out in Investigation 1.

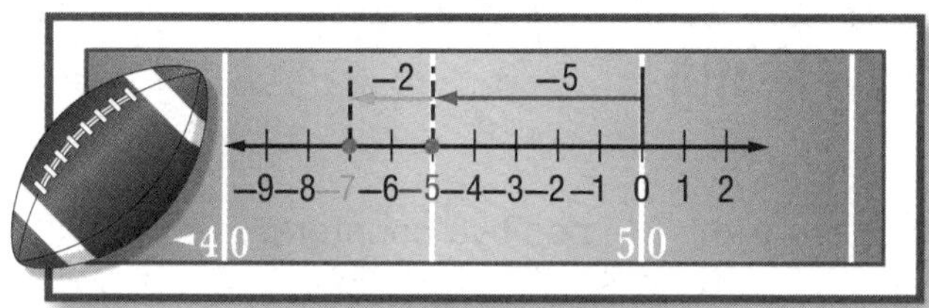

Investigation 1

Use counters to find the total yardage.

Step 1 Use negative integers to represent the yards lost on each play.

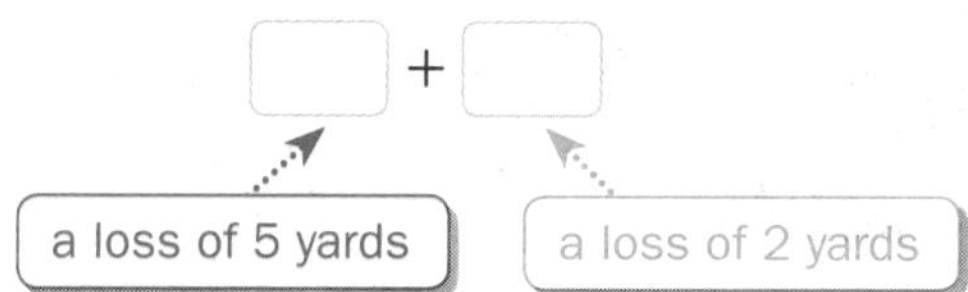

Step 2 Combine a set of 5 negative counters and a set of 2 negative counters.

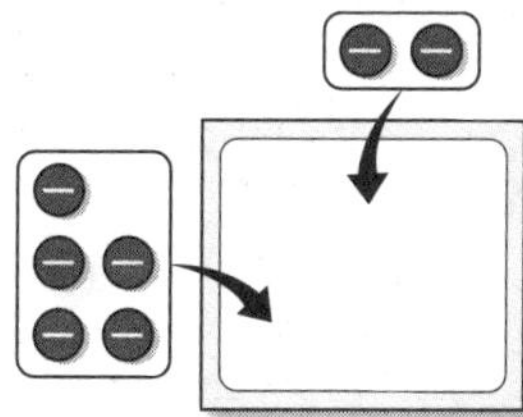

Step 3 There is a total of ☐ negative counters.

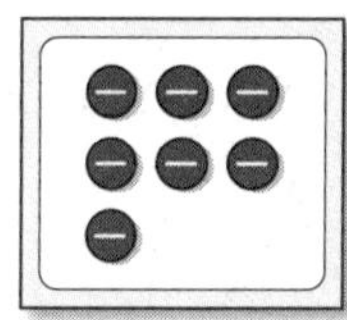

So, $-5 + (-2) =$ ☐. The team lost a total of ☐ yards on the first two plays.

The following two properties are important when modeling operations with integers.

- When one positive counter is paired with one negative counter, the result is called a **zero pair**. The value of a zero pair is 0.

- You can add or remove zero pairs from a mat because adding or removing zero does not change the value of the counters on the mat.

Investigation 2

Use counters to find −4 + 2.

Step 1 Combine ☐ negative counters with ☐ positive counters.

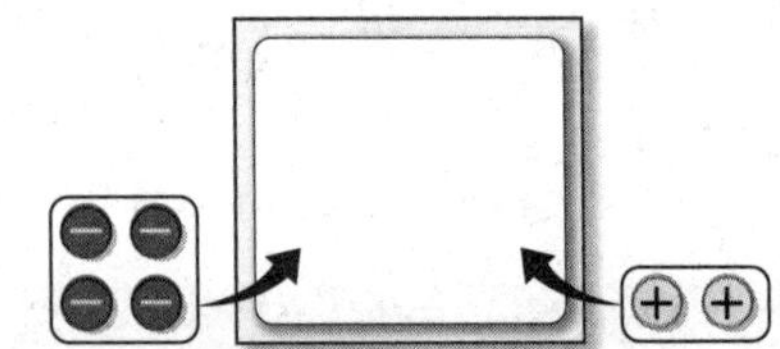

Step 2 Remove all zero pairs.

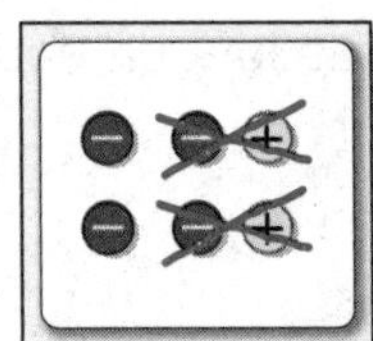

Step 3 Find the number of counters remaining.

There are ☐ negative counters remaining.

So, −4 + 2 = ☐.

Work with a partner. Find each sum. Show your work using drawings.

1. $5 + 6 =$ ______

2. $-3 + (-5) =$ ______

3. $-5 + (-4) =$ ______

4. $7 + 3 =$ ______

5. $-6 + 5 =$ ______

6. $-2 + 7 =$ ______

7. $8 + (-3) =$ ______

8. $3 + (-6) =$ ______

Analyze

Work with a partner to complete the table. The first one is done for you.

	Addition Expression	Sum	Sign of Addend with Greater Absolute Value	Sign of Sum
	$5 + (-2)$	3	Positive	Positive
9.	$-6 + 2$			
10.	$7 + (-12)$			
11.	$-4 + 9$			
12.	$-12 + 20$			
13.	$15 + (-18)$			

14. CCSS **Identify Structure** Write two addition sentences where the sum is positive. In each sentence, one addend should be positive and the other negative.

15. Write two addition sentences where the sum is negative. In each sentence, one addend should be positive and the other negative.

16. Write two addition sentences where the sum is zero. Describe the numbers.

Reflect

17. CCSS **Justify Conclusions** A contestant scores −100 points in the first round, −250 points in the second round, and 500 points in the third round. Find the contestant's total number of points. Explain your reasoning.

18. Inquiry WHEN is the sum of two integers a negative number?

Lesson 2

Add Integers

What You'll Learn

Scan the lesson. List two headings you would use to make an outline of the lesson.

-
-

Essential Question

WHAT happens when you add, subtract, multiply, and divide integers?

Vocabulary

opposites

additive inverse

Common Core State Standards

Content Standards

7.NS.1, 7.NS.1a, 7.NS.1b, 7.NS.1d, 7.NS.3, 7.EE.3

Mathematical Practices

1, 3, 4, 7

Vocabulary Start-Up

Integers like 2 and −2 are called **opposites** because they are the same distance from 0, but on opposite sides. Complete the graphic organizer about opposites.

Picture It

−3 −2 −1 0 1 2 3

Real-World Example	Math Example

Two integers that are opposites are also called **additive inverses**. The Additive Inverse Property states that the sum of any number and its additive inverse is zero. You can show 2 + (−2) on a number line.

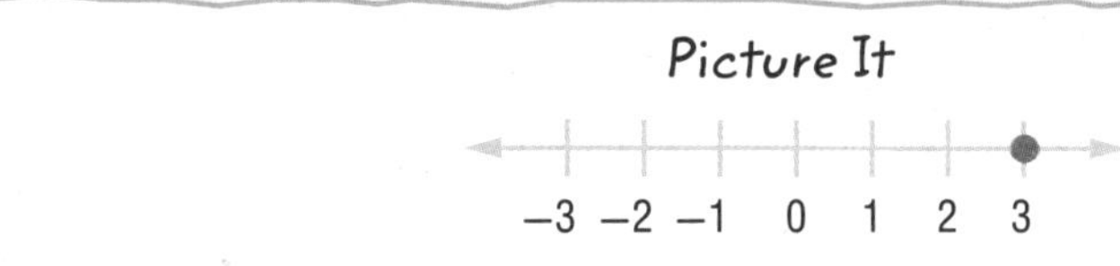

Start at zero.
Move 2 units to the right to show 2.
Then move 2 units to the left to show −2.

So, 2 + (−2) = ☐.

Real-World Link

The temperature outside is −5°. Name the temperature that would make the sum of the two temperatures 0°. ☐

Key Concept Add Integers with the Same Sign

Work Zone

Words To add integers with the same sign, add their absolute values. The sum is:

- positive if both integers are positive.
- negative if both integers are negative.

Examples $7 + 4 = 11$ $-7 + (-4) = -11$

Examples

1. Find $-3 + (-2)$.

Start at 0. Move 3 units down to show -3.

From there, move 2 units down to show -2.

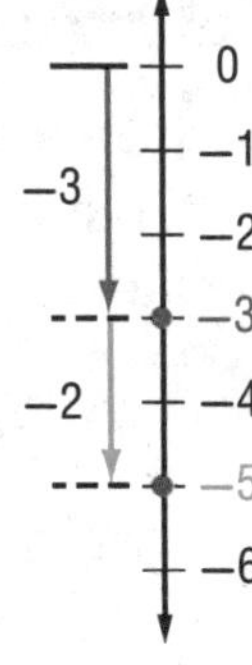

So, $-3 + (-2) = -5$.

Show your work.

2. Find $-26 + (-17)$.

$-26 + (-17) = -43$ Both integers are negative, so the sum is negative.

Got It? Do these problems to find out.

a. $-5 + (-7)$ **b.** $-10 + (-4)$ **c.** $-14 + (-16)$

a. ____________

b. ____________

c. ____________

Key Concept Add Integers with Different Signs

Words To add integers with different signs, subtract their absolute values. The sum is:

- positive if the positive integer's absolute value is greater.
- negative if the negative integer's absolute value is greater.

Examples $9 + (-4) = 5$ $-9 + 4 = -5$

When you add integers with different signs, start at zero. Move right for positive integers. Move left for negative integers. So, the sum of $p + q$ is located a distance $|q|$ from p.

Examples

3. Find 5 + (–3).

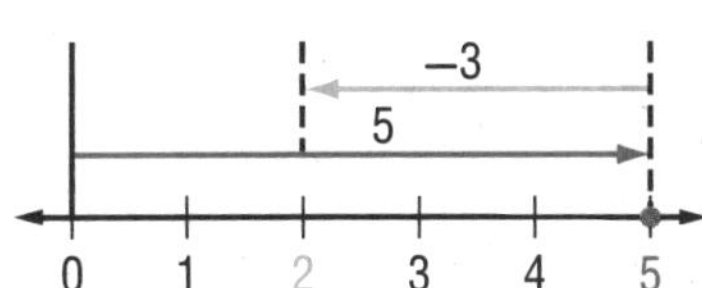

So, $5 + (-3) = 2$.

4. Find –3 + 2.

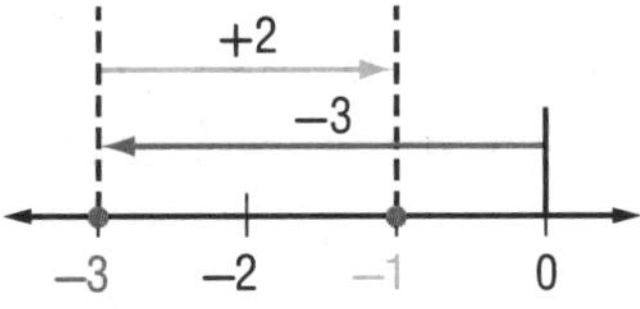

So, $-3 + 2 = -1$.

Got It? Do these problems to find out.

d. $6 + (-7)$ **e.** $-15 + 19$

d. ______

e. ______

Examples

5. Find 7 + (–7).

$7 + (-7) = 0$ Subtract absolute values; $7 - 7 = 0$. 7 and (-7) are opposites. The sum of any number and its opposite is always zero.

6. Find –8 + 3.

$-8 + 3 = -5$ Subtract absolute values; $8 - 3 = 5$. Since -8 has the greater absolute value, the sum is negative.

7. Find 2 + (–15) + (–2).

$2 + (-15) + (-2) = 2 + (-2) + (-15)$	Commutative Property (+)
$= [2 + (-2)] + (-15)$	Associative Property (+)
$= 0 + (-15)$	Additive Inverse Property
$= -15$	Additive Identity Property

Commutative Properties
$a + b = b + a$
$a \cdot b = b \cdot a$

Associative Properties
$a + (b + c) = (a + b) + c$
$a \cdot (b \cdot c) = (a \cdot b) \cdot c$

Identity Properties
$a + 0 = a$
$a \cdot 1 = a$

Got It? Do these problems to find out.

f. $10 + (-12)$ **g.** $-13 + 18$ **h.** $(-14) + (-6) + 6$

f. ______

g. ______

h. ______

Example

8. A roller coaster starts at point *A*. It goes up 20 feet, down 32 feet, and then up 16 feet to point *B*. Write an addition sentence to find the height at point *B* in relation to point *A*. Then find the sum and explain its meaning.

$20 + (-32) + 16 = 20 + 16 + (-32)$ Commutative Property (+)

$= 36 + (-32)$ $20 + 16 = 36$

$= 4$ Subtract absolute values.

Point *B* is 4 feet higher than point *A*.

Got It? Do this problem to find out.

Show your work.

i. The temperature is −3°. An hour later, it drops 6° and 2 hours later, it rises 4°. Write an addition expression to describe this situation. Then find the sum and explain its meaning.

i. ____________

Guided Practice

Add. (Examples 1–7)

1. $-6 + (-8) =$ ______

2. $-3 + 10 =$ ______

3. $-8 + (-4) + 12 =$ ______

4. Sofia owes her brother \$25. She gives her brother the \$18 she earned dog-sitting. Write an addition expression to describe this situation. Then find the sum and explain its meaning. (Example 8) ____________

5. **Building on the Essential Question** Explain how you know whether a sum is positive, negative, or zero without actually adding. ____________

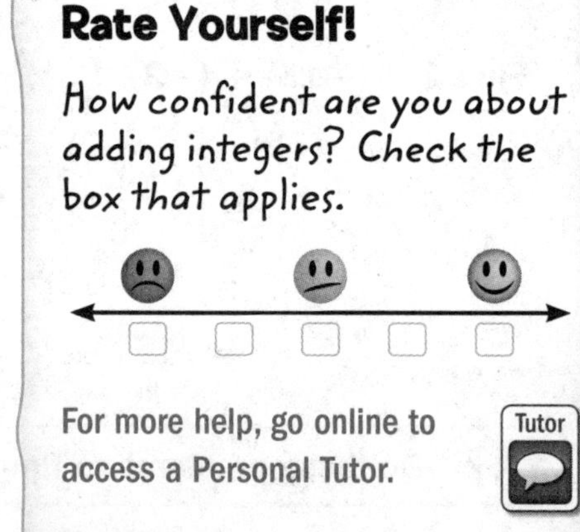

Rate Yourself!
How confident are you about adding integers? Check the box that applies.
For more help, go online to access a Personal Tutor.
Tutor
FOLDABLES Time to update your Foldable!

Name ______ My Homework ______

Independent Practice

Go online for Step-by-Step Solutions 

Add. (Examples 1–7)

1. $-22 + (-16) =$ ______

2. $-10 + (-15) =$ ______

3. $6 + 10 =$ ______

4. $21 + (-21) + (-4) =$ ______

5. $-17 + 20 + (-3) =$ ______

6. $-34 + 25 + (-25) =$ ______

7. $4 + 5 =$ ______

8. $-15 + 8 =$ ______

9. $7 + (-11) =$ ______

10. **Financial Literacy** Stephanie has $152 in the bank. She withdraws $20. Then she deposits $84. Write an addition expression to represent this situation. Then find the sum and explain its meaning. (Example 8)

______

11. CCSS **Model with Mathematics** Find the total profit or loss for each color of T-shirt. ______

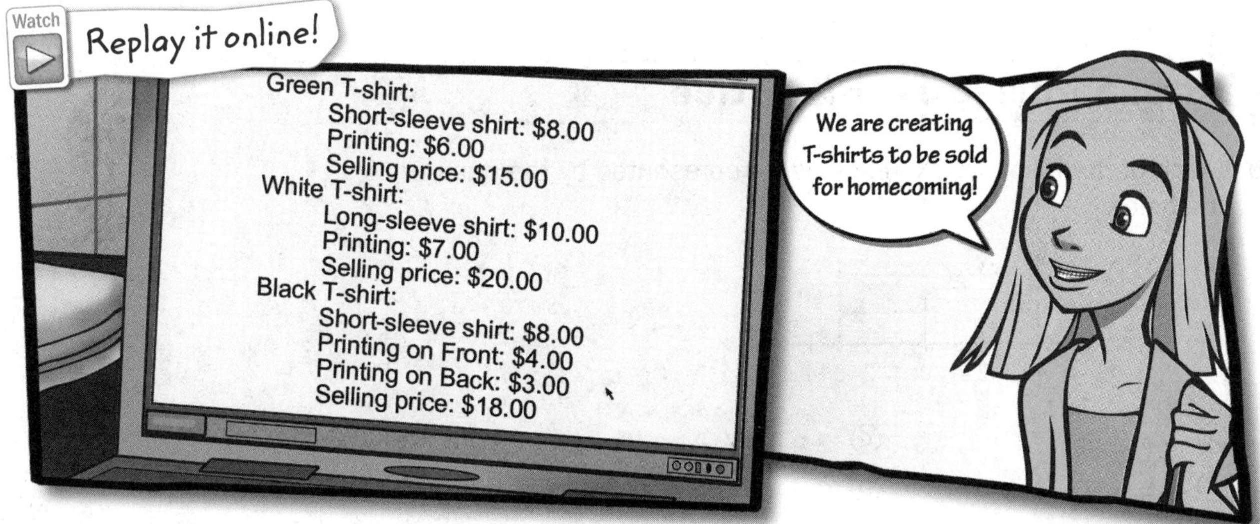

Watch
Replay it online!
Green T-shirt:
Short-sleeve shirt: $8.00
Printing: $6.00
Selling price: $15.00
White T-shirt:
Long-sleeve shirt: $10.00
Printing: $7.00
Selling price: $20.00
Black T-shirt:
Short-sleeve shirt: $8.00
Printing on Front: $4.00
Printing on Back: $3.00
Selling price: $18.00
We are creating T-shirts to be sold for homecoming!

12. **Reason Abstractly** Lena deposits and withdraws money from a bank account. The table shows her transactions for March. Write an addition expression to describe her transactions. Then find the sum and explain its meaning.

March	
Week	**Transaction**
1	deposit \$300
2	withdraw \$50
3	withdraw \$75
4	deposit \$225

H.O.T. Problems Higher Order Thinking

13. **Model with Mathematics** Describe two situations in which opposite quantities combine to make zero.

14. **Identify Structure** Name the property illustrated by the following.

a. $x + (-x) = 0$ ______

b. $x + (-y) = -y + x$ ______

Model with Mathematics Simplify.

15. $8 + (-8) + a$ ______

16. $x + (-5) + 1$ ______

17. $-9 + m + (-6)$ ______

Standardized Test Practice

18. Which of the following expressions is represented by the number line below?

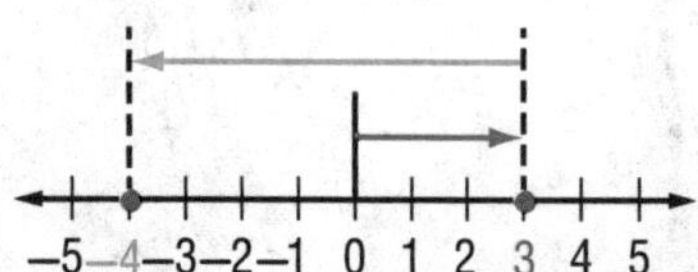

Ⓐ $-4 + 3$

Ⓑ $-4 + 7$

Ⓒ $3 + (-7)$

Ⓓ $0 + (-7)$

Name ______________________ My Homework ______________________

Extra Practice

Add.

19. $18 + (-5) =$ 13

$18 + (-5) = 18 - 5$

$= 13$

20. $-19 + 24 =$ 5

$-19 + 24 = 24 + (-19)$

$= 24 - 19$

$= 5$

21. $13 + (-19) =$ ______

22. $14 + (-6) =$ ______

23. $15 + 9 + (-9) =$ ______

24. $-4 + 12 + (-9) =$ ______

25. $-16 + 16 + 22 =$ ______

26. $25 + 3 + (-25) =$ ______

27. $7 + (-19) + (-7) =$ ______

CCSS Justify Conclusions Write an addition expression to describe each situation. Then find each sum and explain its meaning.

28. Ronnie receives \$40 for his birthday. Then he spends \$15 at the movies.

29. A quarterback is sacked for a loss of 5 yards. On the next play, his team loses 15 yards. Then the team gains 12 yards on the third play.

30. A pelican starts at 60 feet above sea level. It descends 60 feet to catch a fish.

Standardized Test Practice

31. At 8 A.M., the temperature was 3°F below zero. By 1 P.M., the temperature rose 14°F and by 10 P.M., dropped 12°F. What was the temperature at 10 P.M?

Ⓐ 5°F above zero

Ⓑ 5°F below zero

Ⓒ 1°F above zero

Ⓓ 1°F below zero

32. What is the value of $-8 + 7 + (-3)$?

Ⓕ −18

Ⓖ −4

Ⓗ 2

Ⓘ 18

33. **Short Response** Write an addition sentence to represent the number line below. ______

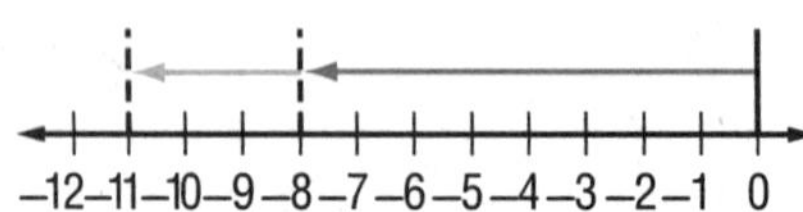

Common Core Review

Write an integer for each situation. 6.NS.5

34. a bank deposit of $75 ______

35. a loss of 8 pounds ______

36. 13° below zero ______

37. a gain of 4 yards ______

38. spending $12 ______

39. a gain of 5 hours ______

Inquiry Lab

Subtract Integers

HOW is the subtraction of integers related to the addition of integers?

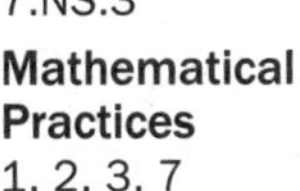

Content Standards
7.NS.1, 7.NS.1c, 7.NS.3

Mathematical Practices
1, 2, 3, 7

Dolphins A dolphin swims 6 meters below the surface of the ocean. Then it jumps to a height of 5 meters above the surface of the water. Determine the difference between the two distances.

Investigation 1

Use counters to find 5 − (−6), the difference between the distances.

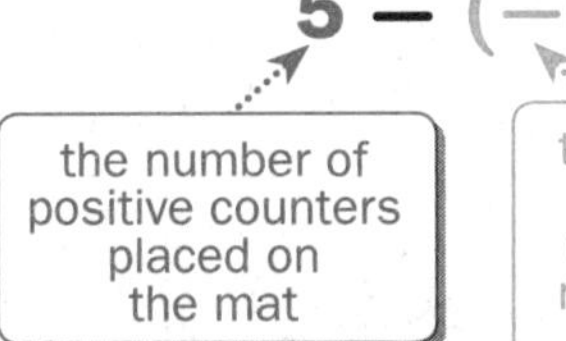

Step 1 Place 5 positive counters on the mat. Remove 6 negative counters. However, there are 0 negative counters.

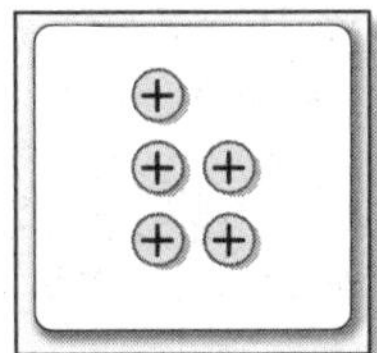

Step 2 Add ☐ zero pairs to the mat.

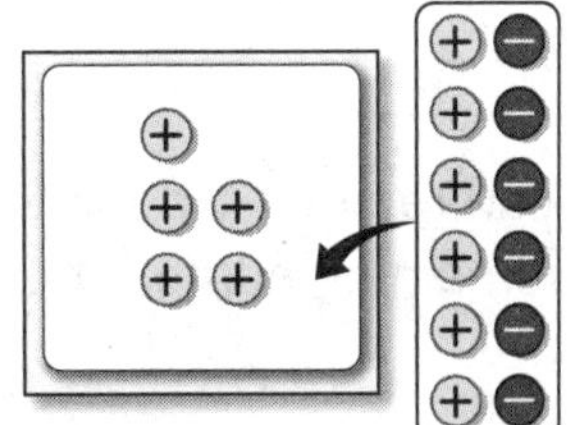

Step 3 Now you can remove ☐ negative counters. Count the remaining positive counters.

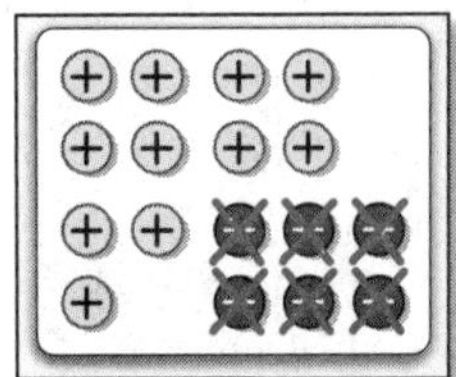

So, 5 − (−6) = ☐. The difference between the distances is ☐ meters.

Investigation 2

Use counters to find –6 – (–3).

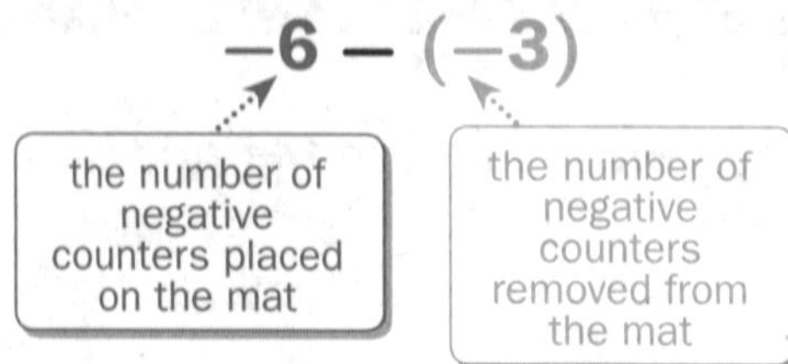

Step 1 Place 6 negative counters on the mat.

Step 2 Remove 3 negative counters.

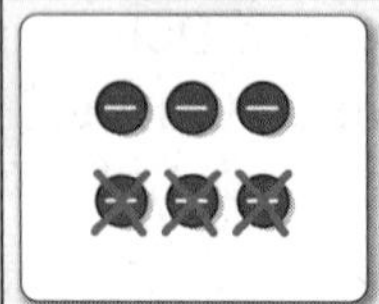

There are ☐ negative counters remaining. So, –6 – (–3) = ☐.

Investigation 3

Use counters to find –5 – 1.

Step 1 Place ☐ negative counters on the mat. You need to remove 1 positive counter. However, there are 0 positive counters.

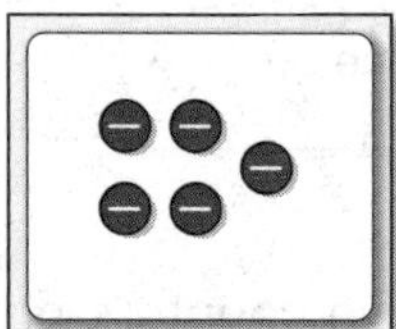

Step 2 Add 1 zero pair to the mat.

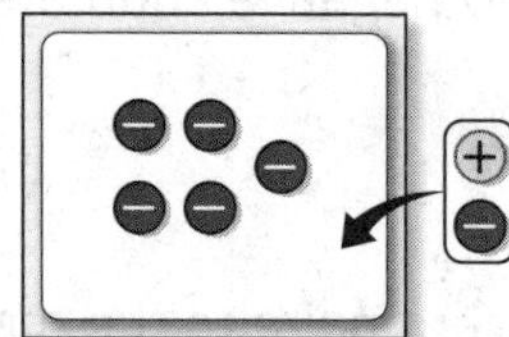

Step 3 Now you can remove 1 positive counter. Find the remaining number of counters.

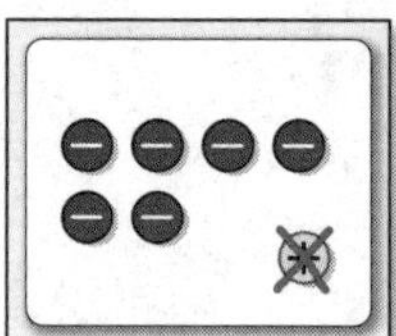

There are ☐ negative counters remaining.

So, –5 – 1 = ☐.

Check Use addition. $-6 + 1 \stackrel{?}{=} -5$

$-5 = -5$ ✓

Work with a partner. Find each difference. Show your work using drawings.

1. $7 - 6 =$ ______

Show your work.

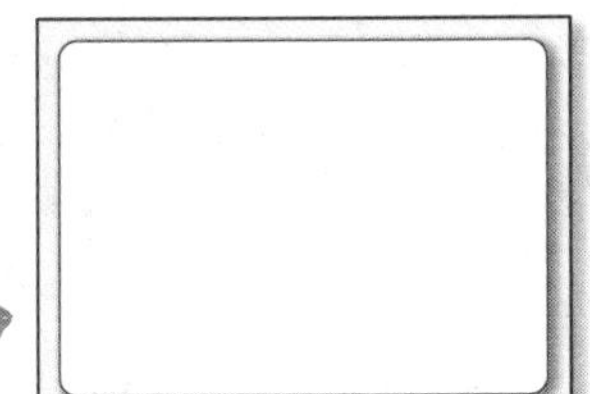

2. $5 - (-3) =$ ______

3. $6 - (-2) =$ ______

4. $5 - 8 =$ ______

5. $-7 - (-2) =$ ______

6. $-7 - 3 =$ ______

7. $-5 - (-7) =$ ______

8. $-2 - (-9) =$ ______

Analyze

Work with a partner. Circle an expression that is equal to the expression in the first column. The first one is done for you.

	$-3-1$	$-3+1$	$-3+(-1)$	$-3-(-1)$
9.	$-2-9$	$-2-(-9)$	$-2+9$	$-2+(-9)$
10.	$-8-4$	$-8+4$	$-8+(-4)$	$-8-(-4)$
11.	$6-(-2)$	$6+2$	$6-2$	$6+(-2)$
12.	$5-(-7)$	$5-7$	$5+(-7)$	$5+7$
13.	$-1-(-3)$	$-1-3$	$-1+3$	$-1+(-3)$
14.	$-3-(-8)$	$-3+8$	$-3-8$	$-3+(-8)$

15. CCSS **Identify Structure** Write a subtraction sentence where the difference is positive. Use a positive and a negative integer.

16. Write a subtraction sentence where the difference is negative. Use a positive and a negative integer.

Reflect

17. CCSS **Reason Abstractly** Jake owes his sister $3. She decides to "take away" his debt. That is, he does not have to pay her back. Write a subtraction sentence for this situation.

18. Inquiry HOW is the subtraction of integers related to the addition of integers?

Lesson 3

Subtract Integers

What You'll Learn

Scan the lesson. List two real-world scenarios in which you would subtract integers.

-
-

Essential Question

WHAT happens when you add, subtract, multiply, and divide integers?

Common Core State Standards

Content Standards
7.NS.1, 7.NS.1c, 7.NS.1d, 7.NS.3

Mathematical Practices
1, 2, 3, 4, 5, 6, 7

Real-World Link

Diving The platform on a diving board is 3 meters high. The actions of a diver climbing up to the diving board platform and diving 1 meter below the water's surface are shown on the number line at the right.

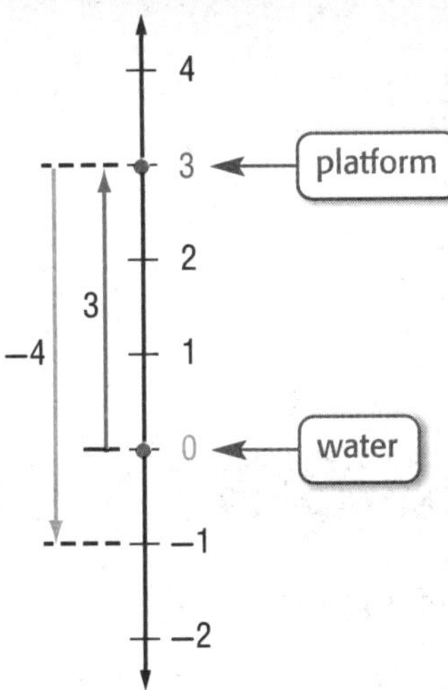

The diver's actions can be represented by the subtraction equation $3 - 4 = -1$.

1. Write a related addition sentence for the subtraction sentence.

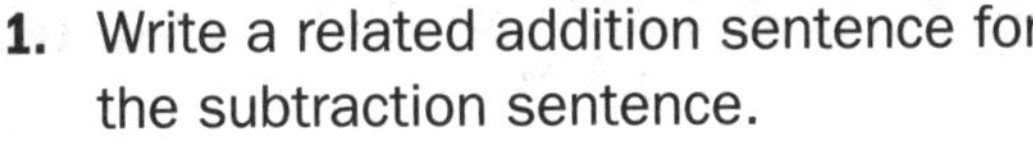

2. Use a number line to find $1 - 5$. Then write a related addition sentence for the subtraction sentence.

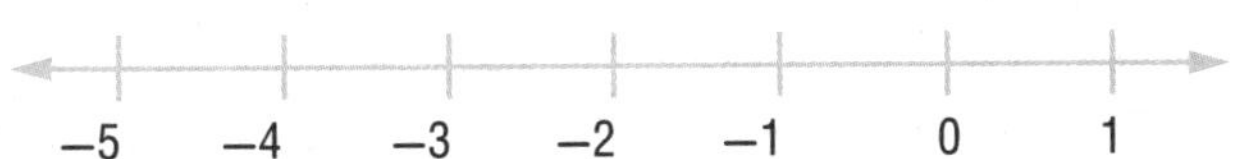

Difference: ______ Addition Sentence: ______

Key Concept

Subtract Integers

Words	To subtract an integer, add its additive inverse.
Symbols	$p - q = p + (-q)$
Examples	$4 - 9 = 4 + (-9) = -5$ $7 - (-10) = 7 + (10) = 17$

Work Zone

When you subtract 7, the result is the same as adding its additive inverse, −7.

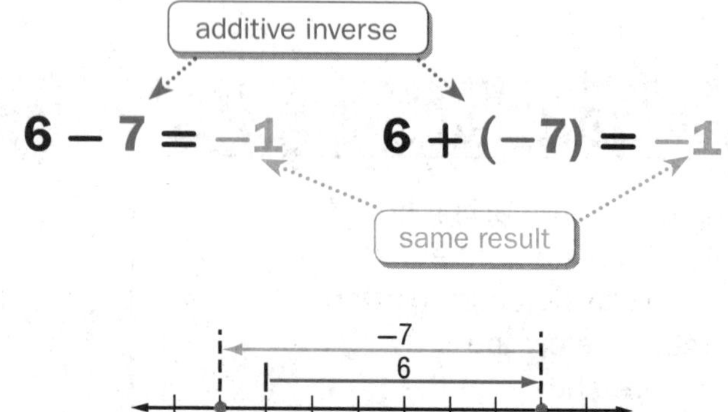

Examples

1. Find 8 − 13.

$8 - 13 = 8 + (-13)$ To subtract 13, add −13.

$= -5$ Simplify.

Check by adding $-5 + 13 \stackrel{?}{=} 8$

$8 = 8$ ✓

2. Find −10 − 7.

$-10 - 7 = -10 + (-7)$ To subtract 7, add −7.

$= -17$ Simplify.

Check by adding $-17 + 7 \stackrel{?}{=} -10$

$-10 = -10$ ✓

Got It? **Do these problems to find out.**

a. $6 - 12$ **b.** $-20 - 15$ **c.** $-22 - 26$

Show your work.

a. ____________

b. ____________

c. ____________

Examples

3. Find $1 - (-2)$.

$1 - (-2) = 1 + 2$ To subtract -2, add 2.

$= 3$ Simplify.

4. Find $-10 - (-7)$.

$-10 - (-7) = -10 + 7$ To subtract -7, add 7.

$= -3$ Simplify.

STOP and Reflect

Circle the integer below that will make this number sentence true.

$-5 - (?) = -3$

-8 -2 2

Got It? Do these problems to find out.

d. $4 - (-12)$ **e.** $-15 - (-5)$ **f.** $18 - (-6)$

d. ____________

e. ____________

f. ____________

Examples

5. Evaluate $x - y$ if $x = -6$ and $y = -5$.

$x - y = -6 - (-5)$ Replace x with -6 and y with -5.

$= -6 + 5$ To subtract -5, add 5.

$= -1$ Simplify.

6. Evaluate $m - n$ if $m = -15$ and $n = 8$.

$m - n = -15 - 8$ Replace m with -15 and n with 8.

$= -15 + (-8)$ To subtract 8, add -8.

$= -23$ Simplify.

Got It? Do these problems to find out.

Evaluate each expression if $a = 5$, $b = -8$, and $c = -9$.

g. $b - 10$ **h.** $a - b$ **i.** $c - a$

g. ____________

h. ____________

i. ____________

Example

7. The temperatures on the Moon vary from –173°C to 127°C. Find the difference between the maximum and minimum temperatures.

Subtract the lower temperature from the higher temperature.

Estimate $100 - (-200) = 300$

$127 - (-173) = 127 + 173$ To subtract –173, add 173.

$= 300$ Simplify.

So, the difference between the temperatures is 300°C.

Got It? **Do this problem to find out.**

Show your work.

j. Brenda had a balance of –$52 in her account. The bank charged her a fee of $10 for having a negative balance. What is her new balance?

j. ______

Guided Practice

Subtract. (Examples 1–4)

1. $14 - 17 =$ ______

2. $14 - (-10) =$ ______

3. $12 - 26 =$ ______

4. Evaluate $q - r$ if $q = -14$ and $r = -6$. (Examples 5 and 6)

______

5. STEM The sea surface temperatures range from –2°C to 31°C. Find the difference between the maximum and minimum temperatures. (Example 7) ______

6.  **Building on the Essential Question** If x and y are positive integers, is $x - y$ always positive? Explain.

______

______

Rate Yourself!

How well do you understand subtracting integers? Circle the image that applies.

Clear Somewhat Clear Not So Clear

For more help, go online to access a Personal Tutor.

FOLDABLES Time to update your Foldable!

Name ______________________ My Homework ______________________

Independent Practice

Go online for Step-by-Step Solutions 

Subtract. (Examples 1–4)

1. $0 - 10 =$ ______

2. $-9 - 5 =$ ______

3. $-4 - 8 =$ ______

4. $31 - 48 =$ ______

5. $-25 - 5 =$ ______

6. $-44 - 41 =$ ______

7. $4 - (-19) =$ ______

8. $-11 - (-42) =$ ______

9. $52 - (-52) =$ ______

Evaluate each expression if $f = -6$, $g = 7$, and $h = 9$. (Examples 5 and 6)

10. $g - 7$ ______

11. $-h - (-9)$ ______

12. $f - g$ ______

 13. **Use Math Tools** Use the information below. (Example 7)

State	Alabama	California	Florida	Louisiana	New Mexico
Lowest Elevation (ft)	0	−282	0	−8	2,842
Highest Elevation (ft)	2,407	14,494	345	535	13,161

a. What is the difference between the highest elevation in Alabama and the lowest elevation in Louisiana? ______

b. Find the difference between the lowest elevation in New Mexico and the lowest elevation in California. ______

c. Find the difference between the highest elevation in Florida and the lowest elevation in California. ______

d. What is the difference between the lowest elevation in Alabama and the lowest elevation in Louisiana? ______

Evaluate each expression if $h = -12$, $j = 4$, and $k = 15$.

14. $-j + h - k$ ________

15. $|h - j|$ ________

16. $k - j - h$ ________

H.O.T. Problems Higher Order Thinking

17. CCSS **Identify Structure** Write a subtraction sentence using integers. Then, write the equivalent addition sentence and explain how to find the sum.

18. CCSS **Identify Structure** Use the properties of operations.

a. The Commutative Property is true for addition. For example, $7 + 2 = 2 + 7$. Does the Commutative Property apply to subtraction. Is $2 - 7$ equal to $7 - 2$? Explain.

b. Using the Associative Property, $9 + (6 + 3) = (9 + 6) + 3$. Is $9 - (6 - 3)$ equal to $(9 - 6) - 3$? Explain.

19. CCSS **Find the Error** Hiroshi is finding $-15 - (-18)$. Find his mistake and correct it.

20. CCSS **Reason Abstractly** *True* or *False*? When n is a negative integer, $n - n = 0$.

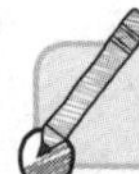

Standardized Test Practice

21. Which of the following expressions is equal to -8?

Ⓐ $15 - 7$ Ⓑ $-15 - 7$ Ⓒ $15 - (-7)$ Ⓓ $-15 - (-7)$

Name ______________________ My Homework ______________________

Extra Practice

Subtract.

22. $13 - 17 =$ -4

Homework Help

$13 - 17 = 13 + (-17)$

$= -4$

23. $27 - (-8) =$ 35

$27 - (-8) = 27 + 8$

$= 35$

24. $-8 - 9 =$ ______

25. $-34 - (-20) =$ ______

26. $15 - (-14) =$ ______

27. $-27 - (-33) =$ ______

Evaluate each expression if $f = -6$, $g = 7$, and $h = 9$.

28. $f - 6$ ______

29. $h - f$ ______

30. $g - h$ ______

31. $5 - f$ ______

32. $4 - (-g)$ ______

33. $-8 - (-h)$ ______

34. CCSS **Be Precise** To find the percent error, you can use this equation.

$$\text{percent error} = \frac{\text{amount of error}}{\text{actual amount}} \times 100$$

Bryan estimates the cost of a vacation to be \$730. The actual cost of the vacation is \$850. Find the percent error. Round to the nearest whole percent if necessary. Is the percent positive or negative? Explain.

____

____

Standardized Test Practice

35. Which sentence about integers is *not* always true?

Ⓐ positive − positive = positive

Ⓑ positive + positive = positive

Ⓒ negative + negative = negative

Ⓓ positive − negative = positive

36. Morgan drove from Los Angeles (elevation 330 feet) to Death Valley (elevation −282 feet). What is the difference in elevation between Los Angeles and Death Valley?

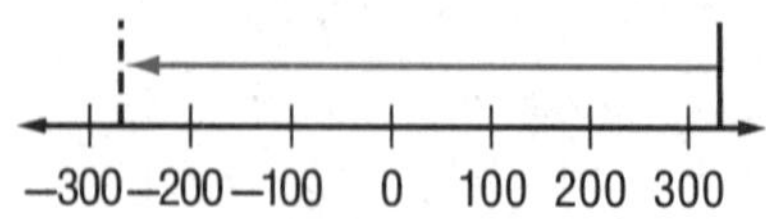

Ⓕ 48 feet

Ⓖ 148 feet

Ⓗ 582 feet

Ⓘ 612 feet

37. **Short Response** In a football game, Landon gained 10 yards on his first carry. He was tackled for a loss of 12 yards on his second carry. Write a subtraction expression to represent Landon's net yardage after the first two plays.

Common Core Review

Multiply. 5.NBT.5

38. 18(10) = ______

39. 15(13) = ______

40. 12(30) = ______

Evaluate each expression. 6.NS.7

41. |−12| = ______

42. |−3| + |−5| = ______

43. |−25| ÷ 5 − |−3| = ______

44. State the quadrant in which the graphed point is located. 6.NS.6

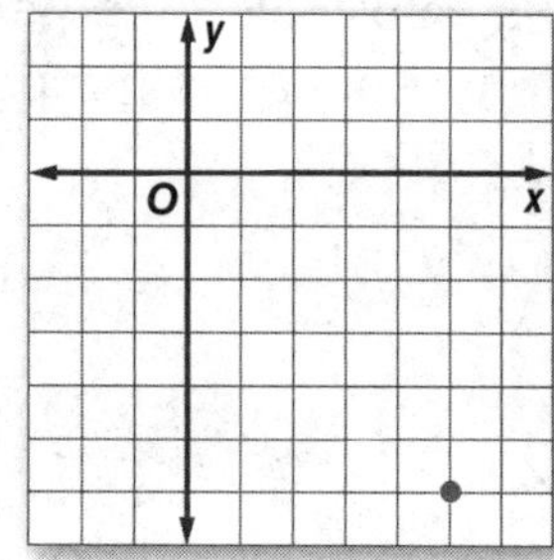

Inquiry Lab

Distance on a Number Line

HOW is the distance between two rational numbers related to their difference?

CCSS **Content Standards** 7.NS.1, 7.NS.1c

Mathematical Practices 1, 2, 3, 8

Weather For a science project, Carmelo recorded the daily low and high temperatures for four days in January. His results are shown in the table below. Find the day that had the greatest difference in temperature readings.

	M	T	W	Th
Low Temperature (°F)	1	−3	−4	−3
High Temperature (°F)	5	0	2	−1

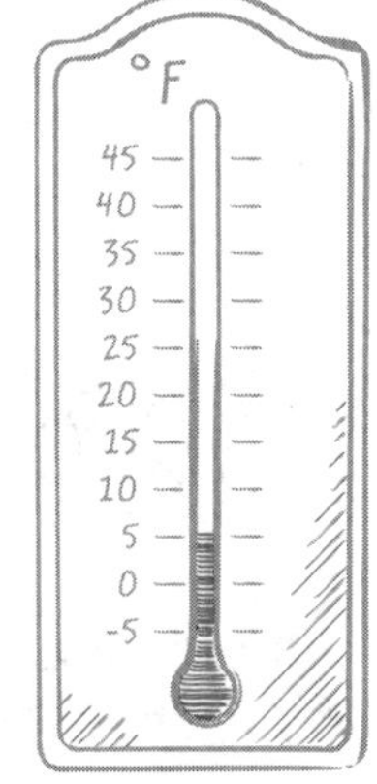

Investigation

Step 1 In the table below, *a* represents each daily low temperature, and *b* represents each daily high temperature. Find $a + b$ and $a - b$. Record your results in the table.

Day	a	b	$a + b$	$a - b$	Distance
M	1	5	6	−4	4 units
T	−3	0			
W	−4	2			
Th	−3	−1			

Step 2 Use a number line to find the distance between each rational number *a* and *b*. For example, the distance between 1 and 5 on the number line below is ☐ units.

Complete the last column in the table.

Step 3 Compare the distances.

So, the day with the greatest difference in temperature readings was ________.

Work with a partner to find the distance between each pair of numbers without using a number line. Then use a number line to check your answer.

1. The distance between −9 and −3 is ________.

2. The distance between −2 and 5 ________.

CCSS Identify Repeated Reasoning Work with a partner to answer the following questions. Refer back to the table in Step 1 of the Investigation.

3. Is there a relationship between the sum of each pair of integers and the distance between them? If so, explain.

4. Is there a relationship between the difference of the integers and the distance between them? If so, explain. ________

Reflect

5. CCSS **Reason Inductively** For each pair of integers in the Investigation, find $b - a$. How does $b - a$ compare to $a - b$? How does it compare to the distance between the points? Use the term *absolute value* in your answer.

6. Inquiry HOW is the distance between two rational numbers related to their difference? ________

Problem-Solving Investigation
Look for a Pattern

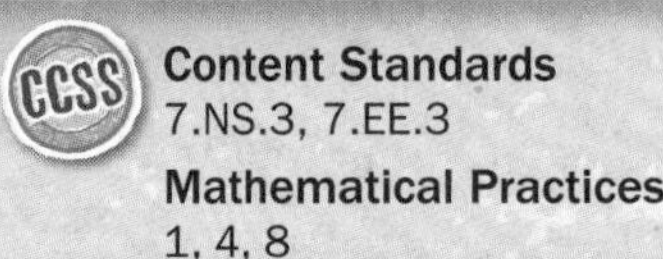

Content Standards
7.NS.3, 7.EE.3
Mathematical Practices
1, 4, 8

Case #1 Shooting Star

Laura wants to make the girls basketball team and knows that making free throws is a skill that impresses the coach. In practice, she makes about 3 out of every 5 free throws she attempts. In tryouts, she has to shoot the ball 30 times from the free throw line.

How many of these can she expect to make?

Understand *What are the facts?*

- Laura can make 3 out of 5 free throw attempts.
- In tryouts, she will have to shoot the ball 30 times from the free throw line.

Plan *What is your strategy to solve this problem?*

Make a table. How can you extend the pattern to solve the problem?

Solve *How can you apply the strategy?*

Make a table to extend the pattern. Complete the table below.

+3 +3 +3 +3 +3

Free Throws	3	6	9	12		
Shots Attempted	5	10	15	20		

+5 +5 +5 +5 +5

If Laura attempts 30 shots, how many should she make? ____

Check *Does the answer make sense?*

She makes free throws a little more than half the time. Since 18 is a little more than 15, the answer is reasonable.

Analyze the Strategy

Identify Repeated Reasoning How would the results have changed if Laura could make 4 out of 5 free throw attempts? ____________________

____

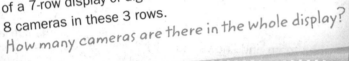

Case #2 Display Dilemma

Tomás looks through a window and sees the top 3 rows of a 7-row display of digital cameras. He sees 4, 6, and 8 cameras in these 3 rows.

How many cameras are there in the whole display?

Understand

Read the problem. What are you being asked to find?

I need to find ______________________________.

Underline key words and values. What information do you know?

The display contains ☐ rows of digital cameras. The problem states that the top three rows have ☐ cameras, ☐ cameras, and ☐ cameras.

Is there any information that you do *not* need to know?

I do not need to know ______________________________.

2 Plan

Choose a problem-solving strategy.

I will use the ______________________________ strategy.

3 Solve

Describe the pattern in the table. Then extend it using your problem-solving strategy. ______________________________

____

Row	7	6	5				
Number of Cameras	4	6	8				

The total number of cameras is ☐.

So, ______________________________.

4 Check

Use information from the problem to check your answer.

____

Collaborate Work with a small group to solve the following cases. Show your work on a separate piece of paper.

Case #3 Nature

A sunflower usually has two different spirals of seeds, one with 34 seeds and the other with 55 seeds. The numbers 34 and 55 are part of the Fibonacci sequence.

1, 1, 2, 3, 5, 8, 13, 21, 34, 55, ...

Find the pattern in the Fibonacci sequence and identify the next two terms.

Case #4 Financial Literacy

Peter is saving money to buy an MP3 player. After one month, he has \$50. After 2 months, he has \$85. After 3 months, he has \$120. After 4 months, he has \$155.

At this rate, how long will it take Peter to save enough money to buy an MP3 player that costs \$295?

Case #5 Geometry

The pattern at the right is made from toothpicks.

How many toothpicks would be needed for the sixth term in the pattern?

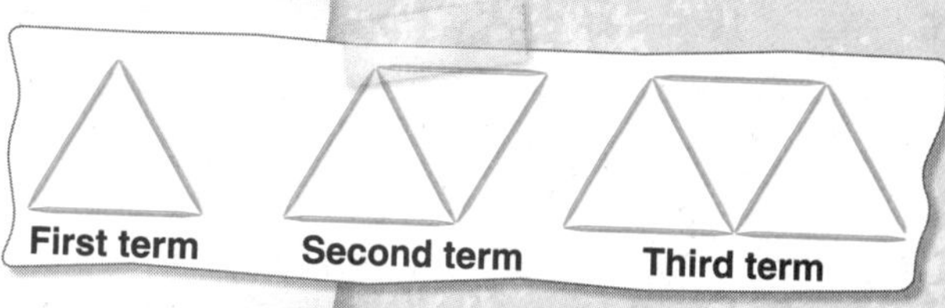

Circle a strategy below to solve the problem.

- Draw a diagram.
- Act it out.
- Make a model.
- Make a table.

Case #6 Diving

A diver descends to −15 feet after 1 minute, −30 feet after 2 minutes, and −45 feet after 3 minutes.

If she keeps descending at this rate, find the diver's position after ten minutes.

Mid-Chapter Check

Vocabulary Check

1. Define *integer*. Give an example of a number that is an integer and a number that is not an integer. (Lesson 1)

____

____

2. Fill in the blank in the sentence below with the correct term. (Lesson 1)

The _______________ of a number is the distance between the number and zero on a number line.

Skills Check and Problem Solving

Evaluate each expression. (Lessons 1, 2, and 3)

3. $|-6| =$ ______

4. $-4 + (-8) =$ ______

5. $3 + 4 + (-5) =$ ______

6. $-3 - 10 =$ ______

7. $8 - (-12) =$ ______

8. $|-5| - |-9| =$ ______

9. The melting point of mercury is −36°F and its boiling point is 672°F. What is the difference between the boiling point and the melting point? (Lesson 3)

____

10. **Standardized Test Practice** Which of the following numerical expressions results in a positive number? (Lessons 1 and 2)

Ⓐ $-4 + (-7)$

Ⓑ $4 + (-7)$

Ⓒ $-4 + 7$

Ⓓ $|-2| + 7 + (-11)$

Inquiry Lab
Multiply Integers

WHEN is the product of two integers a positive number? WHEN is the product a negative number?

Content Standards
7.NS.2, 7.NS.3

Mathematical Practices
1, 3, 4

School The number of students who bring their lunch to Phoenix Middle School had been decreasing at a rate of 4 students each month. What integer represents the total change in the number of students bringing their lunch after three months?

What do you know? ______________________________

What do you need to find? ______________________________

Investigation 1

The integer ☐ represents a decrease of 4 students each month. After three months, the total change will be $3 \times (-4)$.

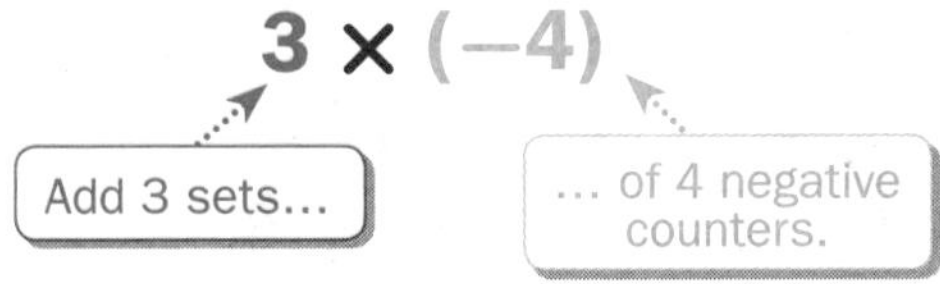

Step 1 Add 3 sets of 4 negative counters to the mat.

Step 2 Count the number of negative counters.

There are ☐ negative counters.

So, $3 \times (-4) =$ ☐. After three months, the total change in the number of students bringing their lunch will be ☐.

Investigation 2

Use counters to find -2×3.

If the first factor is negative, you need to *remove* counters from the mat.

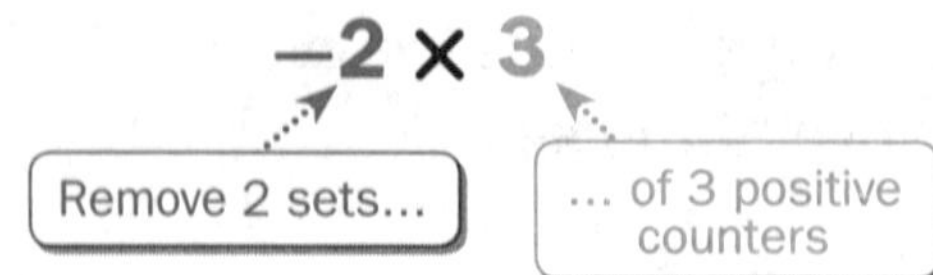

Step 1 Start with 2 sets of 3 zero pairs on the mat. The value on the mat is zero.

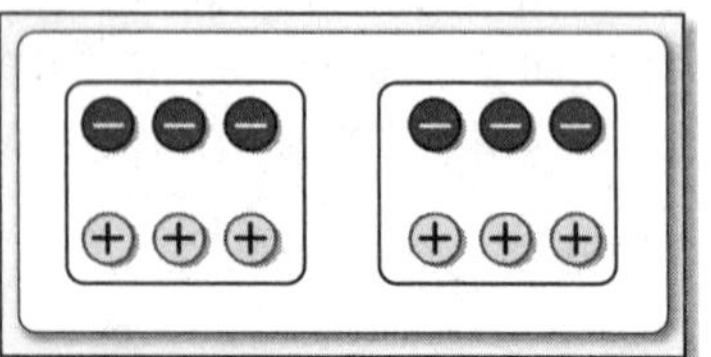

Step 2 Remove 2 sets of 3 positive counters from the mat.

There are ☐ negative counters remaining.

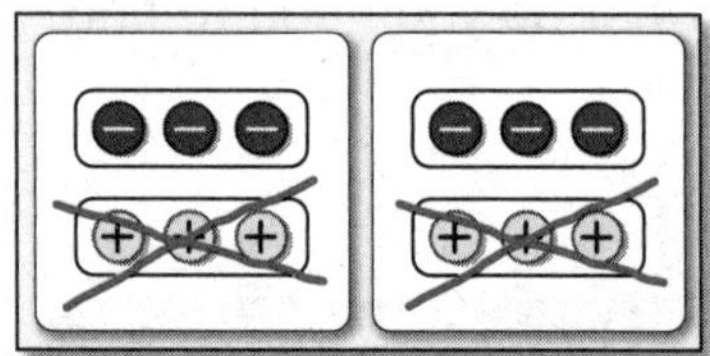

So, $-2 \times 3 =$ ☐.

Investigation 3

Use counters to find $-2 \times (-4)$.

Remove ☐ sets of ☐ negative counters from the mat.

Step 1 Draw 2 sets of 4 zero pairs on the mat.

Step 2 Cross out 2 sets of 4 negative counters from the mat.

There are ☐ positive counters remaining.

So, $-2 \times (-4) =$ ☐.

Work with a partner. Find each product. Show your work using drawings.

1. $2 \times (-3) =$ ______

2. $6 \times (-1) =$ 

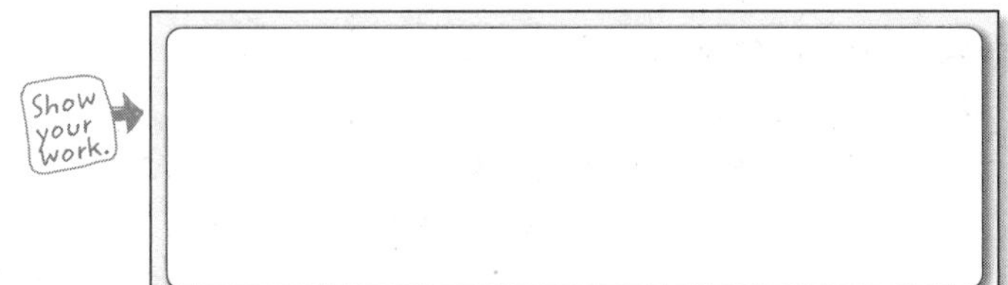

3. $-2 \times 4 =$ ______

4. $-1 \times 5 =$ 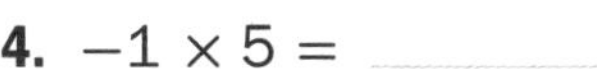

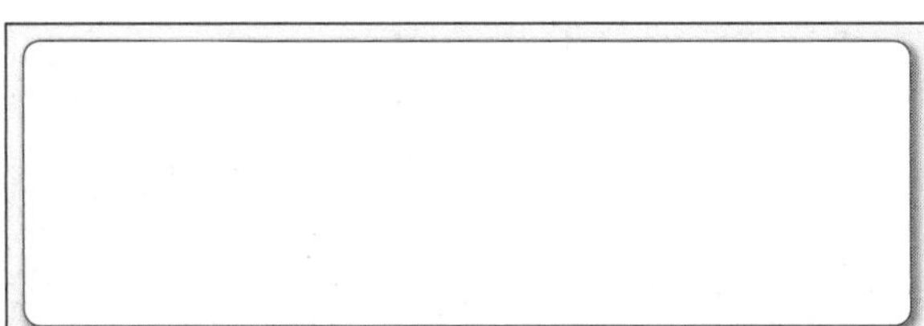

5. $-4 \times 2 =$ ______

6. $-2 \times (-4) =$ ______

7. $-3 \times (-1) =$ ______

8. $-6 \times (-2) =$ 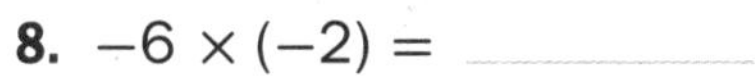

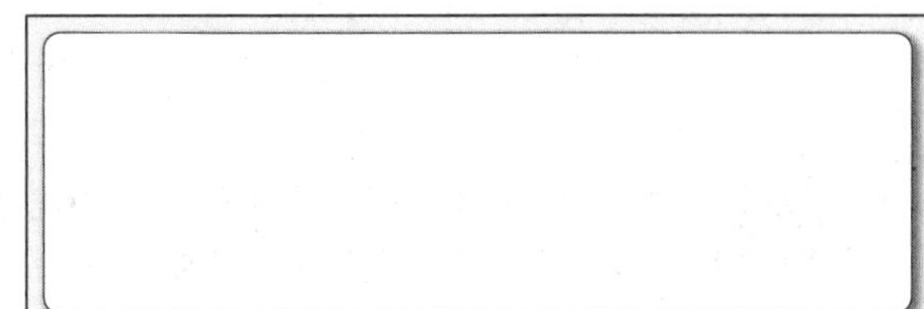

Analyze

Work with a partner to complete the table. Use counters if needed. The first one is already done for you.

	Multiplication Expression	Same Signs or Different Signs?	Product	Positive or Negative?
	2×6	Same signs	12	Positive
9.	$7 \times (-2)$			
10.	$-3 \times (-4)$			
11.	$5 \times (-3)$			
12.	2×8			
13.	$-4 \times (-1)$			
14.	-3×6			
15.	-2×5			

16. CCSS **Reason Inductively** Can you find any patterns in the table? If so, describe them.

Reflect

17. CCSS **Model with Mathematics** Write a real-world problem that could be represented by the expression -5×4.

18. Inquiry WHEN is the product of two integers a positive number? WHEN is the product a negative number?

Lesson 4

Multiply Integers

What You'll Learn

Scan the rest of the lesson. List two headings you would use to make an outline of the lesson.

- ______________________________
- ______________________________

Essential Question

WHAT happens when you add, subtract, multiply, and divide integers?

Common Core State Standards

Content Standards
7.NS.2, 7.NS.2a, 7.NS.2c, 7.NS.3, 7.EE.3

Mathematical Practices
1, 3, 4, 8

Real-World Link

Watch

Skydiving Once a parachute is deployed, a skydiver descends at a rate of about 5 meters per second. Where will the skydiver be in relation to where the parachute deployed after 4 seconds?

1. Descending is usually represented by a negative integer. What integer should you use to represent the position of the skydiver in relation to the parachute's deployment after 1 second? ☐

2. Complete the graphic below. What is the skydiver's position after 2, 3, and 4 seconds?

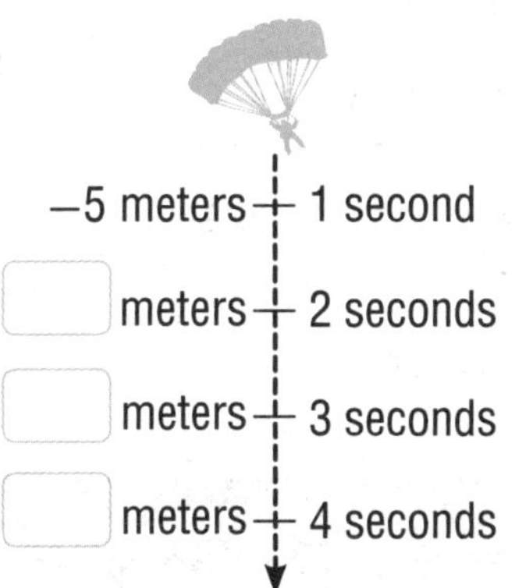

3. Write a multiplication expression to represent the skydiver's position after 5 seconds.

Key Concept Multiply Integers with Different Signs

Words The product of two integers with different signs is negative.

Examples $6(-4) = -24$ $-5(7) = -35$

Work Zone

Remember that multiplication is the same as repeated addition.

$4(-3) = (-3) + (-3) + (-3) + (-3)$ −3 is used as an addend four times.

$= -12$

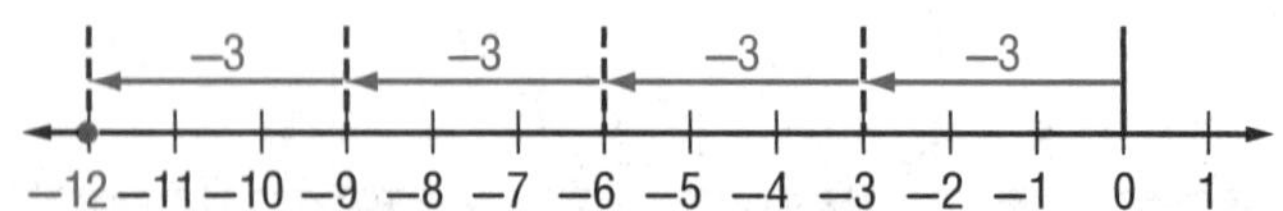

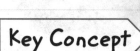

The Commutative Property of Multiplication states that you can multiply in any order. So, $4(-3) = -3(4)$.

Examples

1. Find 3(−5).

$3(-5) = -15$ The integers have different signs. The product is negative.

2. Find −6(8).

$-6(8) = -48$ The integers have different signs. The product is negative.

a. ______

b. ______

Got It? Do these problems to find out.

a. $9(-2)$ **b.** $-7(4)$

Key Concept Multiply Integers with the Same Signs

Words The product of two integers with the same sign is positive.

Examples $2(6) = 12$ $-10(-6) = 60$

The product of two positive integers is positive. You can use a pattern to find the sign of the product of two negative integers. Start with $(2)(-3) = -6$ and $(1)(-3) = -3$.

positive × negative = negative	$(2)(-3) = -6$	+3
	$(1)(-3) = -3$	+3
The Multiplicative Property of Zero	$(0)(-3) = 0$	+3
negative × negative = positive	$(-1)(-3) = 3$	+3
	$(-2)(-3) = 6$	

Each product is 3 more than the previous product. This pattern can also be shown on a number line.

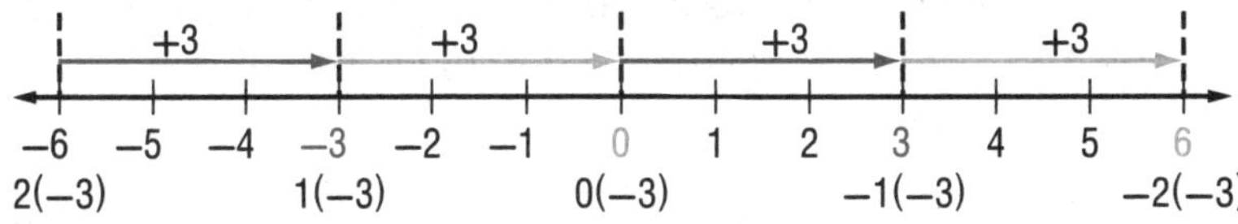

If you extend the pattern, the next two products are $(-3)(-3) = 9$ and $(-4)(-3) = 12$.

Examples

3. Find −11(−9).

$-11(-9) = 99$ The integers have the same sign. The product is positive.

4. Find $(-4)^2$.

$(-4)^2 = (-4)(-4)$ There are two factors of −4.

$= 16$ The product is positive.

5. Find −3(−4)(−2).

$-3(-4)(-2) = [-3(-4)](-2)$ Associative Property

$= 12(-2)$ $-3(-4) = 12$

$= -24$ $12(-2) = -24$

and Reflect

Write three integers with a positive product. At least one of them must be a negative integer. Show your work below.

Got It? Do these problems to find out.

c. $-12(-4)$ **d.** $(-5)^2$ **e.** $-7(-5)(-3)$

c. ______

d. ______

e. ______

Example

6. A submersible is diving from the surface of the water at a rate of 90 feet per minute. What is the depth of the submersible after 7 minutes?

The submersible descends 90 feet per minute. After 7 minutes, the vessel will be at 7(–90) or –630 feet. The submersible will descend to 630 feet below the surface.

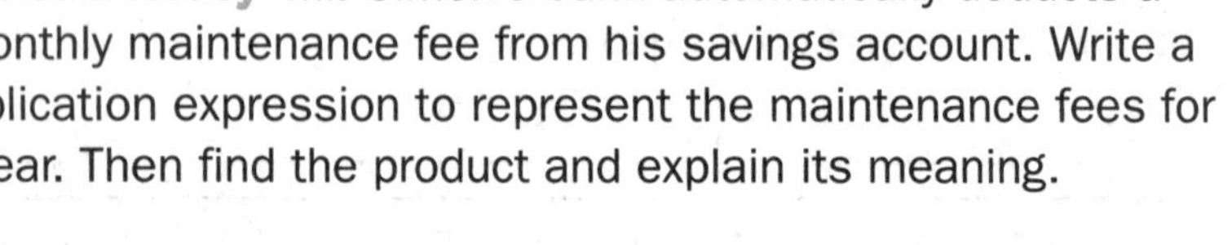

Got It? Do this problem to find out.

f. **Financial Literacy** Mr. Simon's bank automatically deducts a $4 monthly maintenance fee from his savings account. Write a multiplication expression to represent the maintenance fees for one year. Then find the product and explain its meaning.

f. ____________

Guided Practice

Multiply. (Examples 1–5)

1. $6(-10) =$ ________

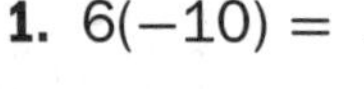

2. $(-3)^3 =$ ________

3. $(-1)(-3)(-4) =$ ________

4. **Financial Literacy** Tamera owns 100 shares of a certain stock. Suppose the price of the stock drops by $3 per share. Write a multiplication expression to find the change in Tamera's investment. Explain your answer. (Example 6)

5. **Building on the Essential Question** When is the product of two or more integers a positive number?

Rate Yourself!
Are you ready to move on? Shade the section that applies.
I have a few questions.
I'm ready to move on.
I have a lot of questions.
For more help, go online to access a Personal Tutor.
Tutor
FOLDABLES Time to update your Foldable!

Name ______________________ My Homework ______________________

Independent Practice

Go online for Step-by-Step Solutions

Multiply. (Examples 1–5)

1. $8(-12) =$ ______

2. $-15(-4) =$ ______

3. $(-6)^2 =$ ______

4. $(-5)^3 =$ ______

5. $-4(-2)(-8) =$ ______

6. $-3(-2)(1) =$ ______

Write a multiplication expression to represent each situation. Then find each product and explain its meaning. (Example 6)

7. Ethan burns 650 Calories when he runs for 1 hour. Suppose he runs 5 hours in one week.

8. Wave erosion causes a certain coastline to recede at a rate of 3 centimeters each year. This occurs uninterrupted for a period of 8 years.

9. CCSS **Model with Mathematics** Refer to the graphic novel frame below. How many black T-shirts would Hannah and Dario need to sell to make up the loss in profit?

Watch
Replay it online!
It looks like we spend $15 on every black T-shirt that is printed on both back and front. We should give one to the mascot for free.
Refer to the graphic novel at the start of the chapter.
HOMECOMING

10. **Multiple Representations** When a movie is rented it has a due date. If the movie is not returned on time, a late fee is assessed. Kaitlyn is charged $5 each day for a movie that is 4 days late.

a. **Words** Explain why $4 \times (-5) = -20$ describes the situation. ______

______

b. **Algebra** Write an expression to represent the fee when the movie is x days late. ______

11. **Identify Repeated Reasoning** When you multiply two positive integers, the product is a positive integer. Complete the graphic organizer to help you remember the other rules for multiplying integers. Describe any patterns in the products.

×	+	−
+		
−		

______

______

______

H.O.T. Problems Higher Order Thinking

12. **Model with Mathematics** Write a multiplication sentence with a product of −18.

______

13. **Justify Conclusions** Explain how to evaluate $(-9)(-6)(15)(-7 + 7)$ as simply as possible.

______

______

14. **Reason Inductively** Evaluate $(-1)^{50}$. Explain your reasoning.

______

______

Standardized Test Practice

15. The temperature drops 2 degrees per hour for 3 hours. Which expression does *not* describe the change in temperature?

Ⓐ $-2(3)$

Ⓑ $-2 + (-2) + (-2)$

Ⓒ $-2 - 2 - 2$

Ⓓ $2(3)$

Name ______________________ My Homework ______________________

Extra Practice

Multiply.

16. $-7(11) =$ -77

$-7(11) = -77$

work → The integers have different signs. The product is negative.

17. $-20(-8) =$ ________

18. $25(-2) =$ ________

19. $(-4)^3 =$ ________

20. $(-9)^2 =$ ________

21. $-9(-1)(-5) =$ ________

Write a multiplication expression to represent each situation. Then find each product and explain its meaning.

22. The average person loses 50 to 80 hairs per day to make way for new growth. Suppose you lose 65 hairs per day for 15 days without growing any.

23. CCSS **Financial Literacy** Lily has a $100 gift card to her favorite pastry shop. She spends $4 a day at the shop for the next 12 days.

Copy and Solve **Evaluate each expression if $a = -6$, $b = -4$, $c = 3$, and $d = 9$. Show your work on a separate sheet of paper.**

24. $-5c =$

25. $b^2 =$

26. $2a =$

27. $bc =$

28. $abc =$

29. $abc^3 =$

30. $-3a^2 =$

31. $-cd^2 =$

32. $-2a + b =$

33. CCSS **Find the Error** Jamar is finding $(-2)(-3)(-4)$. Find his mistake and correct it. Explain your answer.

Standardized Test Practice

34. Which number is the seventh number in the sequence shown?

Position	1	2	3	4	5	6	7
Number	1	−2	4	−8	16	?	?

Ⓐ −64 Ⓒ 32
Ⓑ −32 Ⓓ 64

35. **Short Response** J.J. withdraws $15 from his bank account every week for lunch. Suppose he does not make any additional deposits or withdrawals. What integer represents the change in value of J.J.'s bank account after 8 weeks? Write and evaluate an expression.

36. A camera dropped from a boat descends 9 meters every minute. What will be the change in location of the camera after 5 minutes?

Ⓕ 45 m Ⓗ −14 m
Ⓖ 14 m Ⓘ −45 m

37. The table shows the temperatures on Wednesday night.

Time	Temperature (°F)
3 P.M.	14
5 P.M.	11

At 5 P.M., the temperature began dropping 3°F every hour for 6 hours. What was the temperature after 6 hours?

Ⓐ −7°F Ⓒ 7°F
Ⓑ −29°F Ⓓ 29°F

CCSS Common Core Review

Fill in each ◯ with < or > to make a true sentence. 6.NS.7b

38. 0 ◯ −1

39. −9 ◯ 9

40. −84 ◯ 48

41. 32 ◯ −27

42. Laura's allowance balances over the last three months are shown in the table. Positive values indicate the number of dollars she had left over, and negative values indicate the number of dollars she overspent. Order the allowance balances from least to greatest. 6.NS.7

Month	Allowance Balances ($)
May	−10
June	5
July	−2

43. Graph 1, −4, 3, −2, 0, and 2 on the number line below. 6.NS.6

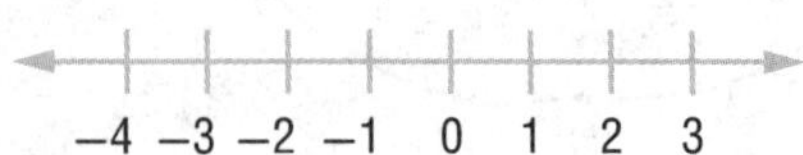

Inquiry Lab

Use Properties to Multiply

HOW can properties be used to prove rules for multiplying integers?

Content Standards
7.NS.2, 7.NS.2a, 7.NS.2c

Mathematical Practices
1, 3

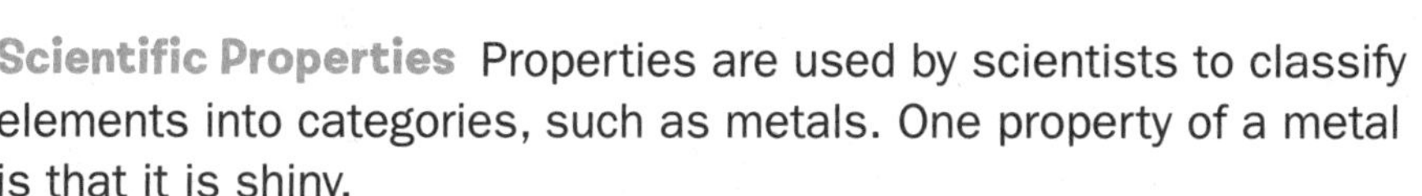

Scientific Properties Properties are used by scientists to classify elements into categories, such as metals. One property of a metal is that it is shiny.

Investigation

You have studied the mathematical properties listed in the table below. In mathematics, properties can be used to justify statements you make while verifying or proving another statement.

Properties of Mathematics	
Additive Inverse	Multiplicative Property of Zero
Distributive Property	Multiplicative Identity

For example, you have used models to show that $2(-1) = -2$.

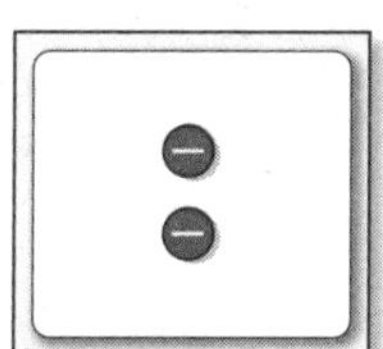

You can *prove* $2(-1) = -2$ by using properties.

Write the correct property from the table above to provide the missing justifications. Use each property name once.

Statements	Properties
$0 = 2(0)$	
$0 = 2[1 + (-1)]$	
$0 = 2(1) + 2(-1)$	
$0 = 2 + 2(-1)$	

Conclusion In the last statement, $0 = 2 + 2(-1)$. In order for this to be true, $2(-1)$ must equal -2. Therefore, $2(-1) =$ ☐.

The sentence $(-2)(-1) = 2$ is an example of the rule that states the product of a negative integer and a negative integer is a positive integer.

Work with a partner to provide the missing information for the statements below.

1. Show that $(-2)(-1) = 2$.

Statements	Properties
$0 = -2(0)$	
$0 = -2[1 + (-1)]$	
$0 = -2(1) + (-2)(-1)$	
$0 = -2 + (-2)(-1)$	

Work with a partner.

2. CCSS **Justify Conclusions** Write a conclusion for Exercise 1.

3. CCSS **Construct an Argument** When you prove a statement mathematically, you must show that the statement is true for all possible values. How could you prove the product of any two negative numbers is a positive number? Explain your reasoning to a classmate.

4. Inquiry HOW can properties be used to prove rules for multiplying integers?

Lesson 5

Divide Integers

What You'll Learn

Scan the lesson. List two headings you would use to make an outline of the lesson.

-
-

Essential Question

WHAT happens when you add, subtract, multiply, and divide integers?

Common Core State Standards

Content Standards
7.NS.2, 7.NS.2b, 7.NS.2c, 7.NS.3

Mathematical Practices
1, 3, 4, 5, 7

Real-World Link

Sharks A Great White Shark has 3,000 teeth! It gains and loses teeth often in its lifetime. Suppose a Great White loses 3 teeth each day for 5 days without gaining any. The shark has lost 15 teeth in all.

1. Write a multiplication sentence for this situation.

2. Division is related to multiplication. Write two division sentences related to the multiplication sentence you wrote for Exercise 1.

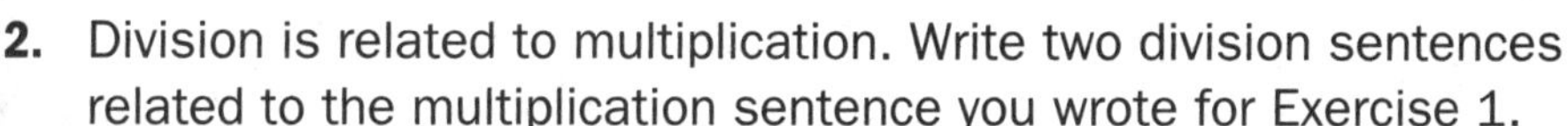

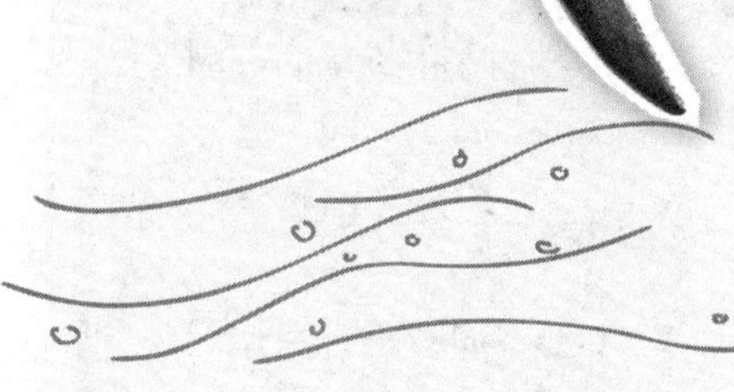

Collaborate Work with a partner to complete the table. The first one is done for you.

	Multiplication Sentence	Division Sentences	Same Signs or Different Signs?	Quotient	Positive or Negative?
	$2 \times 6 = 12$	$12 \div 6 = 2$ $12 \div 2 = 6$	Same signs Same signs	2 6	Positive Positive
3.	$2 \times (-4) = -8$				
4.	$-3 \times 5 = -15$				
5.	$-2 \times (-5) = 10$				

Key Concept — Divide Integers with Different Signs

Words The quotient of two integers with different signs is negative.

Examples $33 \div (-11) = -3$ $\quad -64 \div 8 = -8$

Work Zone

You can divide integers provided that the divisor is not zero. Since multiplication and division sentences are related, you can use them to find the quotient of integers with different signs.

different signs

$2(-6) = -12 \longrightarrow -12 \div 2 = -6$

$-2(-6) = 12 \longrightarrow 12 \div (-2) = -6$

negative quotient

Examples

Tutor

1. Find $80 \div (-10)$. The integers have different signs.

$80 \div (-10) = -8$ The quotient is negative.

2. Find $\frac{-55}{11}$. The integers have different signs.

$\frac{-55}{11} = -5$ The quotient is negative.

Dividing Integers

If p and q are integers and q does not equal 0, then $-\frac{p}{q} = \frac{-p}{q} = \frac{p}{-q}$.

In Example 2, $-\frac{55}{11} = \frac{-55}{11} = \frac{55}{-11}$.

3. Use the table to find the constant rate of change in centimeters per hour.

	Time (h)	Height (cm)	
	1	10	
+1	2	8	−2
+1	3	6	−2
+1	4	4	−2

The height of the candle decreases by 2 centimeters each hour.

$\frac{\text{change in height}}{\text{change in hours}} = \frac{-2}{1}$

So, the constant rate of change is −2 centimeters per hour.

Got It? Do these problems to find out.

a. $20 \div (-4)$ **b.** $\frac{-81}{9}$ **c.** $-45 \div 9$

Show your work.

a. ____________

b. ____________

c. ____________

Divide Integers with the Same Signs

Key Concept

Words The quotient of two integers with the same sign is positive.

Examples $15 \div 5 = 3$ $-64 \div (-8) = 8$

You can also use multiplication and division sentences to find the quotient of integers with the same sign.

same signs

$4(5) = 20 \rightarrow 20 \div 4 = 5$

$-4(5) = -20 \rightarrow -20 \div (-4) = 5$

positive quotient

Examples

Tutor

4. Find $-14 \div (-7)$. The integers have the same sign.

$-14 \div (-7) = 2$ The quotient is positive.

5. Find $\frac{-27}{-3}$. The integers have the same sign.

$\frac{-27}{-3} = 9$ The quotient is positive.

6. Evaluate $-16 \div x$ if $x = -4$.

$-16 \div x = -16 \div (-4)$ Replace x with -4.

$= 4$ Divide. The quotient is positive.

Got It? Do these problems to find out.

d. $-24 \div (-4)$ **e.** $-9 \div (-3)$ **f.** $\frac{-28}{-7}$

g. Evaluate $a \div b$ if $a = -33$ and $b = -3$.

d. ________

e. ________

f. ________

g. ________

Example

7. **STEM** **One year, the estimated Australian koala population was 1,000,000. After 10 years, there were about 100,000 koalas. Find the average change in the koala population per year. Then explain its meaning.**

$$\frac{N - P}{10} = \frac{100{,}000 - 1{,}000{,}000}{10}$$ *N* is the new population, 100,000. *P* is the previous population, 1,000,000.

$$= \frac{-900{,}000}{10} \text{ or } -90{,}000$$ Divide.

The koala population has changed by −90,000 per year.

Got It? Do this problem to find out.

h. **STEM** The average temperature in January for North Pole, Alaska, is −24°C. Use the expression $\frac{9C + 160}{5}$ to find this temperature in degrees Fahrenheit. Round to the nearest degree. Then explain its meaning.

h. ________

Guided Practice

Divide. (Examples 1, 2, 4, and 5)

1. $-16 \div 2 =$ ____

2. $\frac{42}{-7} =$ ____

3. $-30 \div (-5) =$ ____

Evaluate each expression if $x = 8$ and $y = -5$. (Example 6)

4. $15 \div y$ ____

5. $xy \div (-10)$ ____

6. $(x + y) \div (-3)$ ____

7. The lowest recorded temperature in Wisconsin is −55°F on February 4, 1996. Use the expression $\frac{5(F - 32)}{9}$ to find this temperature in degrees Celsius. Round to the nearest tenth. Explain its meaning. (Example 7)

8. **Building on the Essential Question** How is dividing integers similar to multiplying integers?

Rate Yourself!

How confident are you about dividing integers? Check the box that applies.

For more help, go online to access a Personal Tutor.

FOLDABLES *Time to update your Foldable!*

Name ______ My Homework ______

Independent Practice

Go online for Step-by-Step Solutions

Divide. (Examples 1, 2, 4, and 5)

1. $50 \div (-5) =$ ______

2. $-18 \div 9 =$ ______

3. $-15 \div (-3) =$ ______

4. $-100 \div (-10) =$ ______

5. $\frac{22}{-2} =$ ______

6. $\frac{84}{-12} =$ ______

7. $\frac{-26}{13} =$ ______

8. $\frac{-21}{-7} =$ ______

Evaluate each expression if $r = 12$, $s = -4$, and $t = -6$. (Example 6)

9. $r \div s$ ______

10. $rs \div 16$ ______

11. $\frac{t - r}{3}$ ______

12. $\frac{8 - r}{-2}$ ______

13. The distance remaining for a road trip over several hours is shown in the table. Use the information to find the constant rate of change in miles per hour. (Example 3)

______

Time (h)	Distance Remaining (mi)
2	480
4	360
6	240
8	120

14. CCSS **Justify Conclusions** Last year, Mr. Engle's total income was \$52,000, while his total expenses were \$53,800. Use the expression $\frac{I - E}{12}$, where I represents total income and E represents total expenses, to find the average difference between his income and expenses each month. Then explain its meaning. (Example 7) ______

______

Evaluate each expression if $d = -9$, $f = 36$, and $g = -6$.

15. $\frac{-f}{d}$ ______

16. $\frac{12 - (-f)}{-g}$ ______

17. $\frac{f^2}{d^2}$ ______

18. STEM The temperature on Mars ranges widely from −207°F to 80°F.

Find the average of the temperature extremes on Mars. ______

H.O.T. Problems Higher Order Thinking

19. CCSS **Construct an Argument** You know that multiplication is commutative because $9 \times 3 = 3 \times 9$. Is division commutative? Explain.

______

______

CCSS **Identify Structure** Use the graphs shown below to find the slope of each line.

20. ______

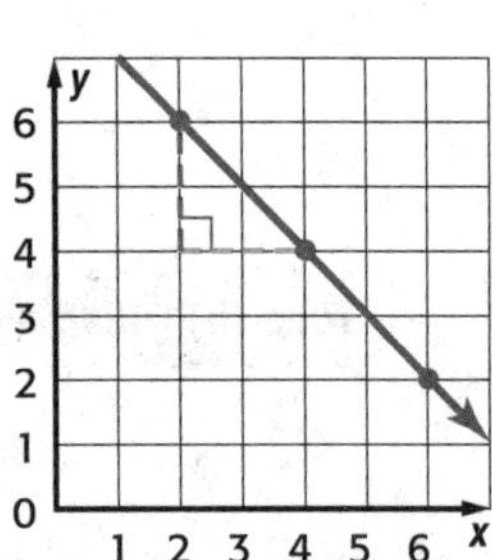

21. ______

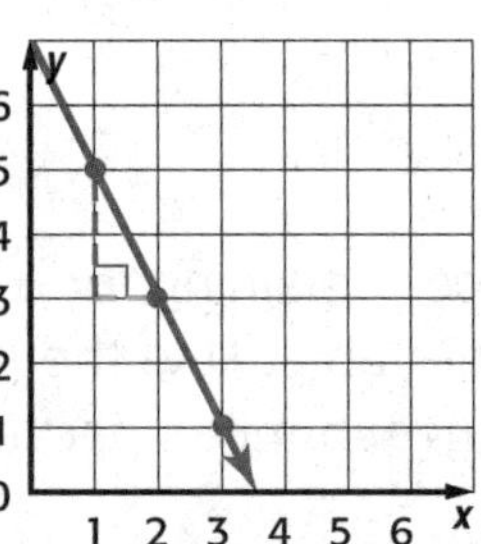

22. CCSS **Identify Structure** Find values for x, y, and z so that all of the following statements are true.

- $y > x$, $z < y$, and $x < 0$
- $x \div z = -z$
- $z \div 2$ and $z \div 3$ are integers
- $x \div y = z$

$x =$ ______ $y =$ ______ $z =$ ______

Standardized Test Practice

23. On December 24, 1924, the temperature in Fairfield, Montana, fell from 63°F at noon to −21°F at midnight. What was the average temperature change per hour?

Ⓐ −3.5°F

Ⓑ −7°F

Ⓒ −42°F

Ⓓ −84°F

Name _______________ My Homework _______________

Extra Practice

Divide.

24. $56 \div (-8) =$ −7

Homework Help → $56 \div (-8) = -7$
The integers have different signs. The quotient is negative.

25. $-36 \div (-4) =$ 9

$-36 \div (-4) = 9$
The integers have the same signs. The quotient is positive.

26. $32 \div (-8) =$ ______

27. $\frac{-16}{-4} =$ ______

28. $\frac{-27}{3} =$ ______

29. $\frac{-54}{-6} =$ ______

Evaluate each expression if $r = 12$, $s = -4$, and $t = -6$.

30. $-12 \div r$ ______

31. $72 \div t$ ______

32. $\frac{s + t}{5}$ ______

33. Divide −200 by −100. ______

34. Find the quotient of −65 and −13. ______

35. STEM The boiling point of water is affected by changes in elevation. Use the expression $\frac{-2A}{1{,}000}$, where A represents the altitude in feet, to find the number of degrees Fahrenheit at which the boiling point of water changes at an altitude of 5,000 feet. Then explain its meaning.

______

36. CCSS **Use Math Tools** The change in altitude over time for several hot air balloons is shown. Find the rate of change in feet per minute for each balloon.

Balloon	Change in Altitude (ft)	Time (min)	Rate of Change (ft/min)
Midnight Express	−2,700	135	
Neon Lights	480	30	
Star Wonder	−1,500	60	

Standardized Test Practice

37. A hang glider flew to an altitude of 10,000 feet. Fifteen minutes later, its altitude was 7,000 feet. What was the average change in elevation per minute?

Ⓐ −300 ft/min
Ⓑ −200 ft/min
Ⓒ 200 ft/min
Ⓓ 300 ft/min

38. **Short Response** During the past week, Mrs. Thorne recorded the following amounts in her checkbook: \$150, −\$75, −\$15, and −\$32. Write and evaluate an expression to find the average of these amounts.

____

39. **Short Response** The table shows the points that each student lost on the first math test. Each question on the test was worth an equal number of points. If Christoper answered 6 questions incorrectly, how many questions did Nythia answer incorrectly? Explain.

Student	Points
Christopher	−24
Nythia	−16
Raul	−4

____

____

____

Common Core Review

Write the opposite of each integer. 6.NS.6a

40. 8 ________

41. 9 ________

42. −7 ________

43. −5 ________

44. A display of cereal boxes has one box in the top row, two boxes in the second row, three boxes in the third row, and so on, as shown. How many rows of boxes will there be in a display of 45 boxes? 5.OA.3

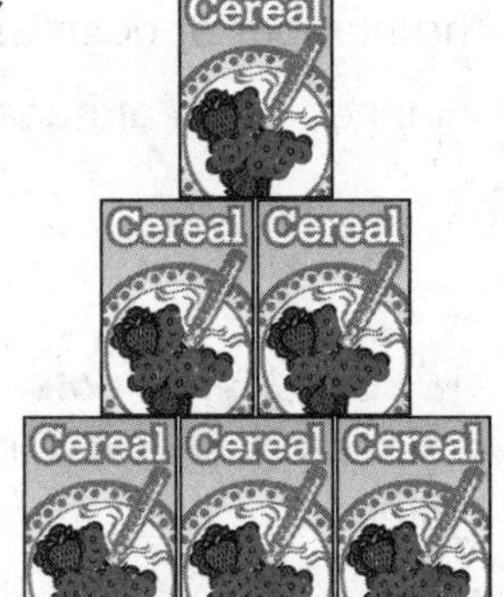

____

45. Name the quadrant in which the point (−4, −3) could be found on the coordinate plane. 6.NS.6b ________________

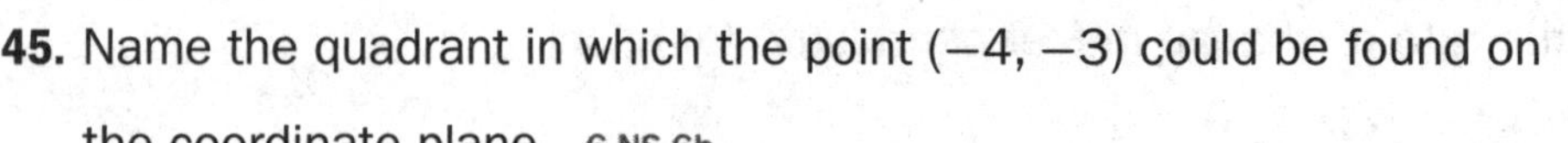

21ST CENTURY CAREER in Astronomy

Space Weather Forecaster

Did you know that space weather, or the conditions on the Sun and in space, can directly affect communication systems and power grids here on Earth? If you enjoy learning about the mysteries of space, then you should consider a career involving space weather. A space weather forecaster uses spacecraft, telescopes, radar, and supercomputers to monitor the sun, solar winds, and the space environment in order to forecast the weather in space.

Is This the Career for You?

Are you interested in a career as a space weather forecaster? Take some of the following courses in high school.

- Astronomy
- Calculus
- Chemistry
- Earth Science
- Physics

Find out how math relates to a career in Astronomy.

Predicting Space Storms!

Use the information in the table to solve each problem.

1. Graph the average temperatures for Earth, Jupiter, Mars, Mercury, Neptune, and Saturn on a number line. Label the points.

2. The temperatures on Mercury range from −279°F to 800°F. What is the difference between the highest and lowest temperatures? ________

3. How much greater is the average temperature on Earth than the average temperature on Jupiter? ________

4. One of Neptune's moons, Triton, has a surface temperature that is 61°F less than Neptune's average temperature. What is Triton's surface temperature? ________

5. The temperature on Mars can reach a low of −187°C. Find the value of the expression $\frac{9(-187) + 160}{5}$ to determine this temperature in degrees Fahrenheit. ________

Average Temperature of Planets

Planet	Average Temperature (°F)	Planet	Average Temperature (°F)
Earth	59	Neptune	−330
Jupiter	−166	Saturn	−220
Mars	−85	Uranus	−320
Mercury	333	Venus	867

Career Project

It's time to update your career portfolio! Investigate the education and training requirements for a career as a space weather forecaster.

List other careers that someone with an interest in astronomy could pursue.

-
-
-
-
-

Chapter Review

Check

Vocabulary Check

Complete each sentence using the vocabulary list at the beginning of the chapter.

1. The sum of an integer and its ________ inverse is 0.
2. A(n) ________ integer is greater than 0.
3. The set of ________ contains all the whole numbers and their opposites.
4. The ________ value of a number is the distance it is from 0 on a number line.
5. 5 and −5 are ________.
6. The result when one positive counter is paired with one negative counter is a ________ pair.

Reconstruct the vocabulary word and definition from the letters under the grid. The letters for each column are scrambled directly under that column.

Key Concept Check

Use Your FOLDABLES

Use your Foldable to help review the chapter.

Tape here

Operations with Integers

How do I add integers with different signs?

How do I subtract integers with different signs?

How do I multiply integers with different signs?

How do I divide integers with different signs?

Got it?

Find the Error **The problems below may or may not contain an error. If the problem is correct, write a "✓" by the answer. If the problem is not correct, write an "X" over the answer and correct the problem.**

1. $|-5| + |2| = -3$ X

$|-5| + |2| = 5 + 2$ or 7

The first one is done for you.

2. $3|-6| = 18$

3. $-24 \div |-2| = 12$

Problem Solving

STEM For Exercises 1–3, use the table that shows the freezing point of various elements.

Element	Freezing Point (°C)
Chlorine	−101
Helium	−272
Krypton	−157
Neon	−249
Nitrogen	−201

1. Graph the temperatures on the number line. (Lesson 1)

−300 −250 −200 −150 −100

2. CCSS **Justify Conclusions** Is the absolute value of the freezing point of chlorine greater than the absolute value of the freezing point of nitrogen? Explain. (Lesson 1)

3. Find the difference between the freezing point of chlorine and the freezing point of nitrogen. (Lesson 3) ______

4. Alicia was rock climbing. She climbed to a height of 22 feet. Next, she descended 8 feet. Then, she climbed up another 34 feet. What was Alicia's elevation? (Lesson 2) ______

5. **Financial Literacy** The price of a certain stock fell \$2 each day for 4 consecutive days. Write an expression that you could use to find the change in the stock's price after 4 days. Suppose the original price of the stock was \$41. What was the price of the stock after 4 days? (Lesson 4)

   Expression: ______ Price: ______

6. The daily high temperature readings for four days in January are shown in the table. Find the average daily temperature for the four days. (Lessons 1 and 5) ______

Day	Temperature (°F)
1	−19
2	−21
3	−22
4	−22

Reflect

 Answering the Essential Question

Use what you learned about integers to complete the graphic organizer. Explain how to determine the sign of the result when performing each operation.

Addition and Subtraction

Essential Question

WHAT happens when you add, subtract, multiply, and divide integers?

Multiply and Divide

Answer the Essential Question. WHAT happens when you add, subtract, multiply, and divide integers?

Chapter 4
Rational Numbers

Essential Question

WHAT happens when you add, subtract, multiply, and divide fractions?

Common Core State Standards

Content Standards
7.NS.1, 7.NS.1b, 7.NS.1c, 7.NS.1d, 7.NS.2, 7.NS.2a, 7.NS.2b, 7.NS.2c, 7.NS.2d, 7.NS.3, 7.RP.3, 7.EE.3

Mathematical Practices
1, 3, 4, 5, 6, 7, 8

Math in the Real World

Tennis About 70,000 tennis balls are used at the U.S. Open tennis tournament each year. This is only a small fraction of the 300,000,000 tennis balls produced each year. Write a fraction in simplest form that compares the number of tennis balls used at the U.S. Open to the number produced per year.

FOLDABLES® Study Organizer

Cut out the Foldable on page FL9 of this book.

Place your Foldable on page 338.

Use the Foldable throughout this chapter to help you learn about rational numbers.

What Tools Do You Need?

Vocabulary

bar notation
common denominator
least common denominator
like fractions
rational numbers
repeating decimal
terminating decimal
unlike fractions

Review Vocabulary

An *improper fraction* is a fraction in which the numerator is greater than or equal to the denominator, such as $\frac{21}{4}$. A *mixed number* is a number composed of a whole number and a fraction, such as $5\frac{1}{4}$.

In the organizer below, write each mixed number as an improper fraction and each improper fraction as a mixed number. The first one in each column is done for you.

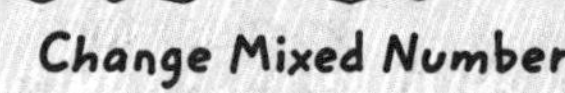

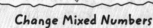

Mixed Numbers and Improper Fractions

Change Mixed Numbers	Change Improper Fractions
$3\frac{1}{2} = \frac{7}{2}$	$\frac{41}{4} = 10\frac{1}{4}$
$5\frac{1}{3} =$	$\frac{16}{3} =$
$8\frac{2}{5} =$	$\frac{23}{5} =$
$6\frac{4}{9} =$	$\frac{90}{11} =$
$10\frac{3}{8} =$	$\frac{66}{7} =$
$7\frac{3}{4} =$	$\frac{101}{2} =$
$15\frac{5}{6} =$	$\frac{87}{20} =$

When Will You Use This?

Watch Play it online!

Caitlyn, Theresa, and Aisha in
Get Organized
Thanks for coming over to help organize my closet!
Sure.
No problem. I live to organize.
I bought a really great new closet organizer! I can't wait to install it!
Uh oh.
I'm not sure this will fit in your closet. It seems a bit big.
CLOSET ORGANIZERS
It says that the size of the storage cube is $18\frac{3}{4}$" and there are three of them.
I'll measure the space, then we can figure it out.
I just measured the closet rod, and it is $12\frac{7}{8}$". I wrote it all down.
$18\frac{3}{4} + 18\frac{3}{4} + 18\frac{3}{4} + 12\frac{7}{8}$
We are doomed!!
YIKES!! Fractions! How are we going to add them up??
Theresa, just breathe! We can figure it out!

Your Turn! **You will solve this problem in the chapter.**

Are You Ready?

Try the Quick Check below.
Or, take the Online Readiness Quiz.

Quick Review

Common Core Review 5.NF.3, 6.NS.6c

Example 1

Write $\frac{25}{100}$ in simplest form.

$\frac{25}{100} = \frac{1}{4}$ (÷25, ÷25)

Divide the numerator and denominator by the GCF, 25.

Since the GCF of 1 and 4 is 1, the fraction $\frac{1}{4}$ is in simplest form.

Example 2

Graph $3\frac{2}{3}$ on a number line.

Find the two whole numbers between which $3\frac{2}{3}$ lies.

$3 < 3\frac{2}{3} < 4$

Since the denominator is 3, divide each space into 3 sections.

Draw a dot at $3\frac{2}{3}$.

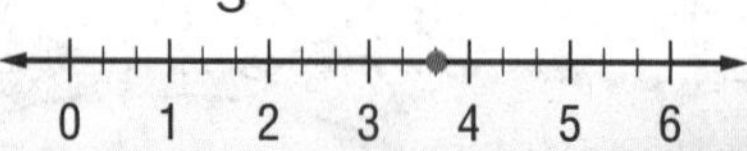

Quick Check

Fractions **Write each fraction in simplest form.**

1. $\frac{24}{36} =$ ______

2. $\frac{45}{50} =$ ______

3. $\frac{88}{121} =$ ______

Graphing **Graph each fraction or mixed number on the number line below.**

4. $\frac{1}{2}$

5. $\frac{3}{4}$

6. $1\frac{1}{4}$

7. $2\frac{1}{2}$

How Did You Do?

Which problems did you answer correctly in the Quick Check? Shade those exercise numbers below.

1 2 3 4 5 6 7

Inquiry Lab

Rational Numbers on the Number Line

HOW can you graph negative fractions on the number line?

CCSS Content Standards
Preparation for 7.NS.1

Mathematical Practices
1, 3, 8

Evaporation Water evaporates from Earth at an average of about $-\frac{3}{4}$ inch per week.

Investigation

Graph $-\frac{3}{4}$ on a number line.

Step 1 Use the bar diagram below that is divided in fourths above a number line.

Mark a 0 on the right side and a −1 on the left side.

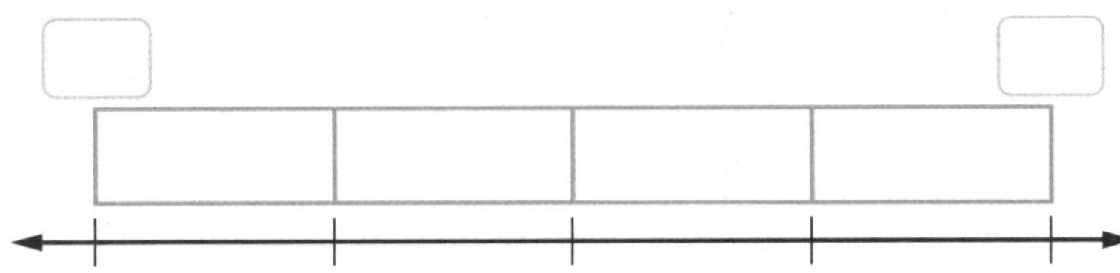

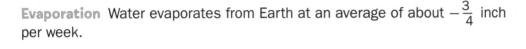

Step 2 Starting from the right, shade three fourths. Label the number line with $-\frac{1}{4}$, $-\frac{2}{4}$, and $-\frac{3}{4}$.

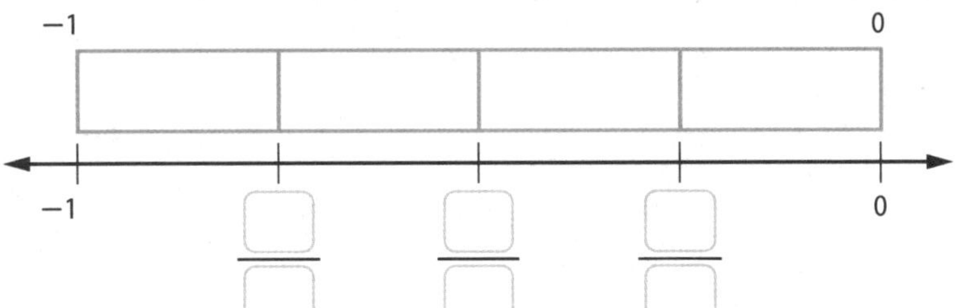

Step 3 Draw the number line portion of the model in Step 2.

Place a dot on the number line to represent $-\frac{3}{4}$.

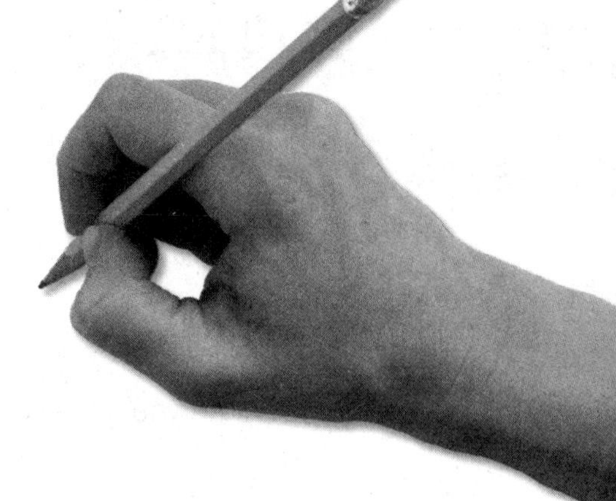

So, on a number line, $-\frac{3}{4}$ is between ☐ and $\frac{☐}{☐}$ or $\frac{☐}{☐}$.

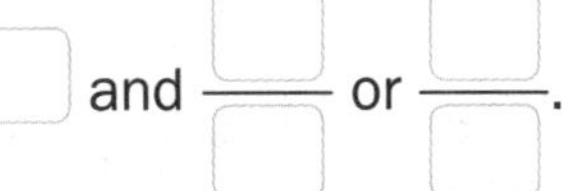

Collaborate

CCSS Model with Mathematics Work with a partner. Graph each fraction on a number line. Use a bar diagram if needed.

1. $-\frac{3}{8}$

2. $-1\frac{2}{5}$

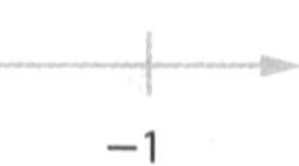

Analyze

Work with a partner to complete each table. Use a number line if needed.

		< or >	
	$\frac{7}{8}$	>	$\frac{3}{8}$
3.	$\frac{9}{8}$		$\frac{5}{8}$
4.	$\frac{13}{8}$		$\frac{3}{8}$
5.	$\frac{15}{8}$		$\frac{13}{8}$

		< or >	
	$-\frac{7}{8}$	<	$-\frac{3}{8}$
6.	$-\frac{9}{8}$		$-\frac{5}{8}$
7.	$-\frac{13}{8}$		$-\frac{3}{8}$
8.	$-\frac{15}{8}$		$-\frac{13}{8}$

9. **CCSS Identify Repeated Reasoning** Compare and contrast the information in the tables.

Reflect

10. **CCSS Reason Inductively** How does graphing $-\frac{3}{4}$ differ from graphing $\frac{3}{4}$?

11. **Inquiry** HOW can you graph negative fractions on the number line?

Lesson 1

Terminating and Repeating Decimals

What You'll Learn

Scan the lesson. List two headings you would use to make an outline of the lesson.

-
-

Essential Question

WHAT happens when you add, subtract, multiply, and divide fractions?

Vocabulary

repeating decimal
bar notation
terminating decimal

Common Core State Standards

Content Standards
7.NS.2, 7.NS.2d, 7.EE.3

Mathematical Practices
1, 3, 4, 6, 7

Vocabulary Start-Up

Any fraction can be expressed as a decimal by dividing the numerator by the denominator.

The decimal form of a fraction is called a **repeating decimal**. Repeating decimals can be represented using **bar notation**. In bar notation, a bar is drawn only over the digit(s) that repeat.

$0.3333... = 0.\overline{3}$ $\quad 0.1212... = 0.\overline{12}$ $\quad 11.38585... = 11.3\overline{85}$

If the repeating digit is zero, the decimal is a **terminating decimal**. The terminating decimal $0.25\overline{0}$ is typically written as 0.25.

Match each repeating decimal to the correct bar notation.

0.1111...	$0.6\overline{1}$
0.61111...	$0.\overline{1}$
0.616161...	$0.\overline{61}$

Real-World Link

Jamie had two hits on her first nine times at bat. To find her batting "average," she divided 2 by 9.

$2 \div 9 = 0.2222...$

Write 0.2222... using bar notation. ______

Round 0.2222... to the nearest thousandth. ______

Work Zone

Write Fractions as Decimals

Our decimal system is based on powers of 10 such as 10, 100, and 1,000. If the denominator of a fraction is a power of 10, you can use place value to write the fraction as a decimal.

Complete the table below. Write fractions in simplest form.

Words	Fraction	Decimal
seven tenths	$\frac{7}{10}$	0.7
nineteen hundredths		
one-hundred five thousandths		

If the denominator of a fraction is a *factor* of 10, 100, 1,000, or any greater power of ten, you can use mental math and place value.

Examples

Tutor

Write each fraction or mixed number as a decimal.

1. $\frac{74}{100}$

Use place value to write the equivalent decimal.

$\frac{74}{100} = 0.74$ Read $\frac{74}{100}$ as *seventy-four hundredths.*

So, $\frac{74}{100} = 0.74$.

2. $\frac{7}{20}$

Think $\frac{7}{20} = \frac{35}{100}$ (× 5, × 5)

So, $\frac{7}{20} = 0.35$.

3. $5\frac{3}{4}$

$5\frac{3}{4} = 5 + \frac{3}{4}$ Think of it as a sum.

$= 5 + 0.75$ You know that $\frac{3}{4} = 0.75$.

$= 5.75$ Add mentally.

So, $5\frac{3}{4} = 5.75$.

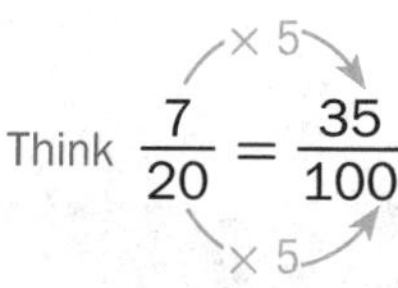

a. ______

b. ______

c. ______

Got It? Do these problems to find out.

a. $\frac{3}{10}$ **b.** $\frac{3}{25}$ **c.** $-6\frac{1}{2}$

Examples

4. Write $\frac{3}{8}$ as a decimal.

$$\begin{array}{r} 0.375 \\ 8\overline{)3.000} \\ -\underline{24} \\ 60 \\ -\underline{56} \\ 40 \\ -\underline{40} \\ 0 \end{array}$$

Divide 3 by 8.

Division ends when the remainder is 0.

So, $\frac{3}{8} = 0.375$.

5. Write $-\frac{1}{40}$ as a decimal.

$$\begin{array}{r} 0.025 \\ 40\overline{)1.000} \\ -\underline{80} \\ 200 \\ -\underline{200} \\ 0 \end{array}$$

Divide 1 by 40.

So, $-\frac{1}{40} = -0.025$.

6. Write $\frac{7}{9}$ as a decimal.

$$\begin{array}{r} 0.777... \\ 9\overline{)7.000} \\ -\underline{63} \\ 70 \\ -\underline{63} \\ 70 \\ -\underline{63} \\ 7 \end{array}$$

Divide 7 by 9.

Notice that the division will never terminate in zero.

So, $\frac{7}{9} = 0.777...$ or $0.\overline{7}$.

Bar Notation
Remember that you can use bar notation to indicate a number pattern that repeats indefinitely. $0.333... = 0.\overline{3}$.

Got It? **Do these problems to find out.**

Write each fraction or mixed number as a decimal. Use bar notation if needed.

d. $-\frac{7}{8}$

e. $2\frac{1}{8}$

f. $-\frac{3}{11}$

g. $8\frac{1}{3}$

d. __________

e. __________

f. __________

g. __________

Write Decimals as Fractions

Every terminating decimal can be written as a fraction with a denominator of 10, 100, 1,000, or a greater power of ten. Use the place value of the final digit as the denominator.

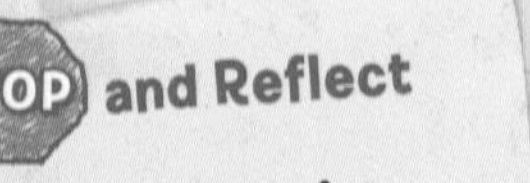

Suppose 0.6 of the fish are goldfish. Write this decimal as a fraction in the space below.

Example

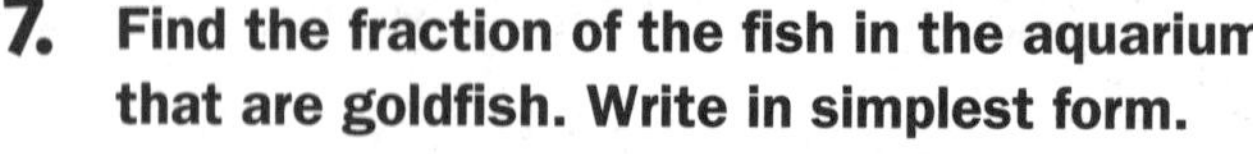

7. Find the fraction of the fish in the aquarium that are goldfish. Write in simplest form.

Fish	Amount
Guppy	0.25
Angelfish	0.4
Goldfish	0.15
Molly	0.2

$0.15 = \frac{15}{100}$ The digit 5 is in the hundredths place.

$= \frac{3}{20}$ Simplify.

So, $\frac{3}{20}$ of the fish are goldfish.

h. ___________

i. ___________

j. ___________

Got It? Do these problems to find out.

Determine the fraction of the aquarium made up by each fish. Write the answer in simplest form.

h. molly **i.** guppy **j.** angelfish

Guided Practice

Write each fraction or mixed number as a decimal. Use bar notation if needed. (Examples 1–6)

1. $\frac{2}{5} =$ ________

2. $-\frac{9}{10} =$ ________

3. $\frac{5}{9} =$ ________

4. During a hockey game, an ice resurfacer travels 0.75 mile. What fraction represents this distance? (Example 7)

____

5. **Building on the Essential Question** How can you write a fraction as a decimal?

____

Rate Yourself!

Are you ready to move on? Shade the section that applies.

For more help, go online to access a Personal Tutor.

Name ______________________ My Homework ______________________

Independent Practice

Go online for Step-by-Step Solutions 

Write each fraction or mixed number as a decimal. Use bar notation if needed. (Examples 1–6)

1. $\frac{1}{2}$ = ______

2. $-4\frac{4}{25}$ = ______

3. $\frac{1}{8}$ = ______

4. $\frac{3}{16}$ = ______

5. $-\frac{33}{50}$ = ______

6. $-\frac{17}{40}$ = ______

7. $5\frac{7}{8}$ = ______

8. $9\frac{3}{8}$ = ______

9. $-\frac{8}{9}$ = ______

10. $-\frac{1}{6}$ = ______

11. $-\frac{8}{11}$ = ______

12. $2\frac{6}{11}$ = ______

Write each decimal as a fraction or mixed number in simplest form. (Example 7)

13. -0.2 = ______

14. 0.55 = ______

15. 5.96 = ______

16. The screen on Brianna's new phone is 2.85 centimeters long. What mixed number represents the length of the phone screen? (Example 7)

______

17. STEM A praying mantis is an interesting insect that can rotate its head 180 degrees. Suppose the praying mantis at the right is 10.5 centimeters long. What mixed number represents this length? (Example 7)

______

18. **CCSS Persevere with Problems** Suppose you buy a 1.25-pound package of ham at $5.20 per pound.

a. What fraction of a pound did you buy?

b. How much money did you spend?

H.O.T. Problems Higher Order Thinking

19. **CCSS Identify Structure** Write a fraction that is equivalent to a terminating decimal between 0.5 and 0.75.

20. **CCSS Persevere with Problems** Fractions in simplest form that have denominators of 2, 4, 8, 16, and 32 produce terminating decimals. Fractions with denominators of 6, 12, 18, and 24 produce repeating decimals. What causes the difference? Explain.

21. **CCSS Persevere with Problems** The value of pi (π) is 3.1415926... . The mathematician Archimedes believed that π was between $3\frac{1}{7}$ and $3\frac{10}{71}$. Was Archimedes correct? Explain your reasoning.

Standardized Test Practice

22. Tanya drew a model for the fraction $\frac{4}{6}$.

Which of the following decimals is equal to $\frac{4}{6}$?

Ⓐ 0.666

Ⓑ $0.\overline{6}$

Ⓒ 0.667

Ⓓ $0.66\overline{7}$

Name ________________ My Homework ________________

Extra Practice

Write each fraction or mixed number as a decimal. Use bar notation if needed.

23. $\frac{4}{5}$ = 0.8

$\frac{4}{5} = \frac{8}{10}$ (×2, ×2)

So, $\frac{4}{5}$ = 0.8.

Homework Help

24. $-7\frac{1}{20}$ = ______

25. $-\frac{4}{9}$ = ______

26. $5\frac{1}{3}$ = ______

27. The fraction of a dime that is made up of copper is $\frac{12}{16}$. Write this fraction as a decimal.

______

Write each decimal as a fraction or mixed number in simplest form.

28. −0.9 = ______

29. 0.34 = ______

30. 2.66 = ______

Write each of the following as an improper fraction.

31. −13 = ______

32. $7\frac{1}{3}$ = ______

33. −3.2 = ______

34. CCSS **Be Precise** Nicolás practiced playing the cello for 2 hours and 18 minutes. Write the time Nicolás spent practicing as a decimal.

______

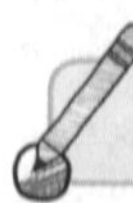

Standardized Test Practice

35. Use the table that shows decimal and fraction equivalents.

Decimal	Fraction
$0.\overline{3}$	$\frac{3}{9}$
$0.\overline{4}$	$\frac{4}{9}$
$0.\overline{5}$	$\frac{5}{9}$
$0.\overline{6}$	$\frac{6}{9}$

Which fraction represents $0.\overline{8}$?

Ⓐ $\frac{4}{5}$
Ⓑ $\frac{80}{99}$
Ⓒ $\frac{5}{6}$
Ⓓ $\frac{8}{9}$

36. The sign shows the lengths of four hiking trails.

HIKING TRAILS
Lakeview .... $1\frac{1}{4}$ mi
Forest Lane .... $1\frac{1}{3}$ mi
Sparrow Stroll .... $1\frac{3}{10}$ mi
Mountain Climb .... $1\frac{2}{3}$ mi

Which trail length is equivalent to $1.\overline{3}$?

Ⓕ Forest Lane
Ⓖ Lakeview
Ⓗ Mountain Climb
Ⓘ Sparrow Stroll

37. Zoe went to lunch with a friend. After tax, her bill was \$12.05. Which mixed number represents this amount in simplest form?

Ⓐ $12\frac{1}{2}$
Ⓑ $12\frac{1}{20}$
Ⓒ $12\frac{5}{10}$
Ⓓ $12\frac{5}{100}$

CCSS Common Core Review

Round each decimal to the tenths place. 5.NBT.4

38. 5.69 ≈ ______

39. 0.05 ≈ ______

40. 98.99 ≈ ______

Graph and label each fraction on the number line below. 6.NS.6

41. $\frac{1}{2}$

42. $\frac{3}{4}$

43. $\frac{2}{3}$

44. The table shows the discount on athletic shoes at two stores selling sporting equipment. Which store is offering the greater discount? Explain. 6.NS.7

Store	Discount
Good Sports	$\frac{1}{5}$
Go Time	25%

______

______

Lesson 2

Compare and Order Rational Numbers

What You'll Learn

Scan the lesson. Write the definitions of common denominator and least common denominator (LCD).

- Common Denominator ______________________________

____

- Least Common Denominator ______________________________

____

Essential Question

WHAT happens when you add, subtract, multiply, and divide fractions?

Vocabulary

rational number
common denominator
least common denominator

Common Core State Standards

Content Standards
7.NS.2, 7.NS.2b, 7.EE.3

Mathematical Practices
1, 3, 4

Vocabulary Start-Up

A **rational number** is a number that can be expressed as a ratio of two integers written as a fraction, in which the denominator is not zero. The Venn diagram below shows that the number 2 can be called many things. It is a whole number, integer, and rational number. The number −1.4444... is only a rational number.

Common fractions, terminating and repeating decimals, percents, and integers are all rational numbers.

Write the numbers from the number bank on the diagram.

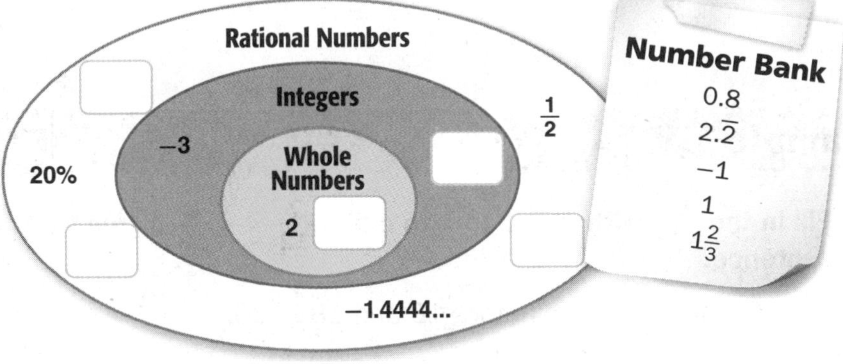

Real-World Link

Not all numbers are rational numbers. The Greek letter π (pi) represents the nonterminating and nonrepeating number whose first few digits are 3.14... . This number is an *irrational number*.

Use the Internet to search for the digits of pi. Describe what you find. ______________________________

____

Work Zone

a. ____________

LCD

To find the least common denominator for $\frac{7}{12}$ and $\frac{8}{18}$, find the LCM of 12 and 18.

$12 = 2 \times 2 \times 3$

$18 = 2 \times 3 \times 3$

$LCM = 2 \times 2 \times 3 \times 3$

$= 36$

Compare Rational Numbers

A **common denominator** is a common multiple of the denominators of two or more fractions. The **least common denominator** or **LCD** is the LCM or least common multiple of the denominators. You can use the LCD to compare fractions. You can also use a number line.

Example

1. Fill in the ◯ with <, >, or = to make $-1\frac{5}{6}$ ◯ $-1\frac{1}{6}$ a true sentence.

Graph each rational number on a number line.
Mark off equal-size increments of $\frac{1}{6}$ between −2 and −1.

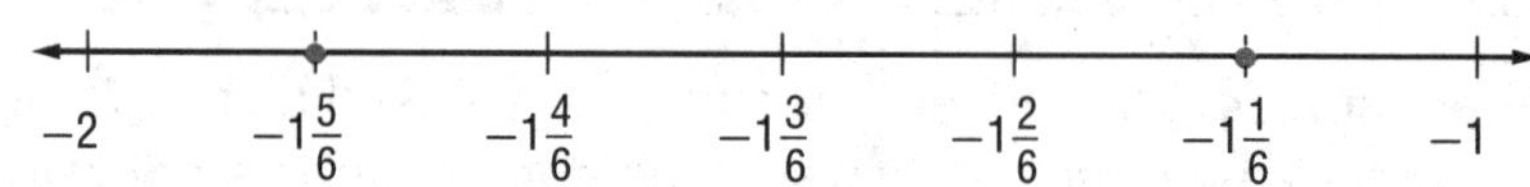

The number line shows that $-1\frac{5}{6} < -1\frac{1}{6}$.

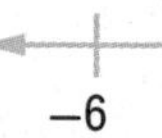

Got It? **Do this problem to find out.**

a. Use the number line to compare $-5\frac{5}{9}$ and $-5\frac{1}{9}$.

−6 −5

Example

2. Fill in the ◯ with <, >, or = to make $\frac{7}{12}$ ◯ $\frac{8}{18}$ a true sentence.

The LCD of the denominators 12 and 18 is 36.

$\frac{7}{12} = \frac{7 \times 3}{12 \times 3} = \frac{21}{36}$ $\qquad$ $\frac{8}{18} = \frac{8 \times 2}{18 \times 2} = \frac{16}{36}$

Since $\frac{21}{36} > \frac{16}{36}$, $\frac{7}{12} > \frac{8}{18}$.

Got It? **Do these problems to find out.**

b. $\frac{5}{6}$ ◯ $\frac{7}{9}$ $\qquad$ **c.** $\frac{1}{5}$ ◯ $\frac{7}{50}$ $\qquad$ **d.** $-\frac{9}{16}$ ◯ $-\frac{7}{10}$

Example

3. In Mr. Huang's class, 20% of students own roller shoes. In Mrs. Trevino's class, 5 out of 29 students own roller shoes. In which class does a greater fraction of students own roller shoes?

Express each number as a decimal and then compare.

$20\% = 0.2 \qquad \frac{5}{29} = 5 \div 29 \approx 0.1724$

Since $0.2 > 0.1724$, $20\% > \frac{5}{29}$.

More students in Mr. Huang's class own roller shoes.

Percents as Decimals

To write a percent as a decimal, remove the percent sign and then move the decimal point two places to the left. Add zeros if necessary.

20% = 20%
= 0.20

Got It? Do this problem to find out.

e. In a second period class, 37.5% of students like to bowl. In a fifth period class, 12 out of 29 students like to bowl. In which class does a greater fraction of the students like to bowl?

e. ____________

Order Rational Numbers

You can order rational numbers using place value.

Example

4. Order the set {3.44, π, 3.14, $3.\overline{4}$} from least to greatest.

Line up the decimal points and compare using place value.

3.140	Annex a zero.	3.440	Annex a zero.
3.1415926...	$\pi \approx 3.1415926...$	3.444...	$3.\overline{4} = 3.444...$
Since $0 < 1$, $3.14 < \pi$.		Since $0 < 4$, $3.44 < 3.\overline{4}$.	

So, the order of the numbers from least to greatest is 3.14, π, 3.44, and $3.\overline{4}$.

Got It? Do this problem to find out.

f. Order the set {23%, 0.21, $\frac{1}{4}$, $\frac{1}{5}$} from least to greatest.

f. ____________

Example

5. Nolan is the quarterback on the football team. He completed 67% of his passes in the first game. He completed 0.64, $\frac{3}{5}$, and 69% of his passes in the next three games. List Nolan's completed passing numbers from least to greatest.

Express each number as a decimal and then compare.

67% = 0.67 0.64 = 64% $\frac{3}{5} = 0.6$ 69% = 0.69

Nolan's completed passing numbers from least to greatest are $\frac{3}{5}$, 0.64, 67%, and 69%.

Guided Practice

Fill in each ◯ with <, >, or = to make a true sentence. Use a number line if necessary. (Examples 1 and 2)

1. $-\frac{4}{5} \bigcirc -\frac{1}{5}$

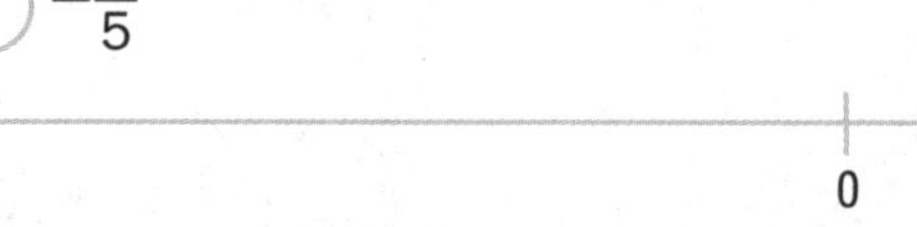

2. $1\frac{3}{4} \bigcirc 1\frac{5}{8}$

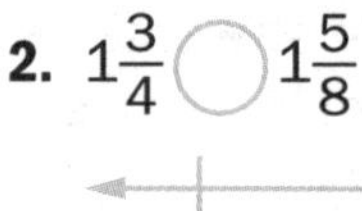

3. Elliot and Shanna are both soccer goalies. Elliot saves 3 goals out of 4. Shanna saves 7 goals out of 11. Who has the better average, Elliot or Shanna? Explain. (Example 3)

4. The lengths of four insects are 0.02 inch, $\frac{1}{8}$ inch, 0.1 inch, and $\frac{2}{3}$ inch. List the lengths in inches from least to greatest. (Examples 4 and 5)

5. **Building on the Essential Question** How can you compare two fractions?

Rate Yourself!

☐ I understand how to compare and order rational numbers.

Great! You're ready to move on!

☐ I still have some questions about comparing and ordering rational numbers.

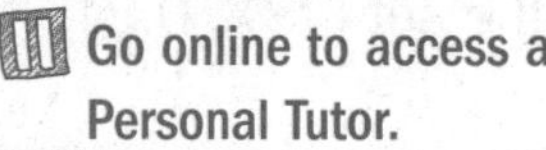

Go online to access a Personal Tutor.

Independent Practice

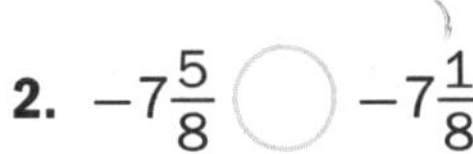

Fill in each ◯ with <, >, or = to make a true sentence. Use a number line if necessary. (Examples 1 and 2)

1. $-\frac{3}{5}$ ◯ $-\frac{4}{5}$

2. $-7\frac{5}{8}$ ◯ $-7\frac{1}{8}$

3. $6\frac{2}{3}$ ◯ $6\frac{1}{2}$

4. $-\frac{17}{24}$ ◯ $-\frac{11}{12}$

5. On her first quiz in social studies, Meg answered 92% of the questions correctly. On her second quiz, she answered 27 out of 30 questions correctly. On which quiz did Meg have the better score? (Example 3)

Order each set of numbers from least to greatest. (Example 4)

6. $\{0.23, 19\%, \frac{1}{5}\}$

7. $\{-0.615, -\frac{5}{8}, -0.62\}$

8. Liberty Middle School is holding a fundraiser. The sixth-graders have raised 52% of their goal amount. The seventh- and eighth-graders have raised 0.57 and $\frac{2}{5}$ of their goal amounts, respectively. List the classes in order from least to greatest of their goal amounts. (Example 5)

Fill in each ◯ with <, >, or = to make a true sentence.

9. $1\frac{7}{12}$ gallons ◯ $1\frac{5}{8}$ gallons

10. $2\frac{5}{6}$ hours ◯ 2.8 hours

11. **Model with Mathematics** Refer to the graphic novel frame below. If the closet organizer has a total width of $69\frac{1}{8}$ inches and the closet is $69\frac{3}{4}$ inches wide, will the organizer fit? Explain.

H.O.T. Problems Higher Order Thinking

12. **Justify Conclusions** Identify the ratio that does not have the same value as the other three. Explain your reasoning.

12 out of 15	0.08	80%	$\frac{4}{5}$

13. **Persevere with Problems** Explain how you know which number, $1\frac{15}{16}$, $\frac{17}{8}$, or $\frac{63}{32}$, is closest to 2.

Standardized Test Practice

14. Which of the following fractions is the least?

Ⓐ $-\frac{7}{8}$

Ⓑ $-\frac{7}{9}$

Ⓒ $-\frac{7}{10}$

Ⓓ $-\frac{7}{11}$

Extra Practice

Fill in each ◯ with <, >, or = to make a true sentence. Use a number line if necessary.

15. $-\frac{5}{7}$ (<) $-\frac{2}{7}$

Mark off equal-size increments of $\frac{1}{7}$ between −1 and 0.

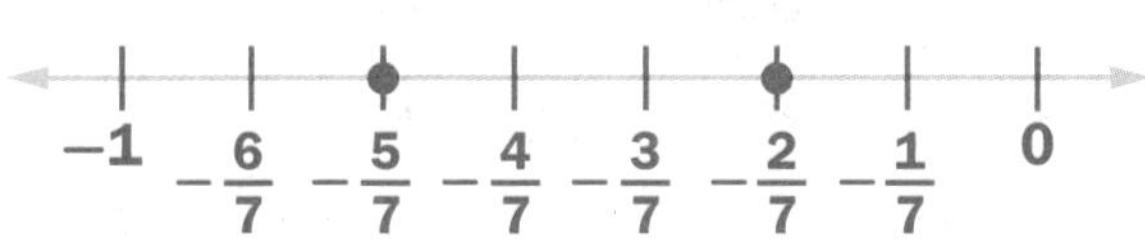

16. $-3\frac{2}{3}$ ◯ $-3\frac{4}{6}$

17. $\frac{4}{7}$ (<) $\frac{5}{8}$

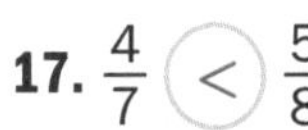

The LCD of the denominators 7 and 8 is 56.

$\frac{4}{7} = \frac{4 \times 8}{7 \times 8} = \frac{32}{56}$ and $\frac{5}{8} = \frac{5 \times 7}{8 \times 7} = \frac{35}{56}$

Since $\frac{32}{56} < \frac{35}{56}$, $\frac{4}{7} < \frac{5}{8}$.

18. $2\frac{3}{4}$ ◯ $2\frac{2}{3}$

19. Gracia and Jim were shooting free throws. Gracia made 4 out of 15 free throws. Jim *missed* 6 out of 16 free throws. Who made the free throw a greater fraction of the time? ______________

Order each set of numbers from least to greatest.

20. $\{7.49, 7\frac{49}{50}, 7.5\%\}$

____

21. $\{-1.4, -1\frac{1}{25}, -1.25\}$

____

22. **STEM** Use the table that shows the lengths of small mammals.

a. Which animal is the smallest mammal?

____

b. Which animal is smaller than the European Mole but larger than the Spiny Pocket Mouse?

____

c. Order the animals from greatest to least size.

____

Animal	Length (ft)
Eastern Chipmunk	$\frac{1}{3}$
European Mole	$\frac{5}{12}$
Masked Shrew	$\frac{1}{6}$
Spiny Pocket Mouse	0.25

Standardized Test Practice

23. Which point shows the location of $\frac{7}{2}$ on the number line?

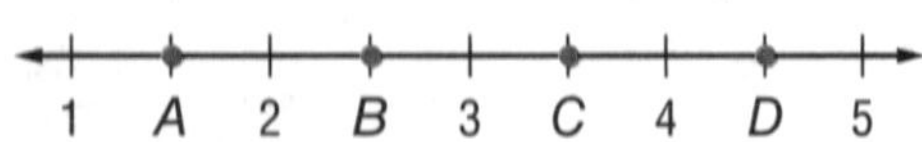

Ⓐ point *A*
Ⓑ point *B*
Ⓒ point *C*
Ⓓ point *D*

24. Which list of numbers is ordered from least to greatest?

Ⓕ $\frac{1}{4}$, $4\frac{1}{4}$, 0.4, 4%
Ⓖ 4%, 0.4, $4\frac{1}{4}$, $\frac{1}{4}$
Ⓗ 4%, $\frac{1}{4}$, 0.4, $4\frac{1}{4}$
Ⓘ 0.4, $\frac{1}{4}$, 4%, $4\frac{1}{4}$

25. The daily price changes for a stock are shown in the table.

Day	Price Change
Monday	−0.21
Tuesday	−1.05
Wednesday	−0.23
Thursday	+0.42
Friday	−1.15

On which day did the price decrease by the greatest amount?

Ⓐ Monday
Ⓑ Tuesday
Ⓒ Wednesday
Ⓓ Friday

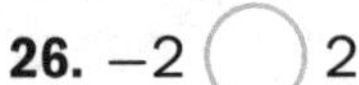

Common Core Review

Fill in each ◯ with < or > to make a true sentence. 6.NS.7

26. −2 ◯ 2

27. −4 ◯ −5

28. −20 ◯ 20

29. −7 ◯ −8

30. −10 ◯ −1

31. 50 ◯ −100

32. Victoria, Cooper, and Diego are reading the same book for their language arts class. The table shows the fraction of the book each student has read. Which student has read the least amount? Explain your reasoning. 6.NS.7

Student	Amount Read
Victoria	$\frac{2}{5}$
Cooper	$\frac{1}{5}$
Diego	$\frac{3}{5}$

Inquiry Lab

Add and Subtract on the Number Line

HOW can you use a number line to add and subtract like fractions?

Content Standards
7.NS.1, 7.NS.1b, 7.NS.3

Mathematical Practices
1, 3, 5

Baseball In eight times at bat, Max hit 2 doubles, 5 singles, and struck out 1 time. Find the fraction of the times that Max hit either a single or a double.

Investigation 1

Step 1 Since there were 8 times at bat, create a vertical number line that is divided into eighths.

Step 2 Graph the fraction of doubles, $\frac{2}{8}$, on the number line.

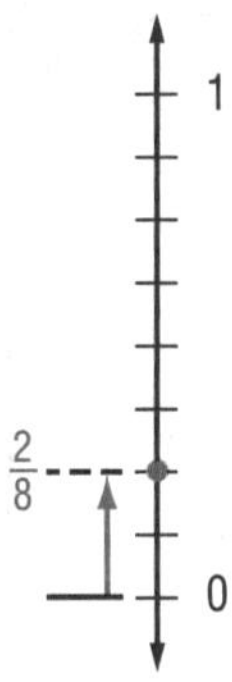

Step 3 From the $\frac{2}{8}$ point, count $\frac{5}{8}$ more on the number line.

So, $\frac{2}{8} + \frac{5}{8} = \frac{\square}{\square}$.

Max got a hit $\frac{\square}{\square}$ of the times he was at bat.

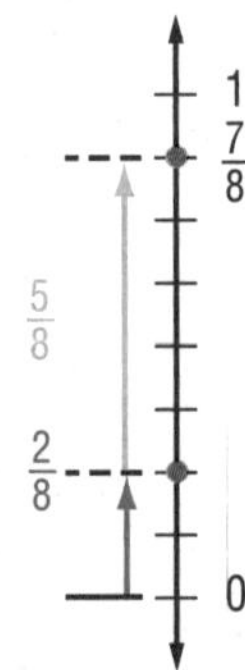

Investigation 2

Find $\frac{3}{6} - \frac{4}{6}$.

Step 1 Divide a number line into sixths. Since we do not know if our answer is negative or positive, include fractions to the left and to the right of zero.

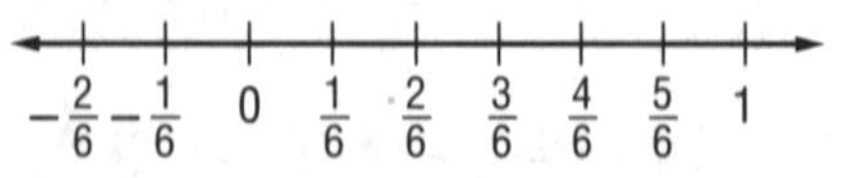

Step 2 Graph $\frac{3}{6}$ on the number line.

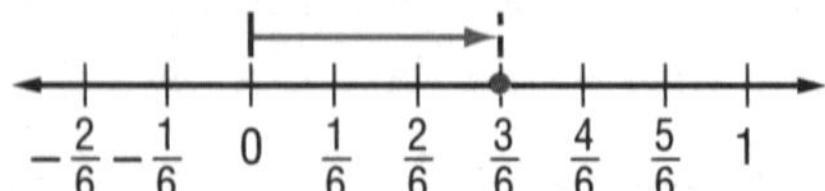

Step 3 Move 4 units to the ________ to show taking away $\frac{4}{6}$.

So, $\frac{3}{6} - \frac{4}{6} = \frac{\square}{\square}$.

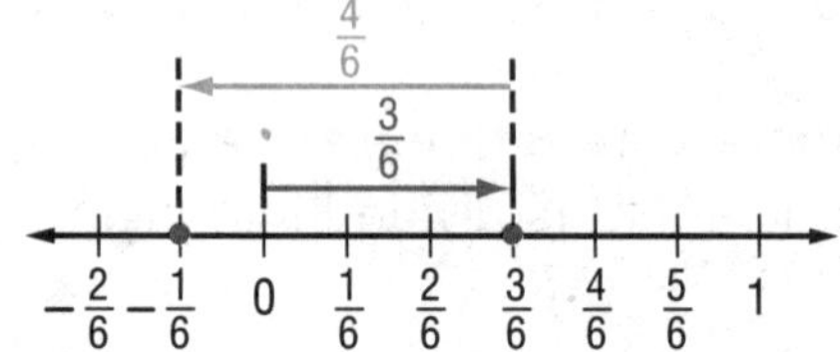

Investigation 3

Find $-\frac{4}{7} - \frac{2}{7}$. Fill in the missing numbers in the diagram below.

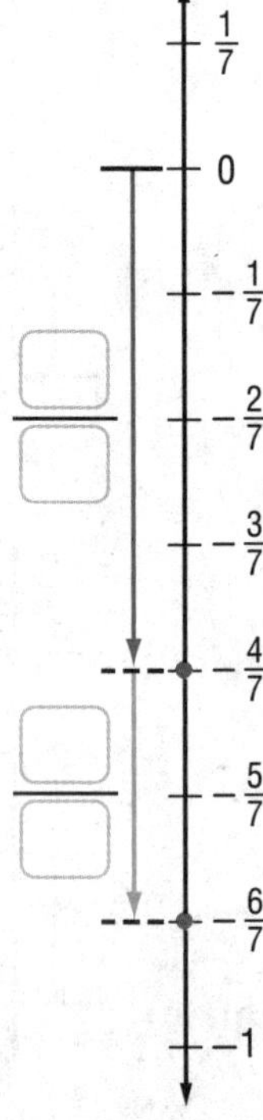

So, $-\frac{4}{7} - \frac{2}{7} = \frac{\square}{\square}$.

Work with a partner. Use a number line to add or subtract. Write in simplest form.

1. $\frac{1}{5} + \frac{2}{5} =$ ______

2. $-\frac{3}{7} + \left(-\frac{1}{7}\right) =$ ______

3. $-\frac{3}{8} + \frac{5}{8} =$ ______

4. $\frac{8}{12} - \frac{4}{12} =$ ______

5. $-\frac{4}{9} + \frac{5}{9} =$ ______

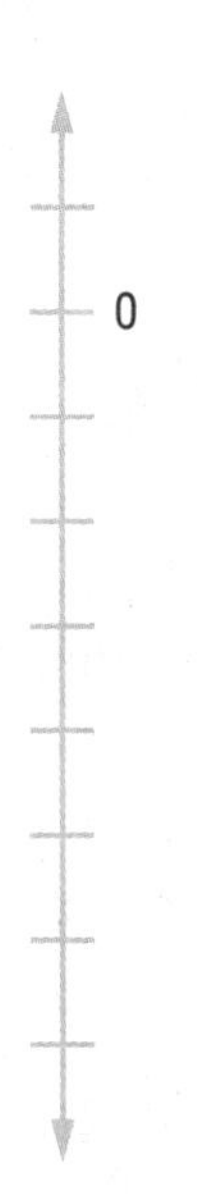

6. $\frac{4}{7} - \frac{6}{7} =$ ______

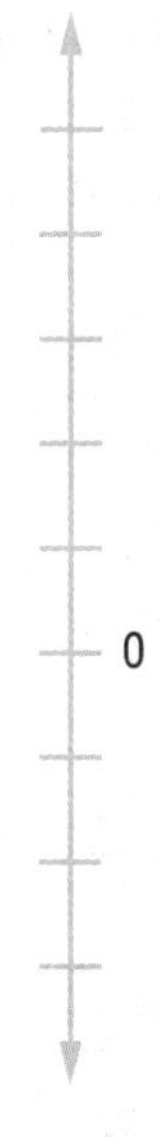

Analyze

Use Math Tools **Work with a partner to complete the table. The first one is done for you.**

	Expression	Use only the Numerators	Use a number line to add or subtract the fractions.
	$-\frac{5}{6} - \left(-\frac{1}{6}\right)$	$-5 - (-1) = -4$	$\frac{1}{6}$, $-\frac{5}{6}$; 1 $-\frac{5}{6}$ $-\frac{4}{6}$ $-\frac{3}{6}$ $-\frac{2}{6}$ $-\frac{1}{6}$ 0 $\frac{1}{6}$ $\frac{2}{6}$
7.	$-\frac{5}{6} - \frac{1}{6}$	$-5 - 1 = -6$	0
8.	$\frac{5}{6} - \frac{1}{6}$	$5 - 1 = 4$	0
9.	$-\frac{5}{6} + \frac{1}{6}$	$-5 + 1 = -4$	0

Reflect

10. **Reason Inductively** Refer back to the Analyze section. Compare your results for using only the numerators with your results for using a number line. Write a rule for adding and subtracting like fractions.

11. **Inquiry** HOW can you use a number line to add and subtract like fractions?

Lesson 3

Add and Subtract Like Fractions

What You'll Learn

Scan the lesson. List two real-world scenarios in which you would add or subtract like fractions.

- ______________________________
- ______________________________

Essential Question

WHAT happens when you add, subtract, multiply, and divide fractions?

Vocabulary

like fractions

Common Core State Standards

Content Standards
7.NS.1, 7.NS.1c, 7.NS.1d, 7.NS.3, 7.EE.3

Mathematical Practices
1, 3, 4, 7

Real-World Link

Shoes Sean surveyed ten classmates to find which type of tennis shoe they like to wear.

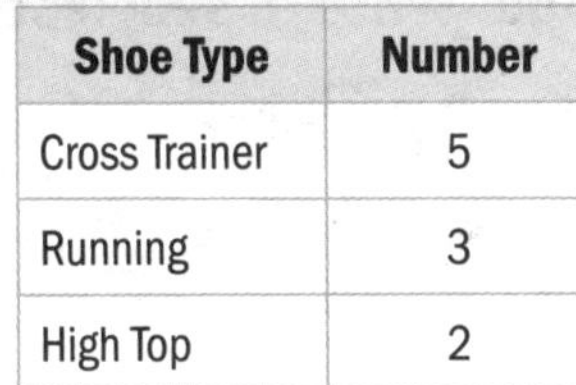

Shoe Type	Number
Cross Trainer	5
Running	3
High Top	2

1. What fraction of students liked to wear cross trainers?

Number of students that wear cross trainers. →

Total number of students surveyed. →

2. What fraction of students liked to wear high tops?

Number of students that wear high tops. →

Total number of students surveyed. →

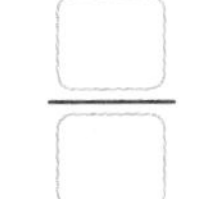

3. What fraction of students liked to wear either cross trainers or high tops?

Fraction of students that wear cross trainers. Fraction of students that wear high tops.

So, ______________ of the students liked to wear either cross trainers or high tops.

4. Explain how to find $\frac{3}{10} + \frac{2}{10}$. Then find the sum.

____

____

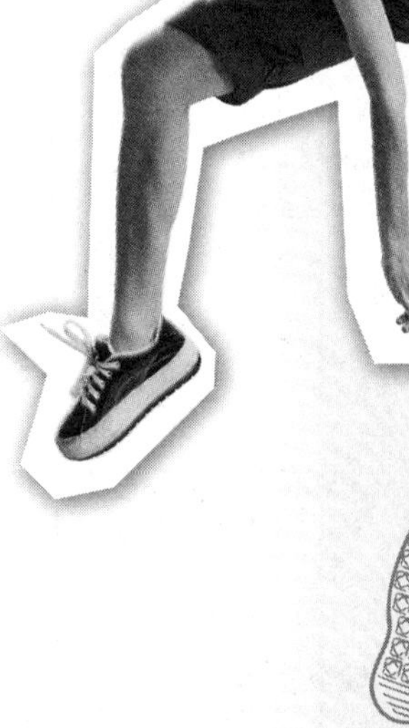

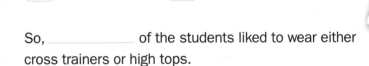

Key Concept Add and Subtract Like Fractions

Words To add or subtract like fractions, add or subtract the numerators and write the result over the denominator.

Examples

Numbers	Algebra
$\frac{5}{10} + \frac{2}{10} = \frac{5+2}{10}$ or $\frac{7}{10}$	$\frac{a}{c} + \frac{b}{c} = \frac{a+b}{c}$, where $c \neq 0$
$\frac{11}{12} - \frac{4}{12} = \frac{11-4}{12}$ or $\frac{7}{12}$	$\frac{a}{c} - \frac{b}{c} = \frac{a-b}{c}$, where $c \neq 0$

Work Zone

Fractions that have the same denominators are called **like fractions.**

Examples

Add. Write in simplest form.

1. $\frac{5}{9} + \frac{2}{9}$

$\frac{5}{9} + \frac{2}{9} = \frac{5+2}{9}$ Add the numerators.

$= \frac{7}{9}$ Simplify.

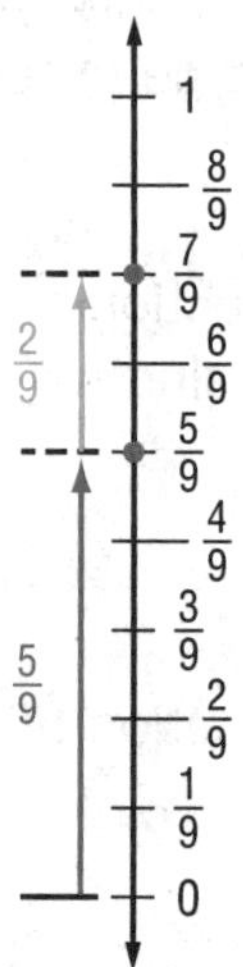

2. $-\frac{3}{5} + \left(-\frac{1}{5}\right)$

$-\frac{3}{5} + \left(-\frac{1}{5}\right) = \frac{-3}{5} + \left(\frac{-1}{5}\right)$

$= \frac{-3 + (-1)}{5}$ Add the numerators.

$= \frac{-4}{5}$ or $-\frac{4}{5}$ Use the rules for adding integers.

Negative Fractions

Remember $-\frac{1}{2} = \frac{-1}{2} = \frac{1}{-2}$. Typically, the form $\frac{-1}{2}$ is used when performing computations.

a. ____

b. ____

c. ____

d. ____

Got It? Do these problems to find out.

a. $\frac{1}{3} + \frac{2}{3}$

b. $-\frac{3}{7} + \frac{1}{7}$

c. $-\frac{2}{5} + \left(-\frac{2}{5}\right)$

d. $-\frac{1}{4} + \frac{1}{4}$

Example

3. **Sofia ate $\frac{3}{5}$ of a cheese pizza. Jack ate $\frac{1}{5}$ of a cheese pizza and $\frac{2}{5}$ of a pepperoni pizza. How much pizza did Sofia and Jack eat altogether?**

$$\frac{3}{5} + \left(\frac{1}{5} + \frac{2}{5}\right) = \frac{3}{5} + \left(\frac{2}{5} + \frac{1}{5}\right) \quad \text{Commutative Property of Addition}$$

$$= \left(\frac{3}{5} + \frac{2}{5}\right) + \frac{1}{5} \quad \text{Associative Property of Addition}$$

$$= 1 + \frac{1}{5} \text{ or } 1\frac{1}{5} \quad \text{Simplify.}$$

So, Sofia and Jack ate $1\frac{1}{5}$ pizzas altogether.

Got It? **Do this problem to find out.**

e. Eduardo used fabric to make three costumes. He used $\frac{1}{4}$ yard for the first, $\frac{2}{4}$ yard for the second, and $\frac{3}{4}$ yard for the third costume. How much fabric did Eduardo use altogether?

e. ______

Examples

4. **Find $-\frac{5}{8} - \frac{3}{8}$.**

$$-\frac{5}{8} - \frac{3}{8} = -\frac{5}{8} + \left(-\frac{3}{8}\right) \quad \text{Add } -\frac{3}{8}.$$

$$= \frac{-5 + (-3)}{8} \quad \text{Add the numerators.}$$

$$= -\frac{8}{8} \text{ or } -1 \quad \text{Simplify.}$$

Subtracting Integers
To subtract an integer, add its opposite.
$-9 - (-4) = -9 + 4$
$= -5$

5. **Find $\frac{5}{8} - \frac{7}{8}$.**

$$\frac{5}{8} - \frac{7}{8} = \frac{5-7}{8} \quad \text{Subtract the numerators.}$$

$$= -\frac{2}{8} \text{ or } -\frac{1}{4} \quad \text{Simplify.}$$

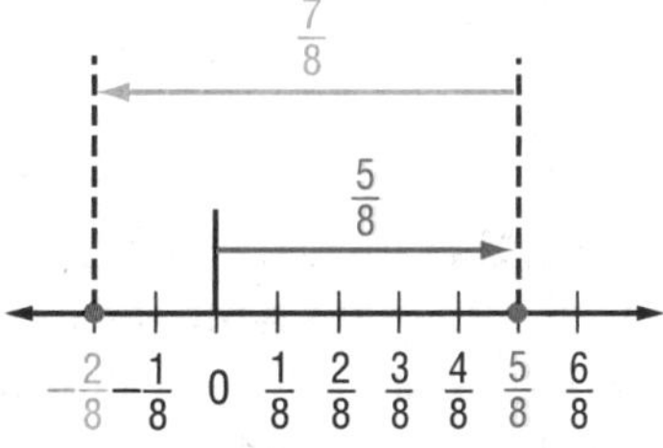

Got It? **Do these problems to find out.**

f. $\frac{5}{9} - \frac{2}{9}$

g. $-\frac{5}{9} - \frac{2}{9}$

h. $-\frac{11}{12} - \left(-\frac{5}{12}\right)$

f. ______

g. ______

h. ______

Choose an Operation

You can add or subtract like fractions to solve real-world problems.

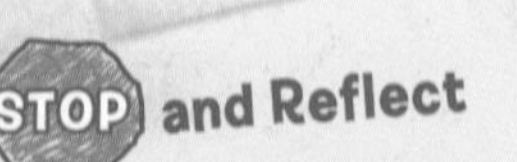

Example

Tutor

6. About $\frac{6}{100}$ of the population of the United States lives in Florida. Another $\frac{4}{100}$ lives in Ohio. About what fraction more of the U.S. population lives in Florida than in Ohio?

$\frac{6}{100} - \frac{4}{100} = \frac{6-4}{100}$ Subtract the numerators.

$= \frac{2}{100}$ or $\frac{1}{50}$ Simplify.

About $\frac{1}{50}$ more of the U.S. population lives in Florida than in Ohio.

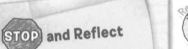

STOP and Reflect

In Example 6, what word or words indicate that you should subtract to solve the problem? Write your answer below.

Guided Practice

Check

Add or subtract. Write in simplest form. (Examples 1–5)

1. $\frac{3}{5} + \frac{1}{5} =$ ______

2. $\frac{2}{7} + \frac{1}{7} =$ ______

3. $\left(\frac{5}{8} + \frac{1}{8}\right) + \frac{3}{8} =$ ______

4. $-\frac{4}{5} - \left(-\frac{1}{5}\right) =$ ______

5. $\frac{5}{14} - \left(-\frac{1}{14}\right) =$ ______

6. $\frac{2}{7} - \frac{6}{7} =$ ______

7. Of the 50 states in the United States, 14 have an Atlantic Ocean coastline and 5 have a Pacific Ocean coastline. What fraction of U.S. states have either an Atlantic Ocean or Pacific Ocean coastline? (Example 6)

______

8. **Building on the Essential Question** What is a simple rule for adding and subtracting like fractions?

______

______

Rate Yourself!

How confident are you about adding and subtracting like fractions? Check the box that applies.

For more help, go online to access a Personal Tutor.

Tutor

FOLDABLES *Time to update your Foldable!*

Name ______ My Homework ______

Independent Practice

Go online for Step-by-Step Solutions

Add or subtract. Write in simplest form. (Examples 1, 2, 4, and, 5)

1. $\frac{5}{7} + \frac{6}{7} =$ ______

2. $\frac{3}{8} + \left(-\frac{7}{8}\right) =$ ______

3. $-\frac{1}{9} + \left(-\frac{5}{9}\right) =$ ______

4. $\frac{9}{10} - \frac{3}{10} =$ ______

5. $-\frac{3}{4} + \left(-\frac{3}{4}\right) =$ ______

6. $-\frac{5}{9} - \frac{2}{9} =$ ______

7. In Mr. Navarro's first period class, $\frac{17}{28}$ of the students got an A on their math test. In his second period class, $\frac{11}{28}$ of the students got an A. What fraction more of the students got an A in Mr. Navarro's first period class than in his second period class? Write in simplest form. (Example 6)

______

8. To make a greeting card, Bryce used $\frac{1}{8}$ sheet of red paper, $\frac{3}{8}$ sheet of green paper, and $\frac{7}{8}$ sheet of white paper. How many sheets of paper did Bryce use? (Example 3)

______

9. The table shows the Instant Messenger abbreviations students at Hillside Middle School use the most.

Instant Messenger Abbreviations	
L8R (Later)	$\frac{48}{100}$
LOL (Laughing out loud)	$\frac{26}{100}$
BRB (Be right back)	$\frac{19}{100}$
CUL8R (See you later)	$\frac{7}{100}$

a. What fraction of these students uses LOL or CUL8R when using Instant Messenger? ______

b. What fraction of these students uses L8R or BRB when using Instant Messenger? ______

c. What fraction more of these students write L8R than CUL8R when using Instant Messenger? ______

10. **Model with Mathematics** Cross out the expression that does not belong. Explain your reasoning.

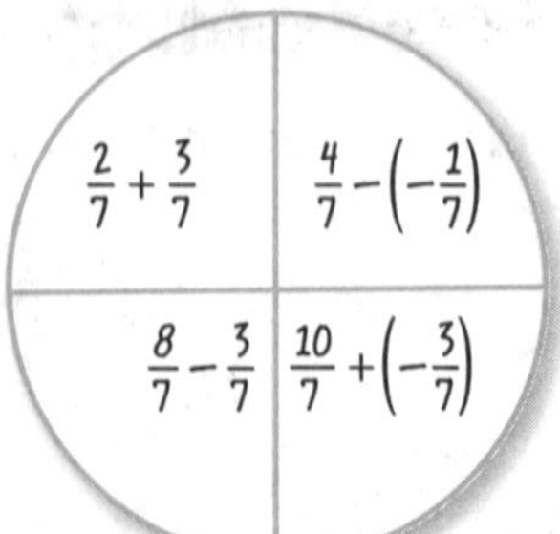

$\frac{2}{7}+\frac{3}{7}$
$\frac{4}{7}-\left(-\frac{1}{7}\right)$
$\frac{8}{7}-\frac{3}{7}$
$\frac{10}{7}+\left(-\frac{3}{7}\right)$

H.O.T. Problems Higher Order Thinking

11. **Justify Conclusions** Select two like fractions with a difference of $\frac{1}{3}$ and with denominators that are *not* 3. Justify your selection.

12. **Persevere with Problems** Simplify the following expression.

$$\frac{14}{15}+\frac{13}{15}-\frac{12}{15}+\frac{11}{15}-\frac{10}{15}+\ldots-\frac{4}{15}+\frac{3}{15}-\frac{2}{15}+\frac{1}{15}$$

Standardized Test Practice

13. The body length of a male Jumping Spider is shown below.

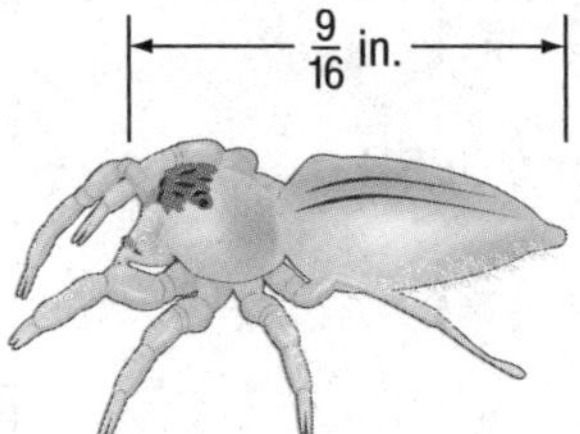

The body length of a female Jumping Spider is about $\frac{5}{16}$ inch.
How much longer is the body length of a male Jumping Spider than the body length of a female Jumping Spider?

Ⓐ $\frac{7}{8}$ inch

Ⓑ $\frac{1}{2}$ inch

Ⓒ $\frac{1}{4}$ inch

Ⓓ $\frac{3}{16}$ inch

Extra Practice

Add or subtract. Write in simplest form.

14. $\frac{4}{5} + \frac{3}{5} = 1\frac{2}{5}$

$\frac{4}{5} + \frac{3}{5} = \frac{4+3}{5}$

$= \frac{7}{5}$ or $1\frac{2}{5}$

15. $-\frac{5}{6} + \left(-\frac{5}{6}\right) =$ ______

16. $-\frac{15}{16} + \left(-\frac{7}{16}\right) =$ ______

17. $\frac{5}{8} - \frac{3}{8} =$ ______

18. $\frac{7}{12} - \frac{2}{12} =$ ______

19. $\frac{15}{18} - \frac{13}{18} =$ ______

20. Two nails are $\frac{5}{16}$ inch and $\frac{13}{16}$ inch long. How much shorter is the $\frac{5}{16}$-inch nail? ______

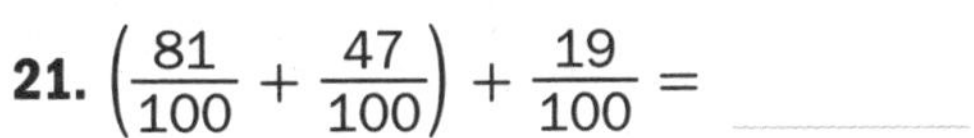

Identify Structure **Add. Write in simplest form.**

21. $\left(\frac{81}{100} + \frac{47}{100}\right) + \frac{19}{100} =$ ______

22. $\frac{\frac{1}{3}}{6} + \frac{\frac{2}{3}}{6} =$ ______

23. A recipe for Michigan blueberry pancakes calls for $\frac{3}{4}$ cup flour, $\frac{1}{4}$ cup milk, and $\frac{1}{4}$ cup blueberries. How much more flour is needed than milk? Write in simplest form.

______

24. The graph shows the location of volcanic eruptions.

a. What fraction represents the volcanic eruptions for both North and South America?

______

b. How much larger is the section for Asia and South Pacific than for Europe? Write in simplest form.

______

Worldwide Volcano Eruptions

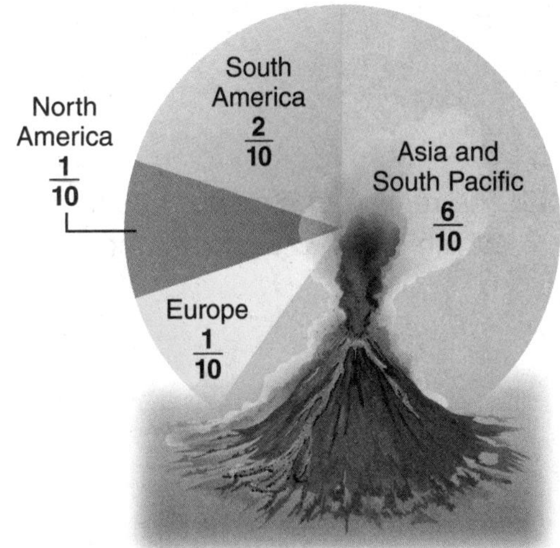

Standardized Test Practice

25. A group of friends bought two large pizzas and ate only part of each pizza. The picture shows how much was left.

How many pizzas did they eat?

Ⓐ $\frac{3}{8}$

Ⓑ $\frac{5}{8}$

Ⓒ $1\frac{1}{4}$

Ⓓ $1\frac{3}{8}$

26. At a school carnival, homemade pies were cut into 8 equal-size pieces. Eric sold 13 pieces, Elena sold 7 pieces, and Tanya sold 10 pieces. Which expression can be used to find the total number of pies sold by Eric, Elena, and Tanya?

Ⓕ $13 + 7 + 10$

Ⓖ $8(13 + 7 + 10)$

Ⓗ $\frac{13}{8} \times \frac{7}{8} \times \frac{10}{8}$

Ⓘ $\frac{13}{8} + \frac{7}{8} + \frac{10}{8}$

27. Short Response What is the value of x that makes the statement below true? ______

$$\frac{7}{9} - \frac{x}{9} = \frac{1}{3}$$

CCSS Common Core Review

Fill in each ◯ with <, >, or = to make a true sentence. 6.NS.7

28. $\frac{7}{8}$ ◯ $\frac{3}{4}$

29. $\frac{1}{3}$ ◯ $\frac{7}{9}$

30. $\frac{5}{7}$ ◯ $\frac{4}{5}$

31. $\frac{6}{11}$ ◯ $\frac{9}{14}$

Find the least common denominator for each pair of fractions. 6.NS.4

32. $\frac{1}{2}$ and $\frac{1}{3}$ ______

33. $\frac{4}{7}$ and $\frac{3}{28}$ ______

34. $\frac{1}{5}$ and $\frac{7}{6}$ ______

35. $\frac{13}{15}$ and $\frac{7}{12}$ ______

36. The results of a survey about favorite lunch choices are shown. Which lunch was chosen most often? 6.NS.7

______

Favorite Lunch	
Food	**Fraction of Students**
Pizza	$\frac{39}{50}$
Hot Dogs	$\frac{3}{25}$
Grilled Cheese	$\frac{1}{10}$

Lesson 4

Add and Subtract Unlike Fractions

What You'll Learn

Scan the lesson. List two headings you would use to make an outline of the lesson.

-
-

Essential Question

WHAT happens when you add, subtract, multiply, and divide fractions?

Vocabulary

unlike fractions

Common Core State Standards

Content Standards
7.NS.1, 7.NS.1d, 7.NS.3, 7.EE.3

Mathematical Practices
1, 3, 4

Real-World Link

Time The table shows the fractions of one hour for certain minutes.

1. What fraction of one hour is equal to the sum of 15 minutes and 20 minutes?

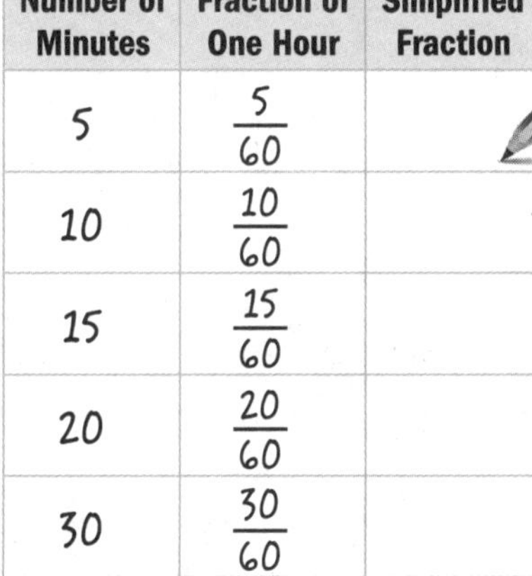

Number of Minutes	Fraction of One Hour	Simplified Fraction
5	$\frac{5}{60}$	
10	$\frac{10}{60}$	
15	$\frac{15}{60}$	
20	$\frac{20}{60}$	
30	$\frac{30}{60}$	

15 minutes   20 minutes

$\frac{\square}{\square} + \frac{\square}{\square} = \frac{\square}{\square}$

2. Write each fraction of an hour in simplest form in the third column of the table.

3. Explain why $\frac{1}{6}$ hour $+ \frac{1}{3}$ hour $= \frac{1}{2}$ hour.

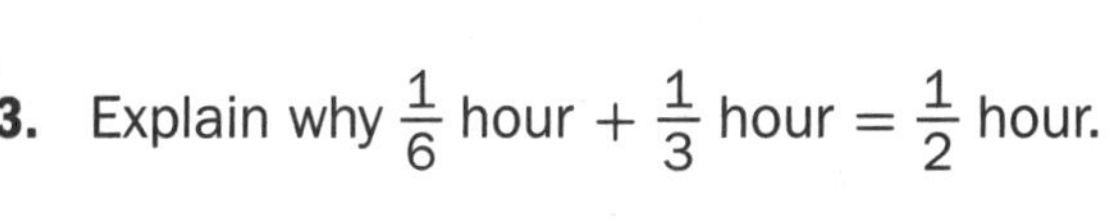

4. Explain why $\frac{1}{12}$ hour $+ \frac{1}{2}$ hour $= \frac{7}{12}$ hour.

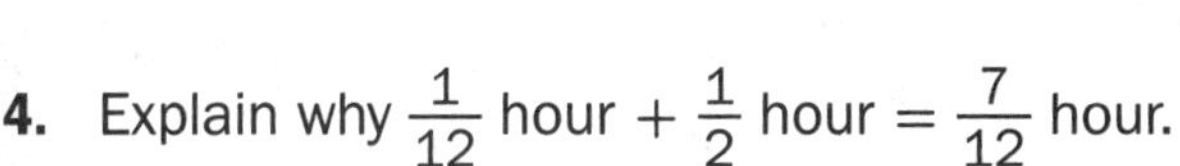

Key Concept Add or Subtract Unlike Fractions

To add or subtract fractions with different denominators,

- Rename the fractions using the least common denominator (LCD).
- Add or subtract as with like fractions.
- If necessary, simplify the sum or difference.

Work Zone

Before you can add two **unlike fractions**, or fractions with different denominators, rename one or both of the fractions so that they have a common denominator.

STOP and Reflect

Circle the pairs of fractions that are unlike fractions.

$\frac{1}{3}$ and $\frac{5}{3}$ $\frac{1}{7}$ and $\frac{1}{5}$ $\frac{5}{9}$ and $\frac{4}{11}$

Example

1. **Find $\frac{1}{2} + \frac{1}{4}$.**

Use a number line.

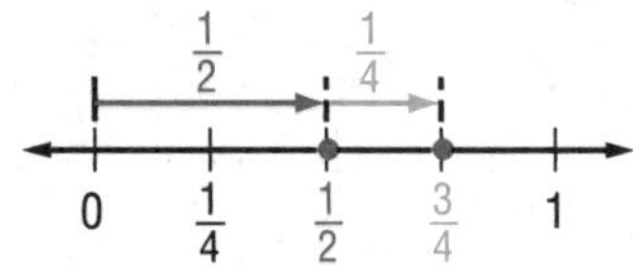

Divide the number line into fourths since the LCD is 4.

Method 2 **Use the LCD.**

The least common denominator of $\frac{1}{2}$ and $\frac{1}{4}$ is 4.

$$\frac{1}{2} + \frac{1}{4} = \frac{1 \times 2}{2 \times 2} + \frac{1 \times 1}{4 \times 1}$$ Rename using the LCD, 4.

$$= \frac{2}{4} + \frac{1}{4}$$ Add the fractions.

$$= \frac{3}{4}$$ Simplify.

Using either method, $\frac{1}{2} + \frac{1}{4} = \frac{3}{4}$.

Got It? **Do these problems to find out.**

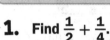

Show your work.

Add. Write in simplest form.

a. $\frac{1}{6} + \frac{2}{3}$

b. $\frac{9}{10} + \left(-\frac{1}{2}\right)$

c. $\frac{1}{4} + \frac{3}{8}$

d. $-\frac{1}{3} + \left(-\frac{1}{4}\right)$

a. ________

b. ________

c. ________

d. ________

Example

2. Find $\left(-\frac{3}{4}+\frac{5}{9}\right)+\frac{7}{4}$.

$$\left(-\frac{3}{4}+\frac{5}{9}\right)+\frac{7}{4}=\left(\frac{5}{9}+\left(-\frac{3}{4}\right)\right)+\frac{7}{4}$$ Commutative Property of Addition

$$=\frac{5}{9}+\left(-\frac{3}{4}+\frac{7}{4}\right)$$ Associative Property of Addition

$$=\frac{5}{9}+1 \text{ or } 1\frac{5}{9}$$ Simplify.

Got It? Do these problems to find out.

e. $\frac{2}{5}+\left(\frac{4}{7}+\frac{3}{5}\right)$

f. $\left(-\frac{3}{10}+\frac{5}{8}\right)+\frac{23}{10}$

e. ______

f. ______

Example

3. Find $-\frac{2}{3}-\frac{1}{2}$.

Method 1 Use a number line.

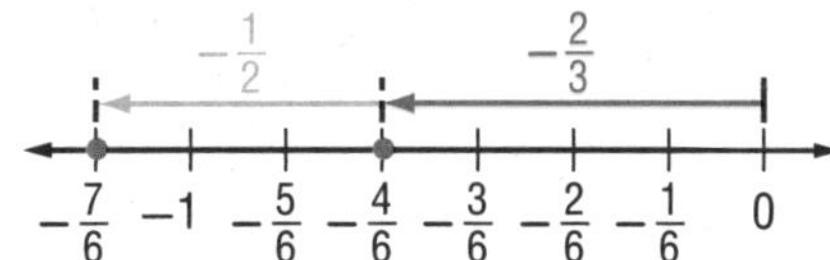

Divide the number line into sixths since the LCD is 6.

Method 2 Use the LCD.

$$-\frac{2}{3}-\frac{1}{2}=-\frac{2\times 2}{3\times 2}-\frac{1\times 3}{2\times 3}$$ Rename using the LCD, 6.

$$=-\frac{4}{6}-\frac{3}{6}$$ Simplify.

$$=\frac{-4}{6}-\frac{3}{6}$$ Rewrite $-\frac{4}{6}$ as $\frac{-4}{6}$.

$$=\frac{-4-3}{6} \text{ or } \frac{-7}{6}$$ Subtract the numerators. Simplify.

Check by adding $-\frac{7}{6}+\frac{1}{2}=-\frac{7}{6}+\frac{3}{6}=-\frac{4}{6}$ or $-\frac{2}{3}$ ✓

Using either method, $-\frac{2}{3}-\frac{1}{2}=-\frac{7}{6}$ or $-1\frac{1}{6}$.

Check for Reasonableness

Estimate the difference.

$-\frac{2}{3}-\frac{1}{2}\approx-\frac{1}{2}-\frac{1}{2}$ or -1

Compare $-\frac{7}{6}$ to the estimate.

$-\frac{7}{6}\approx-1$. So, the answer is reasonable.

Got It? Do these problems to find out.

Subtract. Write in simplest form.

g. $\frac{5}{8}-\frac{1}{4}$

h. $\frac{3}{4}-\frac{1}{3}$

i. $\frac{1}{2}-\left(-\frac{2}{5}\right)$

g. ______

h. ______

i. ______

Choose an Operation

Add or subtract unlike fractions to solve real-world problems.

Example

4. STEM Use the table to find the fraction of the total population that has type A or type B blood.

Blood Type Frequencies				
ABO Type	O	A	B	AB
Fraction	$\frac{11}{25}$	$\frac{21}{50}$	$\frac{1}{10}$	$\frac{1}{25}$

To find the fraction of the total population, add $\frac{21}{50}$ and $\frac{1}{10}$.

$$\frac{21}{50} + \frac{1}{10} = \frac{21 \times 1}{50 \times 1} + \frac{1 \times 5}{10 \times 5}$$ Rename using the LCD, 50.

$$= \frac{21}{50} + \frac{5}{50}$$ Add the fractions.

$$= \frac{26}{50} \text{ or } \frac{13}{25}$$ Simplify.

So, $\frac{13}{25}$ of the population has type A or type B blood.

Guided Practice

Add or subtract. Write in simplest form. (Examples 1–3)

1. $\frac{3}{5} + \frac{1}{10} =$ ______

2. $-\frac{5}{6} + \left(-\frac{4}{9}\right) =$ ______

3. $\left(\frac{7}{8} + \frac{3}{11}\right) + \frac{1}{8} =$ ______

4. $\frac{4}{5} - \frac{3}{10} =$ ______

5. $\frac{3}{8} - \left(-\frac{1}{4}\right) =$ ______

6. $\frac{3}{4} - \frac{1}{3} =$ ______

7. Cassandra cuts $\frac{5}{16}$ inch off the top of a photo and $\frac{3}{8}$ inch off the bottom. How much shorter is the total height of the photo now? Explain. (Example 4)

8. **Building on the Essential Question** Compare adding unlike fractions and adding like fractions.

Rate Yourself!

Are you ready to move on? Shade the section that applies.

YES ? NO

For more help, go online to access a Personal Tutor.

FOLDABLES *Time to update your Foldable!*

Name ______ My Homework ______

Independent Practice

Go online for Step-by-Step Solutions

Add or subtract. Write in simplest form. (Examples 1–3)

1. $\frac{1}{6} + \frac{3}{8} =$ ______

2. $-\frac{1}{15} + \left(-\frac{3}{5}\right) =$ ______

3. $\left(\frac{15}{8} + \frac{2}{5}\right) + \left(-\frac{7}{8}\right) =$ ______

4. $\left(-\frac{7}{10}\right) - \frac{2}{5} =$ ______

5. $\frac{7}{9} - \frac{1}{3} =$ ______

6. $-\frac{7}{12} + \frac{7}{10} =$ ______

7. $-\frac{4}{9} - \frac{2}{15} =$ ______

8. $\frac{5}{8} + \frac{11}{12} =$ ______

9. $\frac{7}{9} + \frac{5}{6} =$ ______

CCSS ***Justify Conclusions*** **Choose an operation to solve each problem. Explain your reasoning. Then solve the problem. Write in simplest form.** (Example 4)

10. Mrs. Escalante was riding a bicycle on a bike path. After riding $\frac{2}{3}$ of a mile, she discovered that she still needed to travel $\frac{3}{4}$ of a mile to reach the end of the path. How long is the bike path?

______

______

11. Four students were scheduled to give book reports in 1 hour. After the first report, $\frac{2}{3}$ hour remained. The next two reports took $\frac{1}{6}$ hour and $\frac{1}{4}$ hour. What fraction of the hour remained?

______

______

12. One hundred sixty cell phone owners were surveyed.

    a. What fraction of owners prefers using their cell phone for text messaging or playing games? Explain.

    ______

    b. What fraction of owners prefers using their phone to take pictures or text message?

    ______

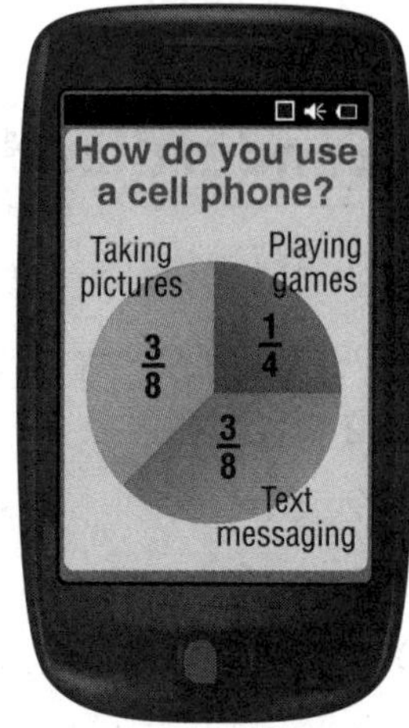

13. Pepita and Francisco each spend an equal amount of time on homework. The table shows the fraction of time they spend on each subject. Complete the table by determining the missing fraction for each student.

Homework	Fraction of Time	
	Pepita	Francisco
Math		$\frac{1}{2}$
English	$\frac{2}{3}$	
Science	$\frac{1}{6}$	$\frac{3}{8}$

14. Chelsie saves $\frac{1}{5}$ of her allowance and spends $\frac{2}{3}$ of her allowance at the mall. What fraction of her allowance remains? Explain.

H.O.T. Problems Higher Order Thinking

15. CCSS **Persevere with Problems** Fractions whose numerators are 1, such as $\frac{1}{2}$ or $\frac{1}{3}$, are called *unit fractions*. Describe a method you can use to add two unit fractions mentally.

16. CCSS **Use a Counterexample** Provide a counterexample to the following statement.

The sum of three fractions with odd numerators is never $\frac{1}{2}$.

Standardized Test Practice

17. Which of the following is the prime factorization of the least common denominator of $\frac{7}{12} + \frac{11}{18}$?

Ⓐ 2×3

Ⓑ 2×3^2

Ⓒ $2^2 \times 3^2$

Ⓓ $2^3 \times 3$

Extra Practice

Add or subtract. Write in simplest form.

18. $\frac{5}{8} + \frac{1}{4} =$ $\frac{7}{8}$

Homework Help

$\frac{5}{8} + \frac{1}{4} = \frac{5}{8} + \frac{1 \times 2}{4 \times 2}$

$= \frac{5}{8} + \frac{2}{8}$

$= \frac{7}{8}$

19. $\frac{4}{5} - \frac{1}{6} =$ ______

20. $\frac{5}{6} - \left(-\frac{2}{3}\right) =$ ______

21. $\frac{3}{10} - \left(-\frac{1}{4}\right) =$ ______

22. $-\frac{2}{3} + \left(\frac{3}{4} + \frac{5}{3}\right) =$ ______

23. $-\frac{7}{8} + \frac{1}{3} =$ ______

Choose an operation to solve each problem. Explain your reasoning. Then solve the problem. Write in simplest form.

24. Ebony is building a shelf to hold the two boxes shown. What is the least width she should make the shelf?

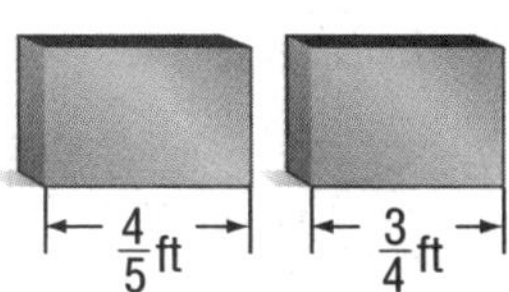

______

______

25. Makayla bought $\frac{1}{4}$ pound of ham and $\frac{5}{8}$ pound of turkey. How much more turkey did she buy? ______

______

26. **Persevere with Problems** Find the sum of $\frac{\frac{3}{4}}{8}$ and $\frac{\frac{1}{3}}{4}$. Write in simplest form.

______

27. **Find the Error** Theresa is finding $\frac{1}{4} + \frac{3}{5}$. Find her mistake and correct it. Explain your answer.

______

______

______

______

Standardized Test Practice

28. The table gives the number of hours Orlando spent at football practice.

Day	Time (h)
Monday	$\frac{1}{2}$
Tuesday	$\frac{3}{4}$
Thursday	$\frac{1}{3}$
Friday	$\frac{5}{6}$

How many more hours did he practice on Friday than on Thursday?

Ⓐ $\frac{1}{3}$
Ⓑ $\frac{1}{2}$
Ⓒ $\frac{5}{6}$
Ⓓ $1\frac{1}{6}$

29. Brett has $\frac{5}{6}$ of his weekly allowance left to spend. He has budgeted $\frac{1}{8}$ of his allowance to save for a new video game. How much of his weekly allowance will he have left after putting the savings away?

Ⓕ $\frac{4}{7}$
Ⓖ $\frac{3}{8}$
Ⓗ $\frac{7}{12}$
Ⓘ $\frac{17}{24}$

30. Felicia needs 1 cup of flour. She only has a $\frac{2}{3}$-cup measure and a $\frac{3}{4}$-cup measure. Which method will bring her closest to having the amount of flour she needs?

Ⓐ Fill the $\frac{2}{3}$-cup measure twice.
Ⓑ Fill the $\frac{3}{4}$-cup measure twice.
Ⓒ Fill the $\frac{2}{3}$-cup measure once.
Ⓓ Fill the $\frac{3}{4}$-cup measure once.

CCSS Common Core Review

Write each improper fraction as a mixed number. 5.NF.3

31. $\frac{7}{5}$ = ______

32. $\frac{14}{3}$ = ______

33. $\frac{101}{100}$ = ______

34. $\frac{22}{9}$ = ______

35. $\frac{77}{10}$ = ______

36. $\frac{23}{8}$ = ______

Divide. 6.NS.2

37. 364 ÷ 14 = ______

38. $4\overline{)5,206}$ = ______

39. $\frac{216}{8}$ = ______

Lesson 5

Add and Subtract Mixed Numbers

What You'll Learn

Scan the lesson. List two real-world scenarios in which you would add or subtract mixed numbers.

-
-

Essential Question

WHAT happens when you add, subtract, multiply, and divide fractions?

Common Core State Standards

Content Standards
7.NS.1, 7.NS.1d, 7.NS.3, 7.EE.3

Mathematical Practices
1, 3, 4

Real-World Link

Hockey Junior and adult hockey sticks are shown below.

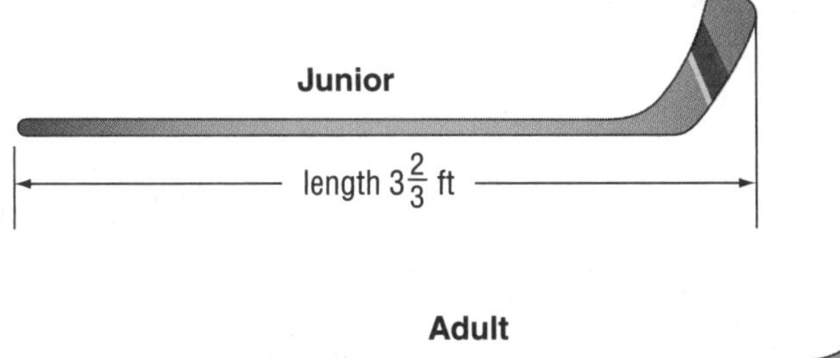

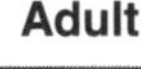

1. Use the expression $4\frac{5}{6} - 3\frac{2}{3}$ to find how much longer the adult hockey stick is than the junior hockey stick.

Rename the fractions using the LCD, 6.

Subtract the fractions. Then subtract the whole numbers.

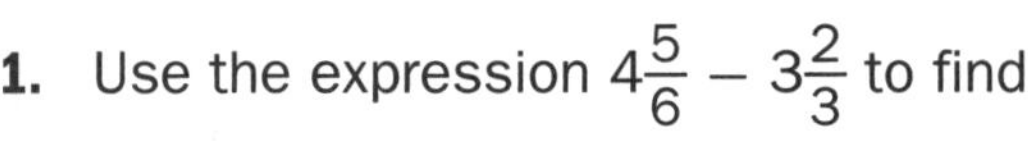

2. Explain how to find $3\frac{7}{10} - 2\frac{2}{5}$. Then use your conjecture to find the difference.

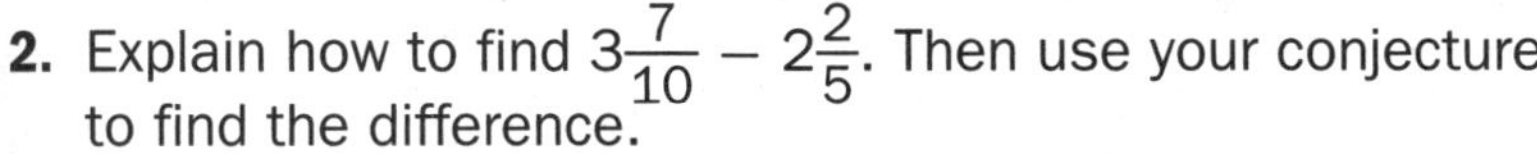

Work Zone

Add and Subtract Mixed Numbers

To add or subtract mixed numbers, first add or subtract the fractions. If necessary, rename them using the LCD. Then add or subtract the whole numbers and simplify if necessary.

Sometimes when you subtract mixed numbers, the fraction in the first mixed number is less than the fraction in the second mixed number. In this case, rename one or both fractions in order to subtract.

Examples

Properties
$120\frac{1}{2} + 40\frac{1}{3}$ can be written as $(120 + \frac{1}{2}) + (40 + \frac{1}{3})$. Then the Commutative and Associative Properties can be used to reorder and regroup the numbers to find the sum.

1. Find $7\frac{4}{9} + 10\frac{2}{9}$. Write in simplest form.

Estimate $7 + 10 = 17$

$$\begin{array}{r} 7\frac{4}{9} \\ +\ 10\frac{2}{9} \\ \hline 17\frac{6}{9} \text{ or } 17\frac{2}{3} \end{array}$$

Add the whole numbers and fractions separately.

Simplify.

Check for Reasonableness $17\frac{2}{3} \approx 17$ ✓

2. Find $8\frac{5}{6} - 2\frac{1}{3}$. Write in simplest form.

Estimate $9 - 2 = 7$

$$\begin{array}{rcr} 8\frac{5}{6} & \rightarrow & 8\frac{5}{6} \\ -2\frac{1}{3} & \rightarrow & -2\frac{2}{6} \\ \hline & & 6\frac{3}{6} \text{ or } 6\frac{1}{2} \end{array}$$

Rename the fraction using the LCD. Then subtract.

Simplify.

Check for Reasonableness $6\frac{1}{2} \approx 7$ ✓

Got It? Do these problems to find out.

Add or subtract. Write in simplest form.

a. $6\frac{1}{8} + 2\frac{5}{8}$ **b.** $5\frac{1}{5} + 2\frac{3}{10}$ **c.** $1\frac{5}{9} + 4\frac{1}{6}$

d. $5\frac{4}{5} - 1\frac{3}{10}$ **e.** $13\frac{7}{8} - 9\frac{3}{4}$ **f.** $8\frac{2}{3} - 2\frac{1}{2}$

Show your work.

a. ______

b. ______

c. ______

d. ______

e. ______

f. ______

Example

3. **Find $2\frac{1}{3} - 1\frac{2}{3}$.**

Method 1 **Rename Mixed Numbers**

Estimate $2 - 1\frac{1}{2} = \frac{1}{2}$

Since $\frac{1}{3}$ is less than $\frac{2}{3}$, rename $2\frac{1}{3}$ before subtracting.

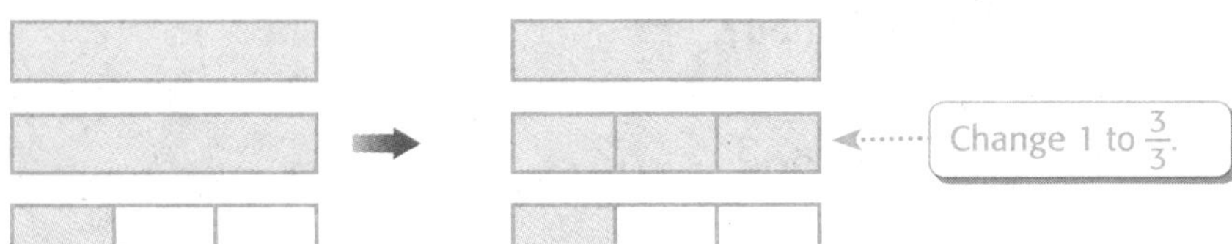

$2\frac{1}{3} = 1\frac{3}{3} + \frac{1}{3}$ or $1\frac{4}{3}$

$$\begin{array}{rcr} 2\frac{1}{3} & \rightarrow & 1\frac{4}{3} \\ -1\frac{2}{3} & \rightarrow & -1\frac{2}{3} \\ \hline & & \frac{2}{3} \end{array}$$

Rename $2\frac{1}{3}$ as $1\frac{4}{3}$.

Subtract the whole numbers and then the fractions.

Check for Reasonableness $\frac{2}{3} \approx \frac{1}{2}$ ✓

Method 2 **Write as Improper Fractions**

$$\begin{array}{rcr} 2\frac{1}{3} & \rightarrow & \frac{7}{3} \\ -1\frac{2}{3} & \rightarrow & -\frac{5}{3} \\ \hline & & \frac{2}{3} \end{array}$$

Write $2\frac{1}{3}$ as $\frac{7}{3}$.

Write $1\frac{2}{3}$ as $\frac{5}{3}$.

Simplify.

So, $2\frac{1}{3} - 1\frac{2}{3} = \frac{2}{3}$.

Using either method, the answer is $\frac{2}{3}$.

Fractions Greater Than One

An improper fraction has a numerator that is greater than or equal to the denominator. Examples of improper fractions are $\frac{5}{4}$ and $2\frac{6}{5}$.

Got It? **Do these problems to find out.**

Subtract. Write in simplest form.

g. $7 - 1\frac{1}{2}$

h. $5\frac{3}{8} - 4\frac{11}{12}$

i. $11\frac{2}{5} - 2\frac{3}{5}$

j. $8 - 3\frac{3}{4}$

k. $3\frac{1}{4} - 1\frac{3}{4}$

l. $16 - 5\frac{5}{6}$

g. ________

h. ________

i. ________

j. ________

k. ________

l. ________

Choose an Operation

Add or subtract unlike fractions to solve real-world problems.

Example

4. An urban planner is designing a skateboard park. The length of the skateboard park is $120\frac{1}{2}$ feet. The length of the parking lot is $40\frac{1}{3}$ feet. What will be the length of the park and the parking lot combined?

$$120\frac{1}{2} + 40\frac{1}{3} = 120\frac{3}{6} + 40\frac{2}{6}$$ Rename $\frac{1}{2}$ as $\frac{3}{6}$ and $\frac{1}{3}$ as $\frac{2}{6}$.

$$= 160 + \frac{5}{6}$$ Add the whole numbers and fractions separately.

$$= 160\frac{5}{6}$$ Simplify.

The total length is $160\frac{5}{6}$ feet.

Guided Practice

Add or subtract. Write in simplest form. (Examples 1–3)

1. $8\frac{1}{2} + 3\frac{4}{5} =$ ______

2. $7\frac{5}{6} - 3\frac{1}{6} =$ ______

3. $11 - 6\frac{3}{8} =$ ______

4. A hybrid car's gas tank can hold $11\frac{9}{10}$ gallons of gasoline. It contains $8\frac{3}{4}$ gallons of gasoline. How much more gasoline is needed to fill the tank? (Example 4) ______

5. 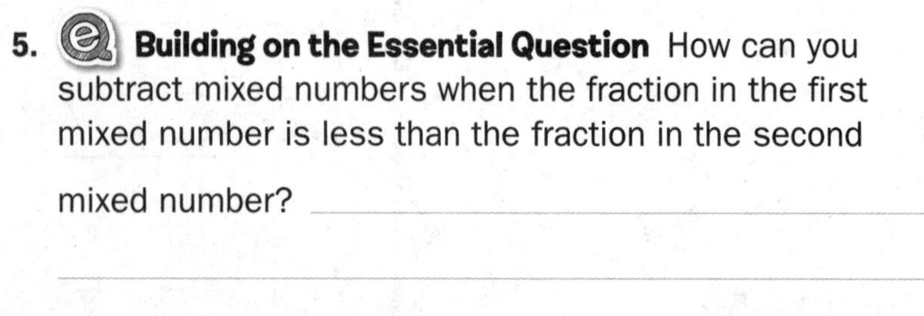 **Building on the Essential Question** How can you subtract mixed numbers when the fraction in the first mixed number is less than the fraction in the second mixed number? ______

Rate Yourself!

How confident are you about adding and subtracting mixed numbers? Shade the ring on the target.

For more help, go online to access a Personal Tutor.

Name ______________________ My Homework ______________________

Independent Practice

Go online for Step-by-Step Solutions eHelp

Add or subtract. Write in simplest form. (Examples 1–3)

1. $2\frac{1}{9} + 7\frac{4}{9} =$ ________

2. $8\frac{5}{12} + 11\frac{1}{4} =$ ________

3. $10\frac{4}{5} - 2\frac{1}{5} =$ ________

Show your work.

4. $9\frac{4}{5} - 2\frac{3}{10} =$ ________

5. $11\frac{3}{4} - 4\frac{1}{3} =$ ________

6. $9\frac{1}{5} - 2\frac{3}{5} =$ ________

7. $6\frac{3}{5} - 1\frac{2}{3} =$ ________

8. $14\frac{1}{6} - 7\frac{1}{3} =$ ________

9. $8 - 3\frac{2}{3} =$ ________

CCSS **Justify Conclusions** **For Exercises 10 and 11, choose an operation to solve. Explain your reasoning. Then solve the problem. Write your answer in simplest form.** (Example 4)

10. If Juliana and Brody hiked both of the trails listed in the table, how far did they hike?

____

Trail	Length (mi)
Woodland Park	$3\frac{2}{3}$
Mill Creek Way	$2\frac{5}{6}$

11 The length of Kasey's garden is $4\frac{5}{8}$ feet. Find the width of Kasey's garden if it is $2\frac{7}{8}$ feet shorter than the length.

____

12. Karen wakes up at 6:00 A.M. It takes her $1\frac{1}{4}$ hours to shower, get dressed, and comb her hair. It takes her $\frac{1}{2}$ hour to eat breakfast, brush her teeth, and make her bed. At what time will she be ready for school? ________________________

Add or subtract. Write in simplest form.

13. $-3\frac{1}{4} + (-1\frac{3}{4}) =$ ______

14. $\dfrac{3\frac{1}{2}}{5} + \dfrac{4\frac{2}{3}}{2} =$ ______

15. $6\frac{1}{3} + 1\frac{2}{3} + 5\frac{5}{9} =$ ______

16. $3\frac{1}{4} + 2\frac{5}{6} - 4\frac{1}{3} =$ ______

H.O.T. Problems Higher Order Thinking

17. CCSS **Model with Mathematics** Write a real-world problem that could be represented by the expression $5\frac{1}{2} - 3\frac{7}{8}$. Then solve your problem.

18. CCSS **Persevere with Problems** A string is cut in half. One of the halves is thrown away. One fifth of the remaining half is cut away and the piece left is 8 feet long. How long was the string initially? Justify your answer.

Standardized Test Practice

19. For a party, Makenna bought $3\frac{1}{3}$ pounds of white grapes. Angelo bought $2\frac{3}{4}$ pounds of red grapes. How many more pounds of grapes did Makenna buy than Angelo?

Ⓐ $\frac{5}{12}$ lb

Ⓑ $\frac{7}{12}$ lb

Ⓒ $5\frac{5}{12}$ lb

Ⓓ $6\frac{1}{12}$ lb

Name ______ My Homework ______

Extra Practice

Add or subtract. Write in simplest form.

20. $6\frac{1}{4} - 2\frac{3}{4} = 3\frac{1}{2}$

$6\frac{1}{4} - 2\frac{3}{4} = 5\frac{5}{4} - 2\frac{3}{4}$

$= 3\frac{2}{4}$

$= 3\frac{1}{2}$

21. $8\frac{3}{8} + 10\frac{1}{3} =$ ______

22. $13 - 5\frac{5}{6} =$ ______

23. $3\frac{2}{7} + 4\frac{3}{7} =$ ______

24. $4\frac{3}{10} - 1\frac{3}{4} =$ ______

25. $12\frac{1}{2} - 6\frac{5}{8} =$ ______

Justify Conclusions **Choose an operation to solve. Explain your reasoning. Then solve the problem. Write your answer in simplest form.**

26. The length of Alana's hair was $9\frac{3}{4}$ inches. After her haircut, the length was $6\frac{1}{2}$ inches. How many inches did she have cut?

______

27. Emeril used a total of $7\frac{1}{4}$ cups of flour to make three pastries. He used $2\frac{1}{4}$ cups of flour for the first and $2\frac{1}{3}$ cups for the second. How much flour did Emeril use for the third pastry?

______

28. Margarite made the jewelry shown. If the necklace is $10\frac{5}{8}$ inches longer than the bracelet, how long is the necklace?

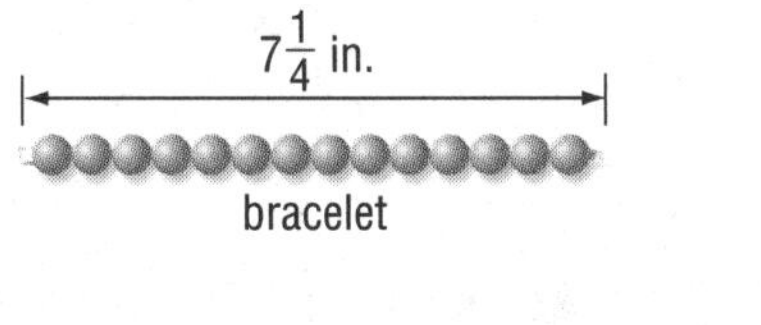

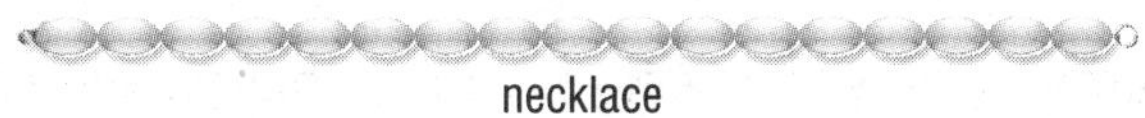

______

29. Find the perimeter of the figure. Write your answer in simplest form.

______

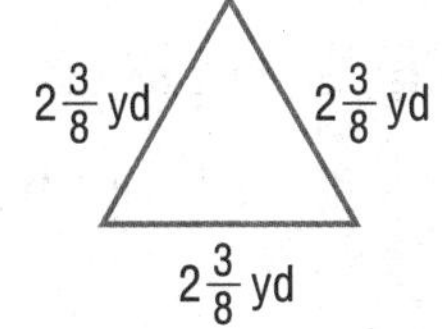

30. Suppose you want to place a shelf that is $30\frac{1}{3}$ inches long in the center of a wall that is $45\frac{3}{4}$ inches wide. About how far from each edge of the wall should you place the shelf? ______

Standardized Test Practice

31. The distance from home plate to the pitcher's mound is 60 feet 6 inches and from home plate to second base is 127 feet $3\frac{3}{8}$ inches. Find the distance from the pitcher's mound to second base.

Ⓐ 68 ft $3\frac{1}{4}$ in.

Ⓑ 67 ft $8\frac{3}{4}$ in.

Ⓒ 67 ft $2\frac{5}{8}$ in.

Ⓓ 66 ft $9\frac{3}{8}$ in.

32. A recipe for party mix calls for $4\frac{3}{4}$ cups of cereal. The amount of peanuts needed is $1\frac{2}{3}$ cups less than the amount of cereal needed. How many cups of peanuts and cereal are needed?

Ⓕ $3\frac{1}{12}$ cups

Ⓖ $6\frac{1}{2}$ cups

Ⓗ $7\frac{5}{6}$ cups

Ⓘ $8\frac{1}{2}$ cups

Common Core Review

Round each mixed number to its nearest whole number. Then estimate each product. 5.NF.4

33. $5\frac{1}{4} \times 7\frac{2}{3} \approx$ ☐ × ☐ ≈ ☐

34. $1\frac{1}{11} \times 8\frac{14}{15} \approx$ ☐ × ☐ ≈ ☐

35. $4\frac{3}{4} \times 11\frac{2}{9} \approx$ ☐ × ☐ ≈ ☐

36. $\frac{1}{20} \times \frac{19}{20} \approx$ ☐ × ☐ ≈ ☐

37. Zoe's average running speed is about $6\frac{4}{5}$ miles per hour. Suppose Zoe runs for $1\frac{3}{4}$ hours. About how far will she have run? Explain. 5.NF.4

____

38. Sam ate about $3\frac{5}{6}$ slices of pizza. There were 12 slices of pizza in the box. About how many slices are left? 5.NF.1 ______________

Problem-Solving Investigation

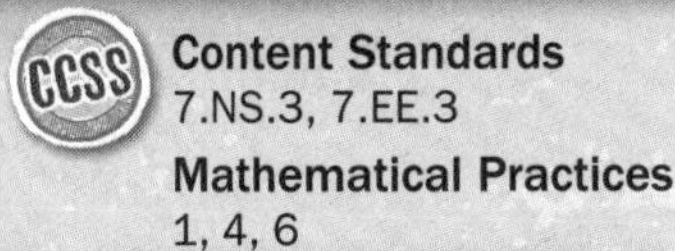

Content Standards
7.NS.3, 7.EE.3
Mathematical Practices
1, 4, 6

Case #1 Science Experiment

Casey drops a ball from a height of 12 feet. It hits the ground and bounces up half as high as it fell. This is true for each successive bounce.

What is the height the ball reaches after the fourth bounce?

Understand *What are the facts?*

Casey dropped the ball from a height of 12 feet. It bounces up half as high for each successive bounce.

Plan *What is your strategy to solve this problem?*

Draw a diagram to show the height of the ball after each bounce.

Solve *How can you apply the strategy?*

The ball reaches a height of ________ foot after the fourth bounce.

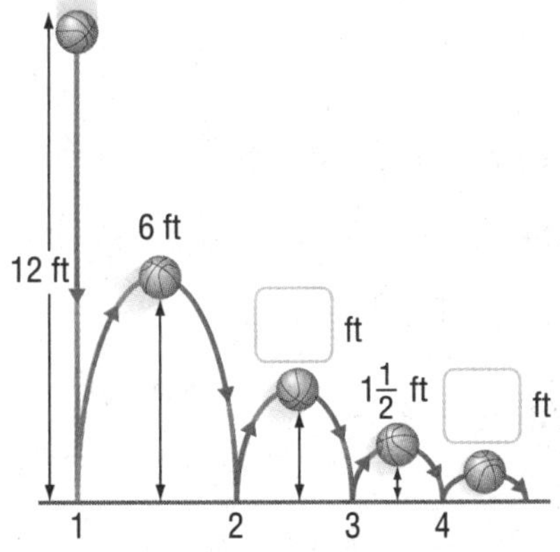

Check *Does the answer make sense?*

Use division to check. $12 \div 2 = 6$, $6 \div 2 = 3$, $3 \div 2 = 1.5$, $1.5 \div 2 = 0.75$.

Analyze the Strategy

Be Precise If the ball is dropped from 12 feet and bounces up $\frac{2}{3}$ as high on each successive bounce, what is the height of the fourth bounce?

Case #2 Travel

Mr. Garcia has driven 60 miles, which is $\frac{2}{3}$ of the way to his sister's house.

How much farther does he have to drive to get to his sister's house?

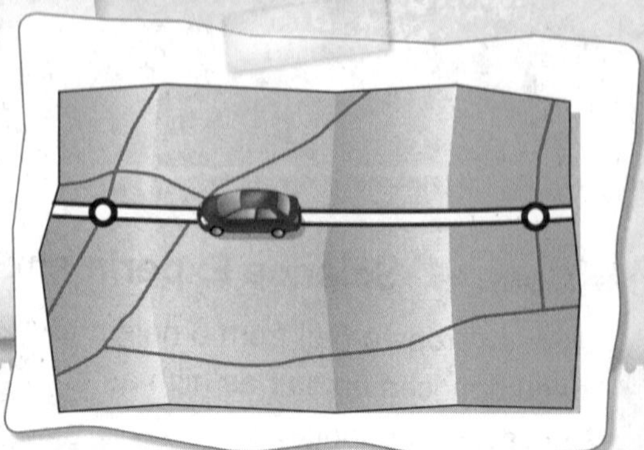

Understand

Read the problem. What are you being asked to find?

I need to find ______________________________.

What information do you know?

Mr. Garcia has driven __________ of the way to his sister's house. This is equal to ______________________________.

Is there any information that you do *not* need to know?

I do not need to know ______________________________.

2 Plan

Choose a problem-solving strategy.

I will use the ______________________________ strategy.

3 Solve

Use your problem-solving strategy to solve the problem.

Use the bar diagram that represents the distance to his sister's house.

Fill in two of the sections to represent $\frac{2}{3}$.

|------- 60 miles -------|

☐ of the 3 parts = 60.

Each part is ☐ miles. The distance to his sister's house is 60 + ☐ = ☐.

So, Mr. Garcia has ________ miles left to drive.

4 Check

Use information from the problem to check your answer.

____

Collaborate Work with a small group to solve the following cases. Show your work on a separate piece of paper.

Case #3 Fractions

Marta ate a quarter of a whole pie. Edwin ate $\frac{1}{4}$ of what was left. Cristina then ate $\frac{1}{3}$ of what was left.

What fraction of the pie remains?

Case #4 Games

Eight members of a chess club are having a tournament. In the first round, every player will play a chess game against every other player.

How many games will be in the first round of the tournament?

Case #5 Distance

Alejandro and Pedro are riding their bikes to school. After 1 mile, they are $\frac{4}{5}$ of the way there.

How much farther do they have to go?

Circle a strategy below to solve the problem.

- Act it out.
- Make a model.
- Look for a pattern.

Case #6 Seats

The number of seats in the first row of a concert hall is 6. The second row has 9 seats, the third row has 12 seats, and the fourth row has 15 seats.

How many seats will be in the eighth row?

Mid-Chapter Check

Vocabulary Check

1. Define *rational number*. Give some examples of rational numbers written in different forms. (Lessons 3 and 4)

2. Fill in the blank in the sentence below with the correct term. (Lesson 1)

   Repeating decimals can be represented using ________.

Skills Check and Problem Solving

Add or subtract. Write in simplest form. (Lessons 3–5)

3. $\frac{5}{8} + \frac{3}{8} =$ ________

4. $-\frac{1}{9} + \frac{2}{9} =$ ________

5. $-\frac{11}{15} - \frac{1}{15} =$ ________

6. $2\frac{5}{9} + 1\frac{2}{3} =$ ________

7. $8\frac{3}{4} - 2\frac{5}{12} =$ ________

8. $5\frac{1}{6} - 1\frac{1}{3} =$ ________

9. The table at the right shows the fraction of each state that is water. Order the states from least to greatest fraction of water. (Lesson 2)

What Part is Water?	
Alaska	$\frac{3}{41}$
Michigan	$\frac{40}{97}$
Wisconsin	$\frac{1}{6}$

10. The maximum height of an Asian elephant is 9.8 feet. What mixed number represents this height? (Lesson 1) ________

11. **Standardized Test Practice** The table shows the weight of a newborn infant for its first year. During which three-month period was the infant's weight gain the greatest? (Lesson 5)

    Ⓐ 0–3 months
    Ⓑ 3–6 months
    Ⓒ 6–9 months
    Ⓓ 9–12 months

Month	Weight (lb)
0	$7\frac{1}{4}$
3	$12\frac{1}{2}$
6	$16\frac{5}{8}$
9	$19\frac{4}{5}$
12	$23\frac{3}{20}$

Lesson 6

Multiply Fractions

What You'll Learn

Scan the lesson. Predict two things you will learn about multiplying fractions.

-
-

Essential Question

WHAT happens when you add, subtract, multiply, and divide fractions?

Common Core State Standards

Content Standards
7.NS.2, 7.NS.2a, 7.NS.2c, 7.NS.3, 7.EE.3

Mathematical Practices
1, 3, 4

Real-World Link

Lunch There are 12 students at the lunch table. Two thirds of the students ordered a hamburger for lunch. One half of those students that ordered a hamburger put cheese on it.

Step 1 Draw an X through the students that did not order a hamburger.

Step 2 Draw a C on the students that ordered cheese on their hamburger.

Didn't I order cheese with that?

1. What fraction of the students at the lunch table ordered a cheeseburger? Write in simplest form. ______

2. What is $\frac{1}{2}$ of $\frac{2}{3}$? Write in simplest form. ______

3. Write your own word problem that involves fractions that can be solved using a diagram like the one above.

Key Concept

Multiply Fractions

Work Zone

Words To multiply fractions, multiply the numerators and multiply the denominators.

Examples

Numbers

$\frac{1}{2} \times \frac{2}{3} = \frac{1 \times 2}{2 \times 3}$ or $\frac{2}{6}$

Algebra

$\frac{a}{b} \cdot \frac{c}{d} = \frac{a \cdot c}{b \cdot d}$ or $\frac{ac}{bd}$, where $b, d \neq 0$

When multiplying two fractions, write the product in simplest form. The numerator and denominator of either fraction may have common factors. If this is the case, you can simplify before multiplying.

Examples

GCF

In Example 3, GCF stands for the greatest of the common factors of two or more numbers.
Example: The GCF of 8 and 2 is 2.

Multiply. Write in simplest form.

1. $\frac{1}{2} \times \frac{1}{3}$

$\frac{1}{2} \times \frac{1}{3} = \frac{1 \times 1}{2 \times 3}$ ← Multiply the numerators. ← Multiply the denominators.

$= \frac{1}{6}$ Simplify.

2. $2 \times \left(-\frac{3}{4}\right)$

$2 \times \left(-\frac{3}{4}\right) = \frac{2}{1} \times \left(\frac{-3}{4}\right)$ Write 2 as $\frac{2}{1}$ and $-\frac{3}{4}$ as $\frac{-3}{4}$.

$= \frac{2 \times (-3)}{1 \times 4}$ ← Multiply the numerators. ← Multiply the denominators.

$= \frac{-6}{4}$ or $-1\frac{1}{2}$ Simplify.

3. $\frac{2}{7} \times \left(-\frac{3}{8}\right)$

$\frac{2}{7} \times \left(-\frac{3}{8}\right) = \frac{\overset{1}{\cancel{2}}}{7} \times \left(-\frac{3}{\underset{4}{\cancel{8}}}\right)$ Divide 2 and 8 by their GCF, 2.

$= \frac{1 \times (-3)}{7 \times 4}$ or $-\frac{3}{28}$ Multiply.

Got It? Do these problems to find out.

Multiply. Write in simplest form.

a. $\frac{3}{5} \times \frac{1}{2}$

b. $\frac{2}{3} \times (-4)$

c. $-\frac{1}{3} \times \left(-\frac{3}{7}\right)$

Show your work.

a. ______

b. ______

c. ______

Multiply Mixed Numbers

When multiplying by a mixed number, you can rename the mixed number as an improper fraction. You can also multiply mixed numbers using the Distributive Property and mental math.

Example

4. Find $\frac{1}{2} \times 4\frac{2}{5}$. Write in simplest form.

Estimate $\frac{1}{2} \times 4 = 2$

Method 1 **Rename the mixed number.**

$\frac{1}{2} \times 4\frac{2}{5} = \frac{1}{\cancel{2}_1} \times \frac{\cancel{22}^{11}}{5}$ — Rename $4\frac{2}{5}$ as an improper fraction, $\frac{22}{5}$. Divide 2 and 22 by their GCF, 2.

$= \frac{1 \times 11}{1 \times 5}$ — Multiply.

$= \frac{11}{5}$ — Simplify.

$= 2\frac{1}{5}$ — Simplify.

Method 2 **Use mental math.**

The mixed number $4\frac{2}{5}$ is equal to $4 + \frac{2}{5}$.

So, $\frac{1}{2} \times 4\frac{2}{5} = \frac{1}{2}\left(4 + \frac{2}{5}\right)$. Use the Distributive Property to multiply, then add mentally.

$\frac{1}{2}\left(4 + \frac{2}{5}\right) = 2 + \frac{1}{5}$ — Think Half of 4 is 2 and half of 2 fifths is 1 fifth.

$= 2\frac{1}{5}$ — Rewrite the sum as a mixed number.

Check for Reasonableness $2\frac{1}{5} \approx 2$ ✓

So, $\frac{1}{2} \times 4\frac{2}{5} = 2\frac{1}{5}$.

Using either method, the answer is $2\frac{1}{5}$.

Simplifying
If you forget to simplify before multiplying, you can always simplify the final answer. However, it is usually easier to simplify before multiplying.

Got It? Do these problems to find out.

Multiply. Write in simplest form.

d. $\frac{1}{4} \times 8\frac{4}{9}$

e. $5\frac{1}{3} \times 3$

f. $-1\frac{7}{8} \times \left(-2\frac{2}{5}\right)$

d. 

e. 

f. ________

Example

5. Humans sleep about $\frac{1}{3}$ of each day. Let each year equal $365\frac{1}{4}$ days. Determine the number of days in a year the average human sleeps.

Find $\frac{1}{3} \times 365\frac{1}{4}$.

Estimate $\frac{1}{3} \times 360 = 120$

$\frac{1}{3} \times 365\frac{1}{4} = \frac{1}{3} \times \frac{1{,}461}{4}$ Rename the mixed number as an improper fraction.

$= \frac{1}{\cancel{3}_1} \times \frac{\cancel{1{,}461}^{487}}{4}$ Divide 3 and 1,461 by their GCF, 3.

$= \frac{487}{4}$ or $121\frac{3}{4}$ Multiply. Then rename as a mixed number.

Check for Reasonableness $121\frac{3}{4} \approx 120$ ✓

The average human sleeps $121\frac{3}{4}$ days each year.

Meaning of Multiplication

Recall that one meaning of 3×4 is three groups with 4 in each group. In Example 5, there are $365\frac{1}{4}$ groups with $\frac{1}{3}$ in each group.

Guided Practice

Multiply. Write in simplest form. (Examples 1–4)

1. $\frac{2}{3} \times \frac{1}{3} =$ ______

2. $-\frac{1}{4} \times \left(-\frac{8}{9}\right) =$ ______

3. $2\frac{1}{4} \times \frac{2}{3} =$ ______

4. **STEM** The weight of an object on Mars is about $\frac{2}{5}$ its weight on Earth. How much would an $80\frac{1}{2}$-pound dog weigh on Mars? (Example 5) ______

5. **Building on the Essential Question** How is the process of multiplying fractions different from the process of adding fractions?

______

______

______

______

Rate Yourself!

How well do you understand multiplying fractions? Circle the image that applies.

Clear

Somewhat Clear

Not So Clear

For more help, go online to access a Personal Tutor.

FOLDABLES Time to update your Foldable!

Name ________________ My Homework ________________

Independent Practice

Go online for Step-by-Step Solutions 

Multiply. Write in simplest form. (Examples 1–4)

1. $\frac{3}{4} \times \frac{1}{8} =$ ______

2. $\frac{2}{5} \times \frac{2}{3} =$ ______

3. $-9 \times \frac{1}{2} =$ ______

4. $-\frac{1}{5} \times \left(-\frac{5}{6}\right) =$ ______

5. $\frac{2}{3} \times \frac{1}{4} =$ ______

6. $-\frac{1}{12} \times \frac{2}{5} =$ ______

7. $\frac{2}{5} \times \frac{15}{16} =$ ______

8. $\frac{4}{7} \times \frac{7}{8} =$ ______

9. $\left(-1\frac{1}{2}\right) \times \frac{2}{3} =$ ______

10. The width of a vegetable garden is $\frac{1}{3}$ times its length. If the length of the garden is $7\frac{3}{4}$ feet, what is the width in simplest form? (Example 5)

______

11. One evening, $\frac{2}{3}$ of the students in Rick's class watched television. Of those students, $\frac{3}{8}$ watched a reality show. Of the students that watched the show, $\frac{1}{4}$ of them recorded the show. What fraction of the students in Rick's class watched and recorded a reality TV show?

______

Write each numerical expression. Then evaluate the expression.

12. one half of negative five eighths

______

13. one third of eleven sixteenths

______

14. **Model with Mathematics** Refer to the graphic novel frame below.

a. The height of the closet is 96 inches, and Aisha would like to have 4 rows of cube organizers. What is the most the height of each cube organizer can be?

b. Aisha would like to stack 3 shoe boxes on top of each other at the bottom of the closet. The height of each shoe box is $4\frac{1}{2}$ inches. What is the total height of the 3 boxes?

H.O.T. Problems Higher Order Thinking

15. **Model with Mathematics** Write a real-world problem that involves finding the product of $\frac{3}{4}$ and $\frac{1}{8}$.

16. **Persevere with Problems** Two positive improper fractions are multiplied. Is the product *sometimes, always*, or *never* less than 1? Explain.

Standardized Test Practice

17. Two-thirds of the students in Levi's homeroom class study Spanish. Of these, one fourth study algebra. What fraction of the students in Levi's homeroom class study both Spanish and algebra?

Ⓐ $\frac{1}{6}$

Ⓑ $\frac{5}{12}$

Ⓒ $\frac{3}{6}$

Ⓓ $\frac{11}{12}$

Extra Practice

Multiply. Write in simplest form.

18. $\frac{4}{5} \times (-6) = -4\frac{4}{5}$

$\frac{4}{5} \times (-6) = \frac{4}{5} \times \left(-\frac{6}{1}\right)$

$= \frac{4 \times (-6)}{5 \times 1}$

$= \frac{-24}{5}$ or $-4\frac{4}{5}$

Homework Help

19. $-\frac{4}{9} \times \left(-\frac{1}{4}\right) =$ ______

20. $3\frac{1}{3} \times \left(-\frac{1}{5}\right) =$ ______

21. $\frac{1}{3} \times \frac{3}{4} =$ ______

22. $\frac{4}{9} \times \left(-\frac{1}{8}\right) =$ ______

23. $\frac{5}{6} \times 2\frac{3}{5} =$ ______

24. Each DVD storage case is about $\frac{1}{5}$ inch thick. What will be the height in simplest form of 12 cases sold together?

______

25. Mark left $\frac{3}{8}$ of a pizza in the refrigerator. On Friday, he ate $\frac{1}{2}$ of what was left of the pizza. What fraction of the entire pizza did he eat on Friday?

______

Multiply. Write in simplest form.

26. $\left(\frac{1}{4}\right)^2 =$ ______

27. $\left(-\frac{2}{3}\right)^3 =$ ______

28. $\frac{1\frac{1}{3}}{\frac{1}{4}} \times \frac{\frac{2}{5}}{\frac{1}{2}} =$ ______

29. CCSS **Justify Conclusions** Alano wants to make one and a half batches of the pasta salad recipe shown at the right. How much of each ingredient will Alano need? Explain how you solved the problem.

______

______

______

Pasta Salad Recipe	
Ingredient	**Amount**
Broccoli	$1\frac{1}{4}$ c
Cooked pasta	$3\frac{3}{4}$ c
Salad dressing	$\frac{2}{3}$ c
Cheese	$1\frac{1}{3}$ c

30. Philip rode his bicycle at $9\frac{1}{2}$ miles per hour. If he rode for $\frac{3}{4}$ of an hour, how many miles in simplest form did he cover? ______

Standardized Test Practice

31. Of the dolls in Marjorie's doll collection, $\frac{1}{5}$ have red hair. Of these, $\frac{3}{4}$ have green eyes. What fraction of Marjorie's doll collection has both red hair and green eyes?

Ⓐ $\frac{2}{9}$

Ⓑ $\frac{3}{20}$

Ⓒ $\frac{4}{9}$

Ⓓ $\frac{19}{20}$

32. Which description gives the relationship between a term and n, its position in the sequence?

Position	1	2	3	4	5	n
Value of Term	$\frac{1}{4}$	$\frac{1}{2}$	$\frac{3}{4}$	1	$1\frac{1}{4}$	

Ⓕ Subtract 4 from n.

Ⓖ Add $\frac{1}{4}$ to n.

Ⓗ Multiply n by $\frac{1}{4}$.

Ⓘ Divide n by $\frac{1}{4}$.

33. **Short Response** Determine if the product of 2 and $\frac{3}{4}$ is greater than or less than 2? ______

Common Core Review

For each multiplication sentence, write two related division sentences. 5.NBT.5

34. $3 \times 4 = 12$

35. $\frac{1}{6} \times \frac{1}{3} = \frac{1}{18}$

36. $2\frac{2}{5} \times 4\frac{1}{2} = 10\frac{4}{5}$

37. $5\frac{5}{8} \times 1\frac{1}{5} = 6\frac{3}{4}$

Solve. 5.MD.1

38. Madelyn is building a computer desk. She has $8\frac{2}{3}$ feet of wood. How many inches of wood does Madelyn have? (*Hint:* 12 inches = 1 foot) ______

39. Victor made punch for a birthday party. He used $10\frac{1}{2}$ cups of soda. How many pints of soda did Victor use? (*Hint:* 2 cups = 1 pint) ______

Lesson 7

Convert Between Systems

What You'll Learn

Scan the lesson. List two real-world scenerios in which you would convert measurments.

-
-

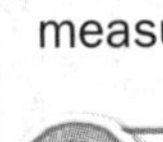

Essential Question

HOW do you convert between measurement systems?

Common Core State Standards

Content Standards
7.RP.3, 7.NS.2, 7.NS.3

Mathematical Practices
1, 3, 4, 5, 6

Real-World Link

5K Race To raise money for a health organization, the Matthews family is participating in a 5K race. A 5K race is 5 kilometers.

1. How many meters long is the race?

5 kilometers = ☐ meters

2. One mile is approximately 1.6 kilometers. About how many miles is the race?

5 kilometers ≈ ☐ miles

3. A kilometer is a unit of length in the metric measurement system. A mile is a measure of length in the customary measurement system. Write the following units of length under the correct measurement system.

centimeter, foot, inch, meter, millimeter, yard

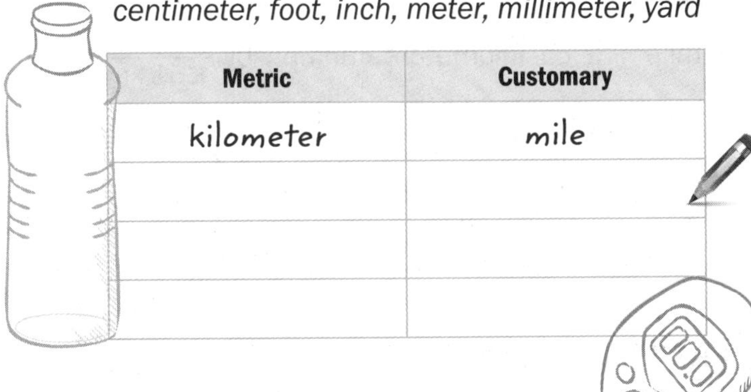

Metric	Customary
kilometer	mile

Work Zone

Convert Between Measurement Systems

You can multiply by fractions to convert between customary and metric units. The table below lists common customary and metric relationships.

Customary and Metric Relationships			
Type of Measure	**Customary**	→	**Metric**
Length	1 inch (in.)	≈	2.54 centimeters (cm)
	1 foot (ft)	≈	0.30 meter (m)
	1 yard (yd)	≈	0.91 meter (m)
	1 mile (mi)	≈	1.61 kilometers (km)
Weight/Mass	1 pound (lb)	≈	453.6 grams (g)
	1 pound (lb)	≈	0.4536 kilogram (kg)
	1 ton (T)	≈	907.2 kilograms (kg)
Capacity	1 cup (c)	≈	236.59 milliliters (mL)
	1 pint (pt)	≈	473.18 milliliters (mL)
	1 quart (qt)	≈	946.35 milliliters (mL)
	1 gallon (gal)	≈	3.79 liters (L)

STOP and Reflect

What metric unit of measure corresponds to miles? to pounds? Write your answers below.

Examples

1. Convert 17.22 inches to centimeters. Round to the nearest hundredth if necessary.

Since 2.54 centimeters ≈ 1 inch, multiply by $\frac{2.54 \text{ cm}}{1 \text{ in.}}$.

$17.22 \approx 17.22 \cancel{\text{in.}} \cdot \frac{2.54 \text{ cm}}{1 \cancel{\text{in.}}}$ Multiply by $\frac{2.54 \text{ cm}}{1 \text{ in.}}$. Divide out common units.

$\approx 43.7388 \text{ cm}$ Simplify.

So, 17.22 inches is approximately 43.74 centimeters.

2. Convert 5 kilometers to miles. Round to the nearest hundredth if necessary.

Since 1 mile ≈ 1.61 kilometers, multiply by $\frac{1 \text{ mi}}{1.61 \text{ km}}$.

$5 \text{ km} \approx 5 \cancel{\text{km}} \cdot \frac{1 \text{ mi}}{1.61 \cancel{\text{km}}}$ Multiply by $\frac{1 \text{ mi}}{1.61 \text{ km}}$. Divide out common units.

$\approx \frac{5 \text{ mi}}{1.61}$ or 3.11 mi Simplify.

So, 5 kilometers is approximately 3.11 miles.

Got It? Do these problems to find out.

Show your work.

Complete. Round to the nearest hundredth if necessary.

a. 6 yd ≈ ■ m **b.** 1.6 cm ≈ ■ in. **c.** 17 m ≈ ■ yd

a. ______

b. ______

c. ______

Examples

3. Convert 828.5 milliliters to cups. Round to the nearest hundredth if necessary.

Since 1 cup ≈ 236.59 milliliters, multiply by $\frac{1 \text{ c}}{236.59 \text{ mL}}$.

$828.5 \text{ mL} \approx 828.5 \cancel{\text{mL}} \cdot \frac{1 \text{ c}}{236.59 \cancel{\text{mL}}}$ — Multiply by $\frac{1 \text{ c}}{236.59 \text{ mL}}$ and divide out common units.

$\approx \frac{828.5 \text{ c}}{236.59}$ or 3.50 c — Simplify.

So, 828.5 milliliters is approximately 3.50 cups.

4. Convert 3.4 quarts to milliliters. Round to the nearest hundredth if necessary.

Since 946.35 milliliters ≈ 1 quart, multiply by $\frac{946.35 \text{ mL}}{1 \text{ qt}}$.

$3.4 \text{ qt} \approx 3.4 \cancel{\text{qt}} \cdot \frac{946.35 \text{ mL}}{1 \cancel{\text{qt}}}$ — Multiply by $\frac{946.35}{1 \text{ qt}}$. Divide out common units.

$\approx 3{,}217.59 \text{ mL}$ — Simplify.

So, 3.4 quarts is approximately 3,217.59 milliliters.

5. Convert 4.25 kilograms to pounds. Round to the nearest hundredth if necessary.

Since 1 pound ≈ 0.4536 kilogram, multiply by $\frac{1 \text{ lb}}{0.4536 \text{ kg}}$.

$4.25 \text{ kg} \approx 4.25 \cancel{\text{kg}} \cdot \frac{1 \text{ lb}}{0.4536 \cancel{\text{kg}}}$ — Multiply by $\frac{1 \text{ lb}}{0.4536 \text{ kg}}$. Divide out common units.

$\approx \frac{4.25 \text{ lb}}{0.4536}$ or 9.37 lb — Simplify.

So, 4.25 kilograms is approximately 9.37 pounds.

Dimensional Analysis

Recall that dimensional analysis is the process of including units of measurement when you compute.

Got It? Do these problems to find out.

Complete. Round to the nearest hundredth if necessary.

d. 7.44 c ≈ ■ mL

e. 22.09 lb ≈ ■ kg

f. 35.85 L ≈ ■ gal

d. ________

e. ________

f. ________

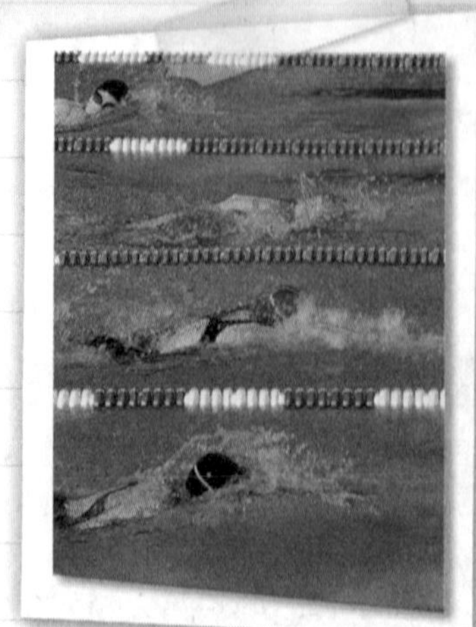

Example

6. An Olympic-size swimming pool is 50 meters long. About how many feet long is the pool?

Since 1 foot ≈ 0.30 meter, use the ratio $\frac{1 \text{ ft}}{0.30 \text{ m}}$.

$50 \text{ m} \approx 50 \text{ m} \cdot \frac{1 \text{ ft}}{0.30 \text{ m}}$ Multiply by $\frac{1 \text{ ft}}{0.30\text{m}}$.

$\approx 50 \not{\text{m}} \cdot \frac{1 \text{ ft}}{0.30 \not{\text{m}}}$ Divide out common units, leaving the desired unit, feet.

$\approx \frac{50 \text{ ft}}{0.30}$ or 166.67 ft Divide.

An Olympic-size swimming pool is about 166.67 feet long.

Guided Practice

Complete. Round to the nearest hundredth if necessary. (Examples 1 – 5)

1. 3.7 yd ≈ ______ m

2. 11.07 pt ≈ ______ mL

3. 650 lb ≈ ______ kg

4. About how many feet does a team of athletes run in a 1,600-meter relay race? (Example 6) ______

5. Raheem bought 3 pounds of bananas. About how many kilograms did he buy? (Example 6) ______

6. **Building on the Essential Question** How can you use dimensional analysis to convert between measurement systems?

______

______

______

______

Rate Yourself!

Are you ready to move on? Shade the section that applies.

YES ? NO

For more help, go online to access a Personal Tutor.

Name ______________________ My Homework ______________________

Independent Practice

Go online for Step-by-Step Solutions 

Complete. Round to the nearest hundredth if necessary. (Examples 1 – 5)

1. 5 in. ≈ __________ cm

2. 2 qt ≈ __________ mL

3. 58.14 kg ≈ __________ lb

4. 4 L ≈ __________ gal

5. 10 mL ≈ __________ c

6. 63.5 T ≈ __________ kg

7. 4.725 m ≈ __________ ft

8. 3 T ≈ __________ kg

9. 680.4 g ≈ __________ lb

10. A notebook computer has a mass of 2.25 kilograms. About how many pounds does the notebook weigh? (Example 6)

____

11. A glass bottle holds 3.75 cups of water. About how many milliliters of water can the bottle hold? (Example 6)

____

12. A Cabbage Palmetto has a height of 80 feet. What is the approximate height of the tree in meters? (Example 6)

____

CCSS Persevere with Problems Determine the greater amount for each situation.

13. Which box is greater, a 1.5-pound box of raisins or a 650-gram box of raisins?

14. Which is greater a 2.75-gallon container of juice or a 12-liter container of juice?

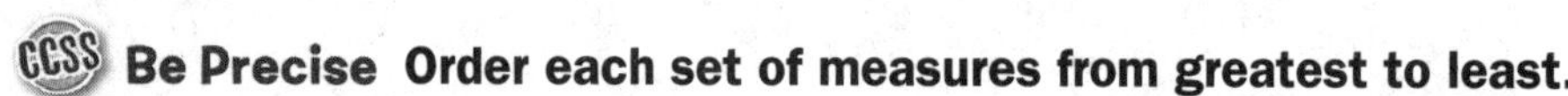

H.O.T. Problems Higher Order Thinking

15. CCSS **Reason Inductively** One gram of water has a volume of 1 milliliter. What is the volume of the water if it has a mass of 1 kilogram?

16. CCSS **Persevere with Problems** The distance from Earth to the Sun is approximately 93 million miles. About how many gigameters is this? Round to the nearest hundredth. *(Hint: In 1 gigameter there are about 621,118.01 miles.)*

CCSS Be Precise Order each set of measures from greatest to least.

17. 1.2 cm, 0.6 in., 0.031 m, 0.1 ft

18. 2 lb, 891 g, 1 kg, 0.02 T

19. $1\frac{1}{4}$ c, 0.4 L, 950 mL, 0.7 gal

20. 4.5 ft, 48 in., 1.3 m, 120 cm

Standardized Test Practice

21. A store sells poster board. The table shows the colors and sizes in-stock. Which of the following metric approximations is the same as the measures of the green poster board?

Ⓐ 2.8 cm by 3.6 cm

Ⓑ 2.8 m by 3.6 m

Ⓒ 28 cm by 36 cm

Ⓓ 28 m by 36 m

Poster Board	
Color	**Size (in.)**
white	11 × 14 16 × 20
green	11 × 14
blue	16 × 20

Name ______________ My Homework ______________

Extra Practice

Complete. Round to the nearest hundredth if necessary.

22. 15 cm ≈ 5.91 in.

Homework Help

$15 \text{ cm} \approx 15 \text{ cm} \cdot \frac{1 \text{ in.}}{2.54 \text{ cm}}$

$\approx 15 \cancel{\text{cm}} \cdot \frac{1 \text{ in.}}{2.54 \cancel{\text{cm}}}$

$\approx \frac{15 \text{ in.}}{2.54} \approx 5.91 \text{ in.}$

23. 350 lb ≈ 158.76 kg

$350 \text{ lb} \approx 350 \text{ lb} \cdot \frac{0.4536 \text{ kg}}{1 \text{ lb}}$

$\approx 350 \cancel{\text{lb}} \cdot \frac{0.4536 \text{ kg}}{1 \cancel{\text{lb}}}$

$\approx 158.76 \text{ kg}$

24. 17 mi ≈ ______ km

25. 32 gal ≈ ______ L

26. 50 mL ≈ ______ fl oz

27. 19 kg ≈ ______ lb

28. The Willis Tower has a height of 1,451 feet. What is the estimated height of the building in meters? ______

29. Which is greater, a bottle containing 64 fluid ounces or a bottle containing 2 liters of water? ______

30. CCSS **Use Math Tools** A bakery uses 900 grams of peaches in a cobbler. About how many pounds of peaches does the bakery use in a cobbler?

______

Determine which quantity is greater.

31. 3 gal, 10 L ______

32. 14 oz, 0.4 kg ______

33. 4 mi, 6.2 km ______

34. Velocity is a rate usually expressed in feet per second or meters per second. How can the units help you calculate velocity using the distance a car traveled and the time recorded? ______

______

Standardized Test Practice

35. The diagram shows the length of a fork from the cafeteria.

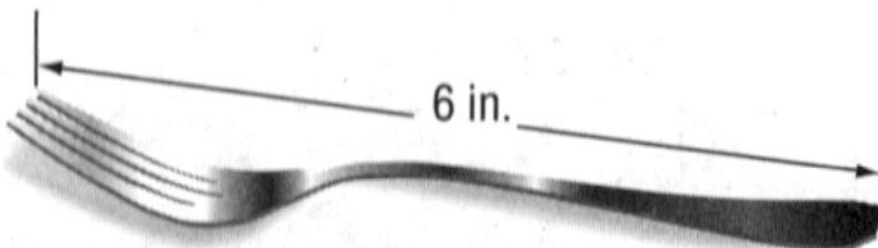

Which of the following measurements is approximately equal to the length of the fork?

Ⓐ 2.4 cm

Ⓑ 15.2 cm

Ⓒ 24 cm

Ⓓ 152 cm

36. The table shows the flying speeds of various birds.

Bird	Speed (km/h)
Spur-winged goose	142
Mallard duck	105

About how fast does the Spur-winged goose travel in miles per hour?

Ⓕ 229 miles per hour

Ⓖ 156 miles per hour

Ⓗ 88 miles per hour

Ⓘ 71 miles per hour

CCSS Common Core Review

Convert. Round to the nearest tenth if necessary. 5.MD.1

37. 17 ft = ________ yd

38. 82 in. = ________ ft

39. 3 mi = ________ ft

40. A skyscraper is 0.484 kilometer tall. What is the height of the skyscraper in meters? 5.MD.1 ________________

Multiply. Write in simplest form. 7.NS.2a

41. $\frac{1}{3} \times \frac{3}{1} =$ ________

42. $\frac{1}{4} \times 8 =$ ________

43. $\frac{4}{13} \times \frac{65}{4} =$ ________

44. $-\frac{7}{8} \times \left(-\frac{8}{7}\right) =$ ________

45. $-6\frac{2}{5} \times \left(-\frac{5}{32}\right) =$ ________

46. $\frac{1}{5} \times (-10) =$ ________

Lesson 8

Divide Fractions

What You'll Learn

Scan the lesson. Predict two things you will learn about dividing fractions.

-
-

Essential Question

WHAT happens when you add, subtract, multiply, and divide fractions?

Common Core State Standards

Content Standards
7.NS.2, 7.NS.2c, 7.NS.3, 7.EE.3

Mathematical Practices
1, 3, 4, 5

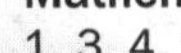

Real-World Link

Oranges Deandre has three oranges and each orange is divided evenly into fourths. Complete the steps below to find $3 \div \frac{1}{4}$.

Step 1 Draw three oranges. The first one is drawn for you.

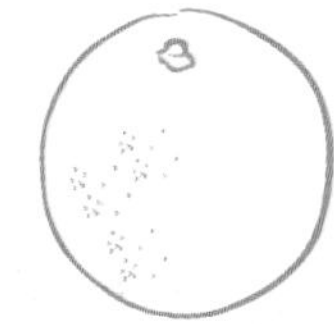

Step 2 Imagine you cut each orange into fourths. Draw the slices for each orange.

So $3 \div \frac{1}{4} = 12$. Deandre will have ☐ orange slices.

1. Find $3 \div \frac{1}{2}$. Use a diagram. ______

2. What is true about $3 \div \frac{1}{2}$ and 3×2? ______

Key Concept: Divide Fractions

Words To divide by a fraction, multiply by its multiplicative inverse, or reciprocal.

Examples

Numbers: $\frac{7}{8} \div \frac{3}{4} = \frac{7}{8} \cdot \frac{4}{3}$

Algebra: $\frac{a}{b} \div \frac{c}{d} = \frac{a}{b} \cdot \frac{d}{c}$, where $b, c, d \neq 0$

Work Zone

Dividing 3 by $\frac{1}{4}$ is the same as multiplying 3 by the reciprocal of $\frac{1}{4}$, which is 4.

reciprocals

$$3 \div \frac{1}{4} = 12 \qquad 3 \cdot 4 = 12$$

same result

STOP and Reflect

What is the reciprocal of $\frac{2}{3}$? of 15? of $-\frac{4}{9}$? Write your answers below.

Is this pattern true for any division expression?

Consider $\frac{7}{8} \div \frac{3}{4}$, which can be rewritten as $\frac{\frac{7}{8}}{\frac{3}{4}}$.

$$\frac{\frac{7}{8}}{\frac{3}{4}} = \frac{\frac{7}{8} \times \frac{4}{3}}{\frac{3}{4} \times \frac{4}{3}}$$ Multiply the numerator and denominator by the reciprocal of $\frac{3}{4}$, which is $\frac{4}{3}$.

$$= \frac{\frac{7}{8} \times \frac{4}{3}}{1}$$ $\frac{3}{4} \times \frac{4}{3} = 1$

$$= \frac{7}{8} \times \frac{4}{3}$$

So, $\frac{7}{8} \div \frac{3}{4} = \frac{7}{8} \times \frac{4}{3}$. The pattern is true in this case.

Examples

1. Find $\frac{1}{3} \div 5$.

$\frac{1}{3} \div 5 = \frac{1}{3} \div \frac{5}{1}$ A whole number can be written as a fraction over 1.

$= \frac{1}{3} \times \frac{1}{5}$ Multiply by the reciprocal of $\frac{5}{1}$, which is $\frac{1}{5}$.

$= \frac{1}{15}$ Multiply.

2. **Find $\frac{3}{4} \div \left(-\frac{1}{2}\right)$. Write in simplest form.**

Estimate $1 \div \left(-\frac{1}{2}\right) = \square$

$$\frac{3}{4} \div \left(-\frac{1}{2}\right) = \frac{3}{4} \cdot \left(-\frac{2}{1}\right)$$ Multiply by the reciprocal of $-\frac{1}{2}$, which is $-\frac{2}{1}$.

$$= \frac{3}{\cancel{4}_2} \cdot \left(-\frac{\cancel{2}^1}{1}\right)$$ Divide 4 and 2 by their GCF, 2.

$$= -\frac{3}{2} \text{ or } -1\frac{1}{2}$$ Multiply.

Check for Reasonableness $-1\frac{1}{2} \approx -2$ ✓

Got It? **Do these problems to find out.**

Divide. Write in simplest form.

a. $\frac{3}{4} \div \frac{1}{4}$ **b.** $-\frac{4}{5} \div \frac{8}{9}$ **c.** $-\frac{5}{6} \div \left(-\frac{2}{3}\right)$

a. ______

b. ______

c. ______

Divide Mixed Numbers

To divide by a mixed number, first rename the mixed number as a fraction greater than one. Then multiply the first fraction by the reciprocal, or multiplicative inverse, of the second fraction.

Example

3. **Find $\frac{2}{3} \div 3\frac{1}{3}$. Write in simplest form.**

$$\frac{2}{3} \div 3\frac{1}{3} = \frac{2}{3} \div \frac{10}{3}$$ Rename $3\frac{1}{3}$ a fraction greater than one.

$$= \frac{2}{3} \cdot \frac{3}{10}$$ Multiply by the reciprocal of $\frac{10}{3}$, which is $\frac{3}{10}$.

$$= \frac{\cancel{2}^1}{\cancel{3}_1} \cdot \frac{\cancel{3}^1}{\cancel{10}_5}$$ Divide out common factors.

$$= \frac{1}{5}$$ Multiply.

Got It? **Do these problems to find out.**

Divide. Write in simplest form.

d. $5 \div 1\frac{1}{3}$ **e.** $-\frac{3}{4} \div 1\frac{1}{2}$ **f.** $2\frac{1}{3} \div 5$

d. ______

e. ______

f. ______

Example

4. The side pieces of a butterfly house are $8\frac{1}{4}$ inches long. How many side pieces can be cut from a board measuring $49\frac{1}{2}$ inches long?

To find how many side pieces can be cut, divide $49\frac{1}{2}$ by $8\frac{1}{4}$.

Estimate Use compatible numbers. $48 \div 8 = 6$

$49\frac{1}{2} \div 8\frac{1}{4} = \frac{99}{2} \div \frac{33}{4}$ Rename the mixed numbers as fractions greater than one.

$= \frac{99}{2} \cdot \frac{4}{33}$ Multiply by the reciprocal of $\frac{33}{4}$, which is $\frac{4}{33}$.

$= \frac{\overset{3}{\cancel{99}}}{\underset{1}{\cancel{2}}} \cdot \frac{\overset{2}{\cancel{4}}}{\underset{1}{\cancel{33}}}$ Divide out common factors.

$= \frac{6}{1}$ or 6 Multiply.

So, 6 side pieces can be cut.

Check for Reasonableness Compare to the estimate. 6 = 6 ✓

Guided Practice

Divide. Write in simplest form. (Examples 1 – 3)

1. $\frac{1}{8} \div \frac{1}{3} =$ ______

2. $-3 \div \left(-\frac{6}{7}\right) =$ ______

3. $-\frac{7}{8} \div \frac{3}{4} =$ ______

4. On Saturday, Lindsay walked $3\frac{1}{2}$ miles in $1\frac{2}{5}$ hours. What was her walking pace in miles per hour? Write in simplest form. (Example 4) ______

5. **Bulding on the Essential Question** How is dividing fractions related to multiplying? ______

Rate Yourself!

Are you ready to move on? Shade the section that applies.

For more help, go online to access a Personal Tutor.

FOLDABLES *Time to update your Foldable!*

Name ______________________ My Homework ______________________

Independent Practice

Go online for Step-by-Step Solutions eHelp 

Divide. Write in simplest form. (Examples 1 – 3)

1. $\frac{3}{8} \div \frac{6}{7} =$ ______

2. $-\frac{2}{3} \div \left(-\frac{1}{2}\right) =$ ______

3. $\frac{1}{2} \div 7\frac{1}{2} =$ ______

4. $6 \div \left(-\frac{1}{2}\right) =$ ______

5. $-\frac{4}{9} \div (-2) =$ ______

6. $\frac{2}{3} \div 2\frac{1}{2} =$ ______

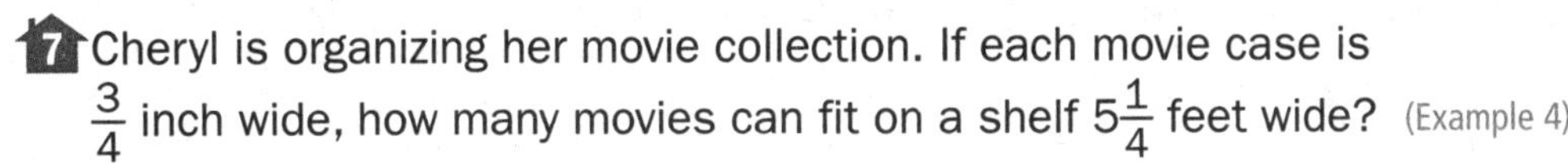

7. Cheryl is organizing her movie collection. If each movie case is $\frac{3}{4}$ inch wide, how many movies can fit on a shelf $5\frac{1}{4}$ feet wide? (Example 4)

______

8. Use the table to solve. Write your answers in simplest form.

a. How many times as heavy is the Golden Eagle as the Red-Tailed Hawk? ______

b. How many times as heavy is the Golden Eagle as the Northern Bald Eagle? ______

Bird	Maximun Weight (lb)
Golden Eagle	$13\frac{9}{10}$
Northern Bald Eagle	$9\frac{9}{10}$
Red-Tailed Hawk	$3\frac{1}{2}$

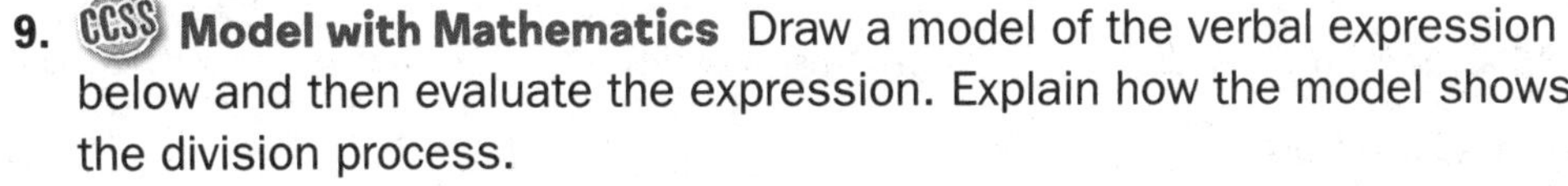

9. CCSS **Model with Mathematics** Draw a model of the verbal expression below and then evaluate the expression. Explain how the model shows the division process.

one half divided by two fifths ______

Show your work.

______

Copy and Solve For Exercises 10 and 11, show your work on a separate piece of paper.

10. **Multiple Representations** Jorge recorded the distance that five of his friends live from his house in the table shown.

Student	Miles
Lucia	$5\frac{1}{2}$
Lon	$8\frac{2}{3}$
Sam	$12\frac{5}{6}$
Jamal	$2\frac{7}{9}$
Tye	$17\frac{13}{18}$

a. **Numbers** Tye lives about how many times farther away than Jamal?

b. **Algebra** The mean is the sum of the data divided by the number of items in the data set. Write and solve an equation to find the mean number of miles that Jorge's friends live from his house. Write your answer in simplest form.

c. **Model** Draw a bar diagram that can be used to find how many more miles Lon travels than Lucia to get to Jorge's house.

11. Tara bought a dozen folders. She took $\frac{1}{3}$ of the dozen and then divided the remaining folders equally among her four friends. What fraction of the dozen did each of her four friends receive? How many folders was this per person?

H.O.T. Problems

12. **Find the Error** Blake is finding $\frac{4}{5} \div \frac{6}{7}$. Find his mistake and correct it.

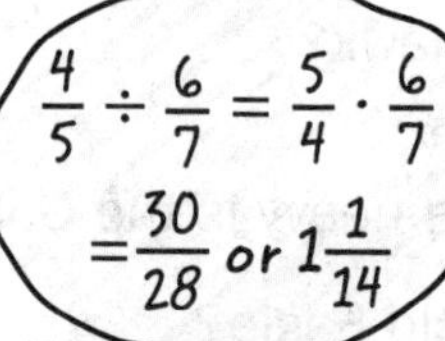

$\frac{4}{5} \div \frac{6}{7} = \frac{5}{4} \cdot \frac{6}{7}$
$= \frac{30}{28}$ or $1\frac{1}{14}$

13. **Persevere with Problems** If $\frac{5}{6}$ is divided by a certain fraction $\frac{a}{b}$, the result is $\frac{1}{4}$. What is the fraction $\frac{a}{b}$?

Standardized Test Practice

14. Which procedure would you use to find $\frac{2}{3} \div \frac{7}{9}$?

Ⓐ Multiply the first fraction by the reciprocal of the second fraction.

Ⓑ Multiply the second fraction by the reciprocal of the first fraction.

Ⓒ Multiply by the least common multiple of 3 and 9.

Ⓓ Multiply by the greatest common factor of 3 and 9.

Name ______________________ My Homework ______________________

Extra Practice

Divide. Write in simplest form.

15. $\frac{5}{9} \div \frac{5}{6} = \frac{2}{3}$

$\frac{5}{9} \div \frac{5}{6} = \frac{5}{9} \times \frac{6}{5}$

$= \frac{\overset{1}{\cancel{5}}}{\underset{3}{\cancel{9}}} \times \frac{\overset{2}{\cancel{6}}}{\underset{1}{\cancel{5}}}$

$= \frac{1 \times 2}{3 \times 1}$

$= \frac{2}{3}$

Homework Help

16. $-5\frac{2}{7} \div \left(-2\frac{1}{7}\right) =$ ______

17. $-5\frac{1}{5} \div \frac{2}{3} =$ ______

18. Vinh bought $4\frac{1}{2}$ gallons of ice cream to serve. If a pint is $\frac{1}{8}$ of a gallon, how many pint-sized servings can be made? ______

19. William has $8\frac{1}{4}$ cups of fruit juice. If he divides the juice into $\frac{3}{4}$-cup servings, how many servings will he have? ______

20. CCSS **Justify Conclusions** So far, a storm has traveled 35 miles in $\frac{1}{2}$ hour. If it is currently 5:00 P.M. and the storm is 105 miles away from you, at what time will the storm reach you? Explain how you solved the problem.

______

______

______

21. Find $\frac{1\frac{2}{3}}{9} \div \frac{1\frac{1}{9}}{3}$. Write in simplest form. ______

22. CCSS **Use Math Tools** Write the letter of each statement below in the section of any operation to which the statement applies.

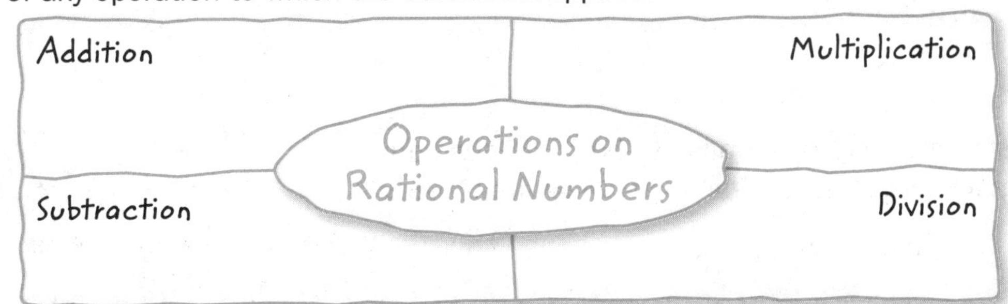

A Use a common denominator.

B Multiply by the multiplicative inverse.

C Write the result in simplest form.

Standardized Test Practice

23. Which expression represents the least value?

Ⓐ $298 + \frac{1}{2}$

Ⓑ $298 - \frac{1}{2}$

Ⓒ $298 \times \frac{1}{2}$

Ⓓ $298 \div \frac{1}{2}$

24. How many times as great is the weight of the large box of peanuts than the small box of peanuts?

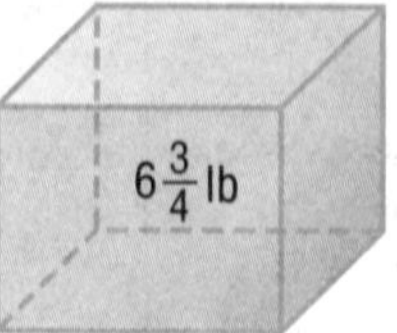

Ⓕ 4

Ⓖ 5

Ⓗ 6

Ⓘ 7

Common Core Review

Add or subtract. Write in simplest form. 5.NF.2

25. $\frac{1}{5} + \frac{1}{4} =$ ______

26. $\frac{1}{3} - \frac{1}{6} =$ ______

27. $\frac{4}{9} + \frac{2}{7} =$ ______

28. $\frac{11}{15} - \frac{3}{20} =$ ______

29. The cheerleaders made spirit buttons for the basketball team. They used blue and red ribbons. How much total ribbon did they use? 5.NF.2

______

Ribbon	
Blue	**Red**
$\frac{3}{8}$ ft	$\frac{3}{8}$ ft

30. How much longer is a $2\frac{1}{2}$-inch-long piece of string than a $\frac{2}{5}$-inch-long piece of string? 5.NF.2 ______

31. The table shows lengths of trails at Sharon Woods Park.

a. How much longer is Oak Trail than Willow Trail? 5.NF.2

Write in simplest form. ______

b. If you walked Maple Trail and Oak Trail, what is the total distance you walked in simplest form? ______

Trail	Length of Trail
Willow	$\frac{1}{8}$ mi
Oak	$\frac{3}{4}$ mi
Maple	$\frac{1}{16}$ mi

21ST CENTURY CAREER in Fashion Design

Fashion Designer

Do you enjoy reading fashion magazines, keeping up with the latest trends, and creating your own unique sense of style? You might want to consider a career in fashion design. Fashion designers create new designs for clothing, accessories, and shoes. In addition to being creative and knowledgeable about current fashion trends, fashion designers need to be able to take accurate measurements and calculate fit by adding, subtracting, and dividing measurements.

Is This the Career for You?

Are you interested in a career as a fashion designer? Take some of the following courses in high school.

- Algebra
- Art
- Digital Design
- Geometry

Find out how math relates to a career in Fashion Design.

A Flair for Fashion!

Use the information in the table to solve each problem. Write in simplest form.

1. For size 8, does Dress Style A or B require more fabric? Explain. ______

2. How many yards of fabric are needed to make Style A in sizes 8 and 14? ______

3. Estimate how many yards of fabric are needed to make Style B in each of the sizes shown. Then find the actual amount of fabric. ______

4. For Style B, how much more fabric is required for size 14 than for size 12? ______

5. A designer has half the amount of fabric needed to make Style A in size 10. How much fabric does she have? ______

6. A bolt has $12\frac{1}{8}$ yards of fabric left on it. How many dresses in Style B size 12 could be made? How much fabric is left over? ______

Amount of Fabric Needed (yards)				
Dress Style	**Size 8**	**Size 10**	**Size 12**	**Size 14**
A	$3\frac{3}{8}$	$3\frac{1}{2}$	$3\frac{3}{4}$	$3\frac{7}{8}$
B	$3\frac{1}{4}$	$3\frac{1}{2}$	$3\frac{7}{8}$	4

Career Project

It's time to update your career portfolio! Use blogs and webpages of fashion designers to answer some of these questions: Where did they go to school? What was their first job? What do they say is the most difficult part about being a fashion designer? What inspires them to create their designs? What advice do they have for new designers?

Suppose you are an employer hiring a fashion designer. What question would you ask a potential employee?

- ______
- ______

Chapter Review

Check

Vocabulary Check

Unscramble each of the clue words. After unscrambling each of the terms, use the numbered letters to find a vocabulary term that relates to all of the other terms.

RAB TONNOTIA ☐☐☐ ☐☐☐☐☐☐☐☐ (1, 7)

TAMTINRINGE ☐☐☐☐☐☐☐☐☐☐☐ (3)

GIEPEATNR ☐☐☐☐☐☐☐☐☐ (4)

KIEL STAFCOIRN ☐☐☐☐ ☐☐☐☐☐☐☐☐☐ (5)

LUKIEN ☐☐☐☐☐☐ (6, 8)

NOMMOC NIOAREOMNDT ☐☐☐☐☐☐ ☐☐☐☐☐☐☐☐☐☐☐ (2)

☐☐☐☐☐☐☐☐
1 2 3 4 5 6 7 8

Complete each sentence using one of the unscrambled words above.

1. The process of using a line over the repeating digits of a decimal is called ________________.
2. Fractions with different denominators are called ________________ fractions.
3. The least common multiple of the denominators is called the least ________________.
4. The decimal form of a fraction is a(n) ________________ decimal.
5. A ________________ decimal is a decimal in which the repeating digit is zero.
6. Fractions with the same denominator are called ________________.

Key Concept Check

Use Your FOLDABLES

Use your Foldable to help review the chapter.

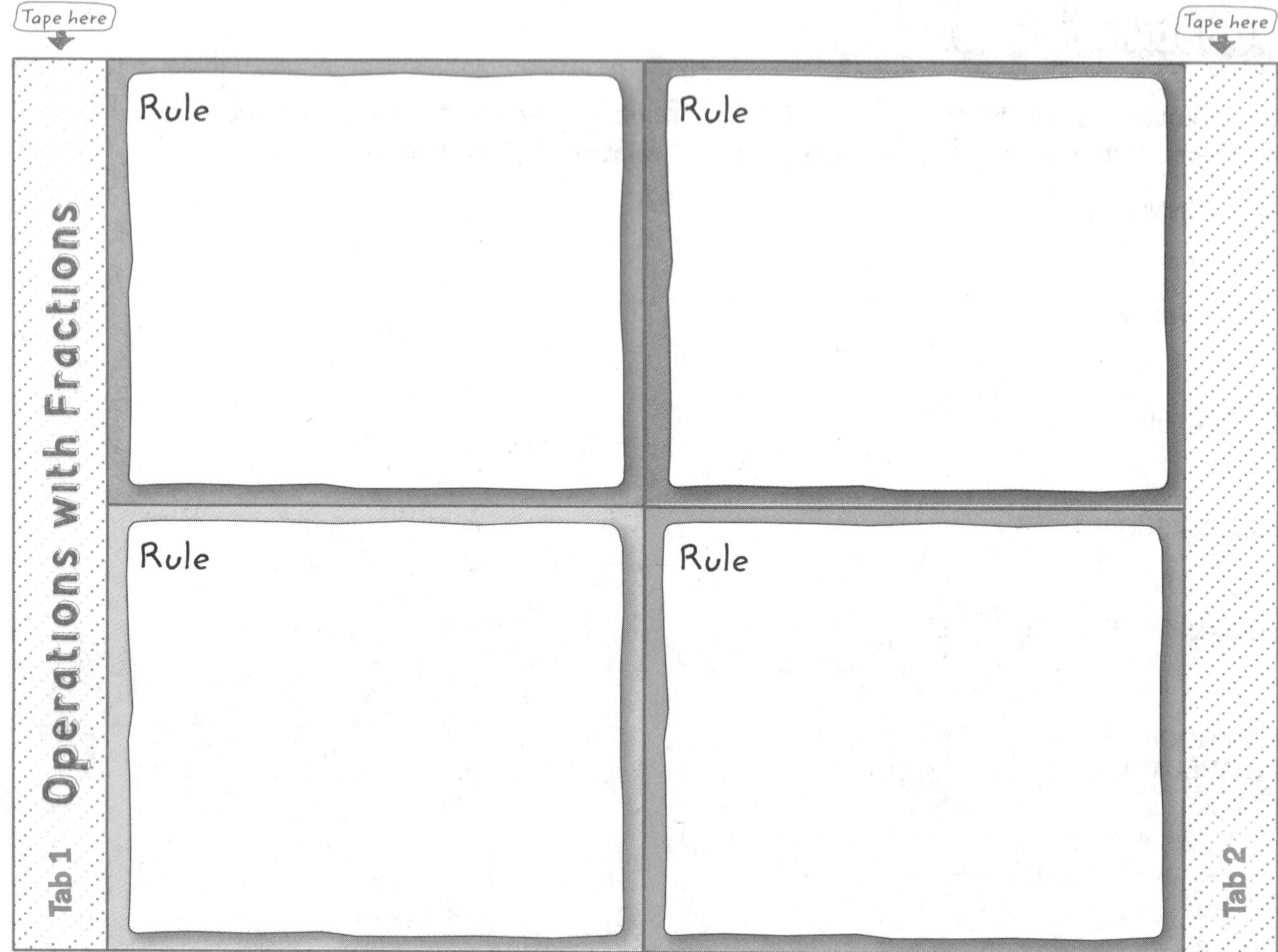

Got it?

Circle the correct term or number to complete each sentence.

1. $\frac{1}{5}$ and $\left(\frac{1}{3}, \frac{3}{5}\right)$ are like fractions.
2. To add like fractions, add the (numerators, denominators).
3. To add unlike fractions, rename the fractions using the least common (numerator, denominator).
4. The reciprocal of $\frac{1}{3}$ is (–3, 3).
5. To divide by a fraction, (multiply, divide) by its reciprocal.
6. The least common denominator of $\frac{1}{5}$ and $\frac{1}{10}$ is (10, 50).

Problem Solving

1. Jeremy ran a mile in 5 minutes and 8 seconds. Write the time in minutes as a decimal. (Lesson 1) ______

2. Amaya received a $\frac{26}{30}$ on her English test and an 81% on her biology test. On which test did she receive the higher score? Explain. (Lesson 2)

______

______

3. CCSS **Model with Mathematics** Mrs. Brown made two desserts. The number line below shows the number of cups of sugar she used to make them. Write an addition sentence that shows the amount of sugar Mrs. Brown used. (Lesson 4) ______

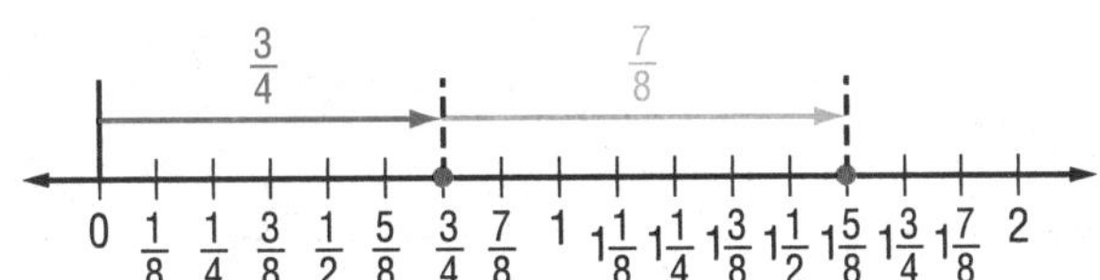

4. Lucas babysat his younger sister $2\frac{1}{2}$ hours on Friday and $3\frac{2}{3}$ hours on Saturday. How much longer did Lucas babysit his younger sister on Saturday than on Friday? (Lesson 5) ______

5. **Financial Literacy** For his birthday, Aiden received a check from his grandmother. The table shows how he spent or saved the money. Two weeks later, Aiden withdrew $\frac{2}{3}$ of the money he had deposited into his savings account. What fraction in simplest form of the original check did Aiden withdraw? Explain. (Lesson 6)

Fraction of Check	How Spent or Saved
$\frac{2}{5}$	Spent on baseball cards
$\frac{1}{4}$	Spent on a CD
$\frac{7}{20}$	Deposited into savings account

______

______

6. STEM The world's largest bird is the ostrich, whose mass can be as much as 156.5 kilograms. What is the approximate weight in pounds? (Lesson 7)

______

7. An ounce is $\frac{1}{16}$ of a pound. How many ounces are in $8\frac{3}{4}$ pounds? (Lesson 8)

______

Reflect

Answering the Essential Question

Use what you learned about operations with rational numbers to complete the graphic organizer. Describe a process to perform each operation.

Add

Subtract

Essential Question

WHAT happens when you add, subtract, multiply, and divide fractions?

Multiply

Divide

Answer the Essential Question. WHAT happens when you add, subtract, multipy, and divide fractions?

Watch

Explore the Ocean Depths For this project, imagine that your dream job is to become an oceanographer. In this project you will:

- **Collaborate** with your classmates as you research information about the ocean.
- **Share** the results of your research in a creative way.
- **Reflect** on how mathematical ideas can be represented.

Collaborate

Go Online **Work with your group to research and complete each activity. Organize the results of each activity in a way that makes sense.**

1. About $\frac{2}{3}$ of Earth is covered by ocean. Research the five oceans of the world and create a table that shows about what fraction each ocean is of that $\frac{2}{3}$.

2. What is the greatest ocean depth? Find out and then display it on a vertical number line along with other facts about what you can find at different ocean depths.

3. Coral reefs are the home of many ocean creatures. Look up some facts about the state of coral reefs in the world today and display them in a creative way.

4. Choose three different types of whales that live in the ocean. Compare things like their size, the amount of food they eat, or the climate in which they live. Organize the information in a table or graph.

5. Research one of the larger icebergs in the Arctic Ocean. Sketch an image of the iceberg next to a vertical number line that shows the approximate top and bottom of the iceberg. Remember, about $\frac{7}{8}$ of an iceberg is under water.

With your group, decide on a way to share what you have learned about ocean depths. Some suggestions are listed below, but you could also think of other creative ways to present your information. Remember to show how you used mathematics in your project!

- Use presentation software to organize what you have learned in this project. Share your presentation with the class.
- Imagine you need to apply for funds to go on a deep sea exploration. Write a persuasive letter or speech that highlights the importance of studying ocean depths.

Check out the note on the right to connect this project with other subjects.

Environmental Literacy Research an animal that lives in the ocean that is on the endangered species list. Give a presentation to your class that answers the following questions:

- What are some of the causes for the animals being on the endangered species list?
- What efforts are currently being made to protect the animal you chose?

6. **Answer the Essential Question** How can mathematical ideas be represented?

   a. How were mathematical ideas involving integers represented in the information you discovered about oceans?

   b. How were mathematical ideas involving rational numbers represented in the information you discovered about oceans?

Glossary/Glosario

Go online for the eGlossary. 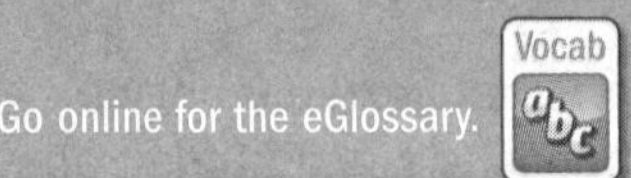

The eGlossary contains words and definitions in the following 13 languages:

Arabic
Bengali
Brazilian Portuguese
Cantonese
English
Haitian Creole
Hmong
Korean
Russian
Spanish
Tagalog
Urdu
Vietnamese

English / Español

Aa

absolute value The distance the number is from zero on a number line.

valor absoluto Distancia a la que se encuentra un número de cero en la recta numérica.

acute angle An angle with a measure greater than 0° and less than 90°.

ángulo agudo Ángulo que mide más de 0° y menos de 90°.

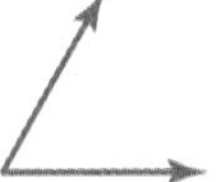

acute triangle A triangle having three acute angles.

triángulo acutángulo Triángulo con tres ángulos agudos.

Addition Property of Equality If you add the same number to each side of an equation, the two sides remain equal.

propiedad de adición de la igualdad Si sumas el mismo número a ambos lados de una ecuación, los dos lados permanecen iguales.

Addition Property of Inequality If you add the same number to each side of an inequality, the inequality remains true.

propiedad de desigualdad en la suma Si se suma el mismo número a cada lado de una desigualdad, la desigualdad sigue siendo verdadera.

Additive Identity Property The sum of any number and zero is the number.

propiedad de identidad de la suma La suma de cualquier número y cero es el mismo número.

additive inverse Two integers that are opposites. The sum of an integer and its additive inverse is zero.

inverso aditivo Dos enteros opuestos.

adjacent angles Angles that have the same vertex, share a common side, and do not overlap.

ángulos adyacentes Ángulos que comparten el mismo vértice y un común lado, pero no se sobreponen.

algebra A branch of mathematics that involves expressions with variables.

álgebra Rama de las matemáticas que trata de las expresiones con variables.

algebraic expression A combination of variables, numbers, and at least one operation.

expresión algebraica Combinación de variables, números y por lo menos una operación.

alternate exterior angles Angles that are on opposite sides of the transversal and outside the parallel lines.

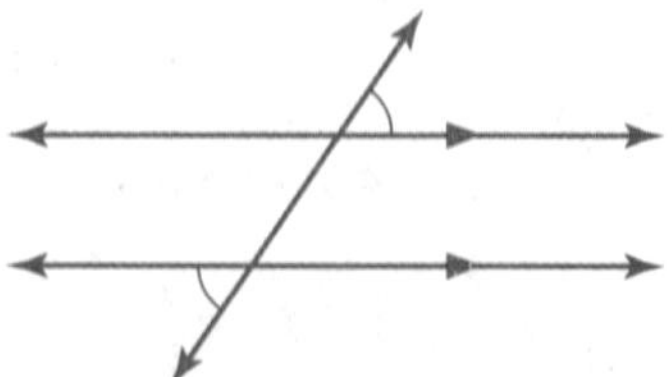

ángulos alternos externos Ángulos en lados opuestos de la trasversal y afuera de las rectas paralelas.

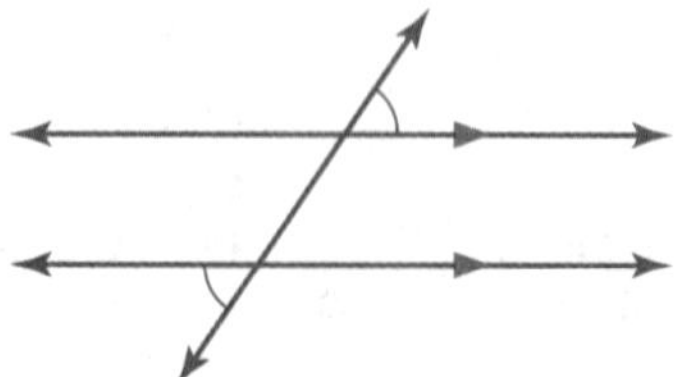

alternate interior angles Angles that are on opposite sides of the transversal and inside the parallel lines.

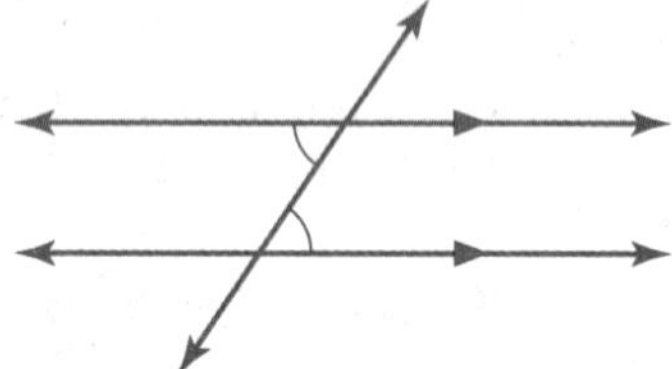

ángulos alternos internos Ángulos en lados opuestos de la trasversal y dentro de las rectas paralelas.

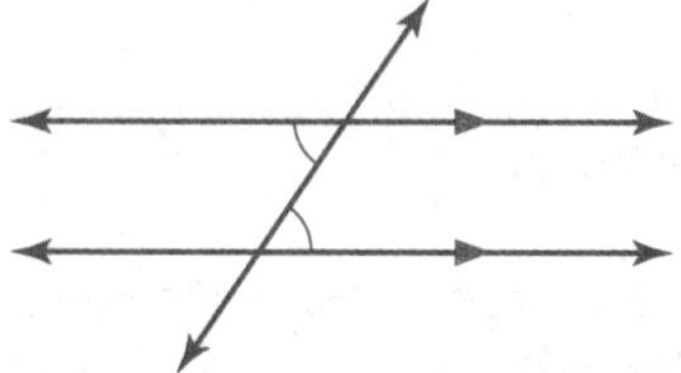

angle Two rays with a common endpoint form an angle. The rays and vertex are used to name the angle.

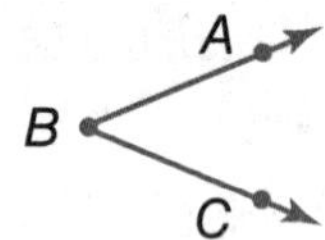

$\angle ABC$, $\angle CBA$, or $\angle B$

ángulo Dos rayos con un extremo común forman un ángulo. Los rayos y el vértice se usan para nombrar el ángulo.

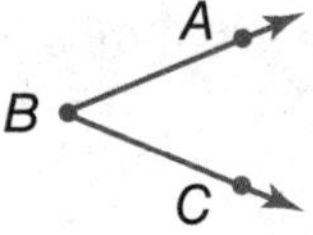

$\angle ABC$, $\angle CBA$ o $\angle B$

arithmetic sequence A sequence in which the difference between any two consecutive terms is the same.

sucesión aritmética Sucesión en la cual la diferencia entre dos términos consecutivos es constante.

Associative Property The way in which numbers are grouped does not change their sum or product.

propiedad asociativa La forma en que se agrupan números al sumarlos o multiplicarlos no altera su suma o producto.

Bb

bar notation In repeating decimals, the line or bar placed over the digits that repeat. For example, $2.\overline{63}$ indicates that the digits 63 repeat.

notación de barra Línea o barra que se coloca sobre los dígitos que se repiten en decimales periódicos. Por ejemplo, $2.\overline{63}$ indica que los dígitos 63 se repiten.

base In a power, the number used as a factor. In 10^3, the base is 10. That is, $10^3 = 10 \times 10 \times 10$.

base En una potencia, el número usado como factor. En 10^3, la base es 10. Es decir, $10^3 = 10 \times 10 \times 10$.

base One of the two parallel congruent faces of a prism.

base Una de las dos caras paralelas congruentes de un prisma.

biased sample A sample drawn in such a way that one or more parts of the population are favored over others.

muestra sesgada Muestra en que se favorece una o más partes de una población.

box plot A method of visually displaying a distribution of data values by using the median, quartiles, and extremes of the data set. A box shows the middle 50% of the data.

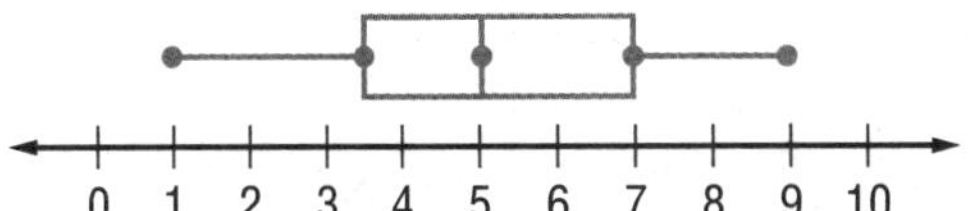

diagrama de caja Un método de mostrar visualmente una distribución de valores usando la mediana, cuartiles y extremos del conjunto de datos. Una caja muestra el 50% del medio de los datos.

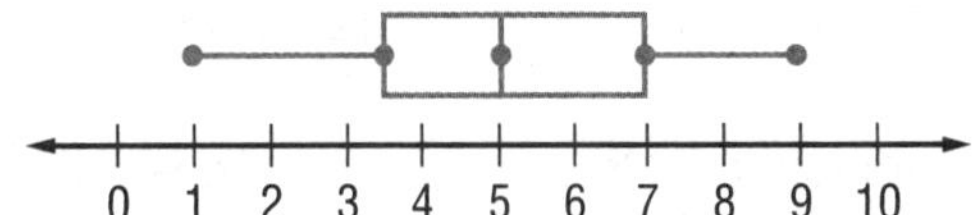

Cc

center The point from which all points on circle are the same distance.

centro El punto desde el cual todos los puntos en una circunferencia están a la misma distancia.

circle The set of all points in a plane that are the same distance from a given point called the center.

círculo Conjunto de todos los puntos de un plano que están a la misma distancia de un punto dado denominado "centro".

circle graph A graph that shows data as parts of a whole. In a circle graph, the percents add up to 100.

Area of Oceans

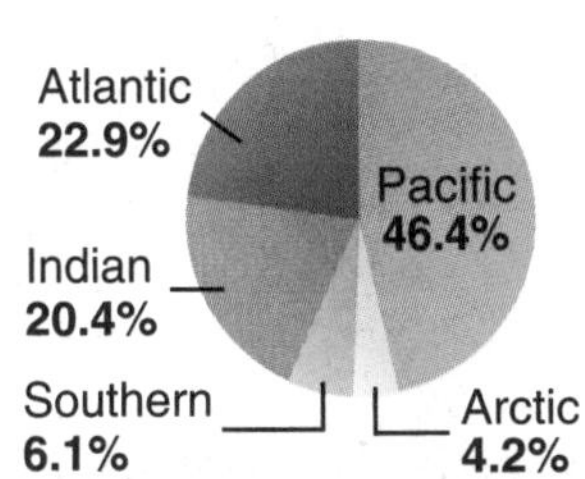

gráfica circular Gráfica que muestra los datos como partes de un todo. En una gráfica circular los porcentajes suman 100.

Área de superficie de los océanos

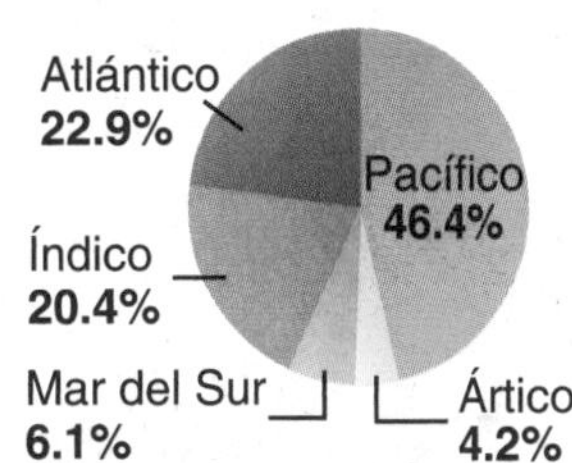

circumference The distance around a circle.

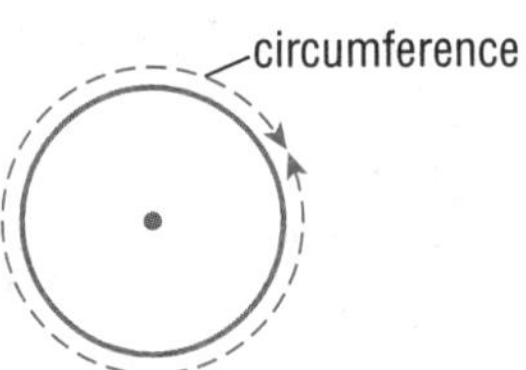

circunferencia Distancia en torno a un círculo.

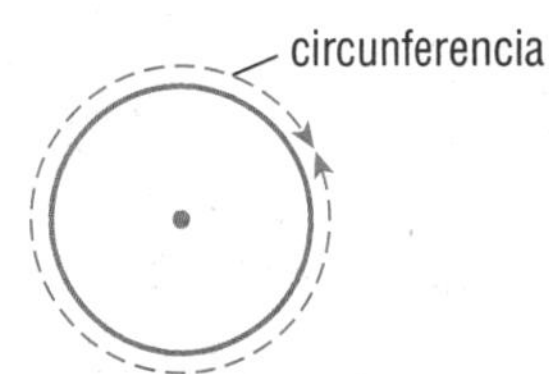

coefficient The numerical factor of a term that contains a variable.

coeficiente El factor numérico de un término que contiene una variable.

common denominator A common multiple of the denominators of two or more fractions. 24 is a common denominator for $\frac{1}{3}$, $\frac{5}{8}$, and $\frac{3}{4}$ because 24 is the LCM of 3, 8, and 4.

común denominador El múltiplo común de los denominadores de dos o más fracciones. 24 es un denominador común para $\frac{1}{3}$, $\frac{5}{8}$ y $\frac{3}{4}$ porque 24 es el mcm de 3, 8 y 4.

Commutative Property The order in which two numbers are added or multiplied does not change their sum or product.

propiedad conmutativa El orden en que se suman o multiplican dos números no altera el resultado.

complementary angles Two angles are complementary if the sum of their measures is 90°.

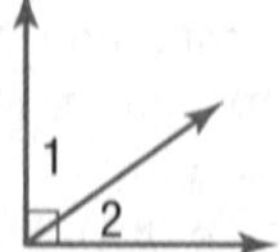

∠1 and ∠2 are complementary angles.

complementary events The events of one outcome happening and that outcome not happening. The sum of the probabilities of an event and its complement is 1 or 100%. In symbols, $P(\text{A}) + P(not\ \text{A}) = 1$.

complex fraction A fraction $\frac{A}{B}$ where A or B are fractions and B does not equal zero.

composite figure A figure that is made up of two or more three-dimensional figures.

compound event An event consisting of two or more simple events.

cone A three-dimensional figure with one circular base connected by a curved surface to a single vertex.

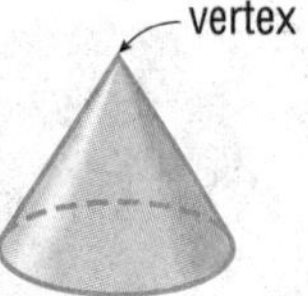

congruent Having the same measure.

congruent angles Angles that have the same measure.

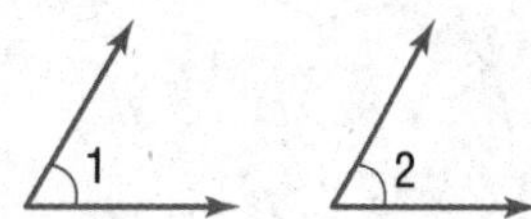

∠1 and ∠2 are congruent angles.

congruent figures Figures that have the same size and same shape and corresponding sides and angles with equal measure.

congruent segments Sides with the same length.

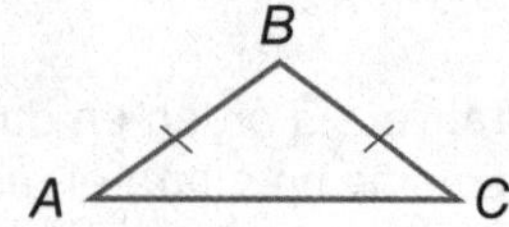

Side $\overline{AB}$ is congruent to side $\overline{BC}$.

ángulos complementarios Dos ángulos son complementarios si la suma de sus medidas es 90°.

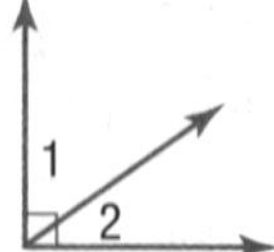

∠1 y ∠2 son complementarios.

eventos complementarios Los eventos de un resultado que ocurre y ese resultado que no ocurre. La suma de las probabilidades de un evento y su complemento es 1 ó 100. En símbolos $P(\text{A}) + P(no\ \text{A}) = 1$.

fracción compleja Una fracción $\frac{A}{B}$ en la cual A o B son fracciones y B no es igual a cero.

figura compuesta Figura formada por dos o más figuras tridimensionales.

evento compuesto Un evento que consiste en dos o más eventos simples.

cono Una figura tridimensional con una base circular conectada por una superficie curva para un solo vértice.

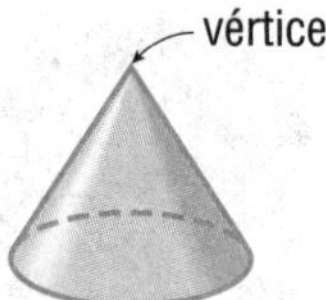

congruente Que tiene la misma medida.

ángulos congruentes Ángulos que tienen la misma medida.

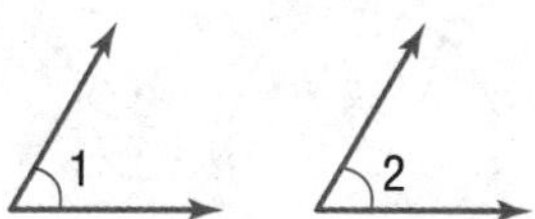

∠1 y ∠2 son congruentes.

figuras congruentes Figuras que tienen el mismo tamaño y la misma forma y los lados y los ángulos correspondientes tienen igual medida.

segmentos congruentes Lados con la misma longitud.

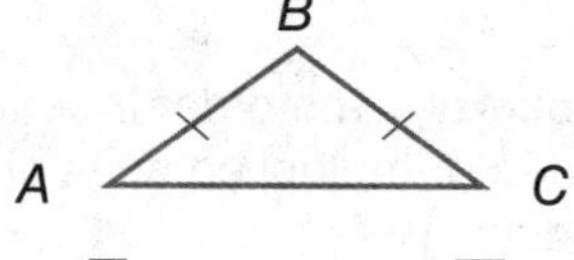

$\overline{AB}$ es congruente a $\overline{BC}$.

constant A term that does not contain a variable.

constante Término que no contiene ninguna variable.

constant of proportionality A constant ratio or unit rate of two variable quantities. It is also called the constant of variation.

constante de proporcionalidad Una razón constante o tasa por unidad de dos cantidades variables. También se llama constante de variación.

constant of variation The constant ratio in a direct variation. It is also called the constant of proportionality.

constante de variación Una razón constante o tasa por unidad de dos cantidades variables. También se llama constante de proporcionalidad.

constant rate of change The rate of change in a linear relationship.

razón constante de cambio Tasa de cambio en una relación lineal.

continuous data Data that take on any real number value. It can be determined by considering what numbers are reasonable as part of the domain.

datos continuos Datos que asumen cualquier valor numérico real. Se pueden determinar al considerar qué números son razonables como parte del dominio.

convenience sample A sample which consists of members of a population that are easily accessed.

muestra de conveniencia Muestra que incluye miembros de una población fácilmente accesibles.

coordinate plane A plane in which a horizontal number line and a vertical number line intersect at their zero points. Also called a coordinate grid.

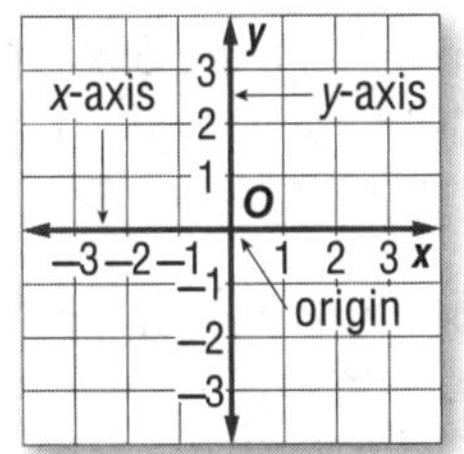

plano de coordenadas Plano en el cual se han trazado dos rectas numéricas, una horizontal y una vertical, que se intersecan en sus puntos cero. También conocido como sistema de coordenadas.

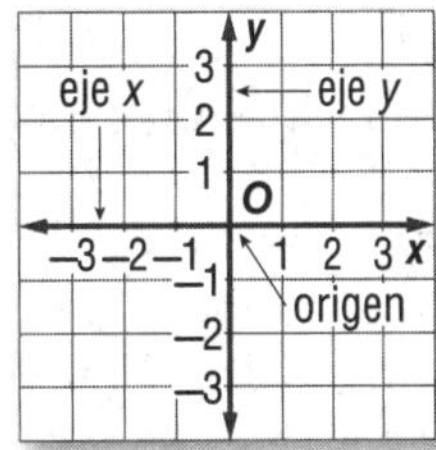

coplanar Lines or points that lie in the same plane.

coplanar Líneas o puntos situados en el mismo plano.

corresponding angles Angles in the same position on parallel lines in relation to a transversal.

ángulos correspondientes Ángulos que están en la misma posición sobre rectas paralelas en relación con la transversal.

corresponding sides The sides of similar figures that are in the same relative postion.

lados correspondientes Lados de figuras semejantes que estan en la misma posición.

counterexample A specific case which proves a statement false.

contraejemplo Caso específico que demuestra la falsedad de un enunciado.

cross product The product of the numerator of one ratio and the denominator of the other ratio. The cross products of any proportion are equal.

producto cruzado Producto del numerador de una razón por el denominador de la otra razón. Los productos cruzados de cualquier proporción son iguales.

cross section The cross section of a solid and a plane.

sección transversal Intersección de un sólido con un plano.

cube root One of three equal factors of a number. If $a^3 = b$, then a is the cube root of b. The cube root of 125 is 5 since $5^3 = 125$.

cubed The product in which a number is a factor three times. Two cubed is 8 because $2 \times 2 \times 2 = 8$.

cylinder A three-dimensional figure with two parallel congruent circular bases connected by a curved surface.

raíz cúbica Uno de tres factores iguales de un número. Si $a^3 = b$, entonces a es la raíz cúbica de b. La raíz cúbica de 125 es 5, dado que $5^3 = 125$.

al cubo El producto de un número por sí mismo, tres veces. Dos al cubo es 8 porque $2 \times 2 \times 2 = 8$.

cilindro Una figura tridimensional con dos paralelas congruentes circulares bases conectados por una superficie curva.

Dd

decagon A polygon having ten sides.

defining a variable Choosing a variable and a quantity for the variable to represent in an expression or equation.

degrees The most common unit of measure for angles. If a circle were divided into 360 equal-sized parts, each part would have an angle measure of 1 degree.

dependent events Two or more events in which the outcome of one event affects the outcome of the other event(s).

dependent variable The variable in a relation with a value that depends on the value of the independent variable.

derived unit A unit that is derived from a measurement system base unit, such as length, mass, or time.

diagonal A line segment that connects two nonconsecutive vertices.

diameter The distance across a circle through its center.

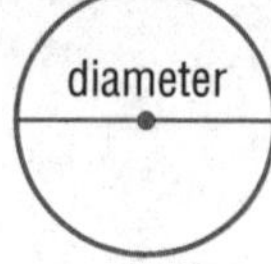

decágono Un polígono con diez lados.

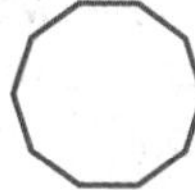

definir una variable El eligir una variable y una cantidad que esté representada por la variable en una expresión o en una ecuacion.

grados La unidad más común para medir ángulos. Si un círculo se divide en 360 partes iguales, cada parte tiene una medida angular de 1 grado.

eventos dependientes Dos o más eventos en que el resultado de un evento afecta el resultado de otro u otros eventos.

variable dependiente La variable en una relación cuyo valor depende del valor de la variable independiente.

unidad derivada Unidad que se deriva de una unidad básica de un sistema de medidas, como la longitud, la masa o el tiempo.

diagonal Segmento de recta que une dos vértices no consecutivos de un polígono.

diámetro Segmento que pasa por el centro de un círculo y lo divide en dos partes iguales.

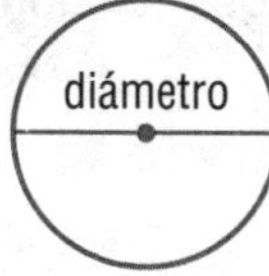

dimensional analysis The process of including units of measurement when you compute.

análisis dimensional Proceso que incluye las unidades de medida al hacer cálculos.

direct variation The relationship between two variable quantities that have a constant ratio.

variación directa Relación entre las cantidades de dos variables que tienen una tasa constante.

discount The amount by which the regular price of an item is reduced.

descuento Cantidad que se le rebaja al precio regular de un artículo.

discrete data When solutions of a function are only integer values. It can be determined by considering what numbers are reasonable as part of the domain.

datos discretos Cuando las soluciones de una función son solo valores enteros. Se pueden determinar considerando qué números son razonables como parte del dominio.

disjoint events Events that cannot happen at the same time.

eventos disjuntos Eventos que no pueden ocurrir al mismo tiempo.

Distributive Property To multiply a sum by a number, multiply each addend of the sum by the number outside the parentheses. For any numbers a, b, and c, $a(b + c) = ab + ac$ and $a(b - c) = ab - ac$.

Example: $2(5 + 3) = (2 \times 5) + (2 \times 3)$ and $2(5 - 3) = (2 \times 5) - (2 \times 3)$

propiedad distributiva Para multiplicar una suma por un número, multiplíquese cada sumando de la suma por el número que está fuera del paréntesis. Sean cuales fuere los números a, b, y c, $a(b + c) = ab + ac$ y $a(b - c) = ab - ac$.

Ejemplo: $2(5 + 3) = (2 \cdot 5) + (2 \cdot 3)$ y $2(5 - 3) = (2 \cdot 5) - (2 \cdot 3)$

Division Property of Equality If you divide each side of an equation by the same nonzero number, the two sides remain equal.

propiedad de igualdad de la división Si divides ambos lados de una ecuación entre el mismo número no nulo, los lados permanecen iguales.

Division Property of Inequality When you divide each side of an inequality by a negative number, the inequality symbol must be reversed for the inequality to remain true.

propiedad de desigualdad en la división Cuando se divide cada lado de una desigualdad entre un número negativo, el símbolo de desigualdad debe invertirse para que la desigualdad siga siendo verdadera.

domain The set of input values for a function.

dominio El conjunto de valores de entrada de una función.

double box plot Two box plots graphed on the same number line.

doble diagrama de caja Dos diagramas de caja sobre la misma recta numérica.

double dot plot A method of visually displaying a distribution of two sets of data values where each value is shown as a dot above a number line.

doble diagrama de puntos Un método de mostrar visualmente una distribución de dos conjuntos de valores donde cada valor se muestra como un punto arriba de una recta numérica.

edge The line segment where two faces of a polyhedron intersect.

borde El segmento de línea donde se cruzan dos caras de un poliedro.

enlargement An image larger than the original.

ampliación Imagen más grande que la original.

equation A mathematical sentence that contains an equals sign, =, stating that two quantities are equal.

ecuación Enunciado matemático que contiene el signo de igualdad = indicando que dos cantidades son iguales.

equiangular In a polygon, all of the angles are congruent.

equiangular En un polígono, todos los ángulos son congruentes.

equilateral In a polygon, all of the sides are congruent.

equilátero En un polígono, todos los lados son congruentes.

equilateral triangle A triangle having three congruent sides.

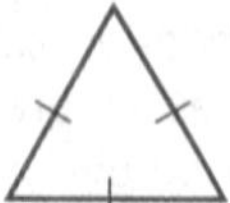

triángulo equilátero Triángulo con tres lados congruentes.

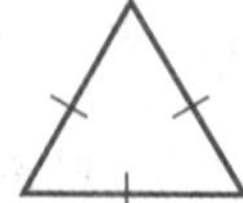

equivalent equations Two or more equations with the same solution.

ecuaciones equivalentes Dos o más ecuaciones con la misma solución.

equivalent expressions Expressions that have the same value.

expresiones equivalentes Expresiones que tienen el mismo valor.

equivalent ratios Two ratios that have the same value.

razones equivalentes Dos razones que tienen el mismo valor.

evaluate To find the value of an expression.

evaluar Calcular el valor de una expresión.

experimental probability An estimated probability based on the relative frequency of positive outcomes occurring during an experiment. It is based on what *actually* occurred during such an experiment.

probabilidad experimental Probabilidad estimada que se basa en la frecuencia relativa de los resultados positivos que ocurren durante un experimento. Se basa en lo que *en realidad* ocurre durante dicho experimento.

exponent In a power, the number that tells how many times the base is used as a factor. In 5^3, the exponent is 3. That is, $5^3 = 5 \times 5 \times 5$.

exponente En una potencia, el número que indica las veces que la base se usa como factor. En 5^3, el exponente es 3. Es decir, $5^3 = 5 \times 5 \times 5$.

exponential form Numbers written with exponents.

forma exponencial Números escritos usando exponentes.

Ff

face A flat surface of a polyhedron.

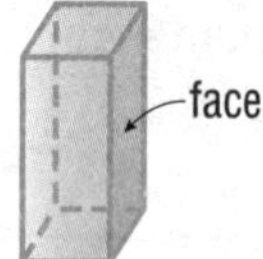

cara Una superficie plana de un poliedro.

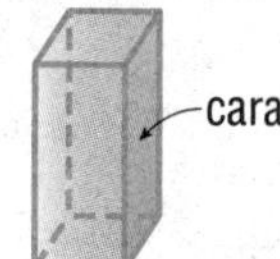

factor To write a number as a product of its factors.

factorizar Escribir un número como el producto de sus factores.

factored form An expression expressed as the product of its factors.

forma factorizada Una expresión expresada como el producto de sus factores.

factors Two or more numbers that are multiplied together to form a product.

factores Dos o más números que se multiplican entre sí para formar un producto.

fair game A game where each player has an equally likely chance of winning.

juego justo Juego donde cada jugador tiene igual posibilidad de ganar.

first quartile For a data set with median M, the first quartile is the median of the data values less than M.

primer cuartil Para un conjunto de datos con la mediana M, el primer cuartil es la mediana de los valores menores que M.

formula An equation that shows the relationship among certain quantities.

fórmula Ecuación que muestra la relación entre ciertas cantidades.

function A relationship which assigns exactly one output value for each input value.

función Relación que asigna exactamente un valor de salida a cada valor de entrada.

function rule The operation performed on the input of a function.

regla de función Operación que se efectúa en el valor de entrada.

function table A table used to organize the input numbers, output numbers, and the function rule.

tabla de funciones Tabla que organiza las entradas, la regla y las salidas de una función.

Fundamental Counting Principle Uses multiplication of the number of ways each event in an experiment can occur to find the number of possible outcomes in a sample space.

Principio Fundamental de Contar Este principio usa la multiplicación del número de veces que puede ocurrir cada evento en un experimento para calcular el número de posibles resultados en un espacio muestral.

Gg

gram A unit of mass in the metric system equivalent to 0.001 kilogram. The amount of matter an object can hold.

gramo Unidad de masa en el sistema métrico que equivale a 0.001 de kilogramo. La cantidad de materia que puede contener un objeto.

graph The process of placing a point on a number line or on a coordinate plane at its proper location.

graficar Proceso de dibujar o trazar un punto en una recta numérica o en un plano de coordenadas en su ubicación correcta.

gratuity Also known as a tip. It is a small amount of money in return for a service.

gratificación También conocida como propina. Es una cantidad pequeña de dinero en retribución por un servicio.

Hh

heptagon A polygon having seven sides.

heptágono Polígono con siete lados.

hexagon A polygon having six sides.

hexágono Polígono con seis lados.

histogram A type of bar graph used to display numerical data that have been organized into equal intervals.

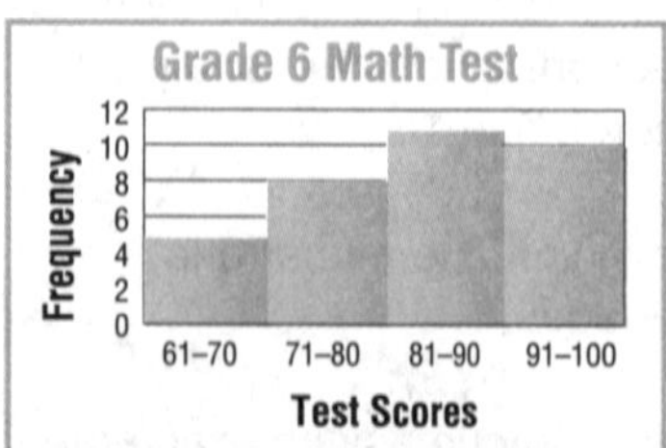

histograma Tipo de gráfica de barras que se usa para exhibir datos que se han organizado en intervalos iguales.

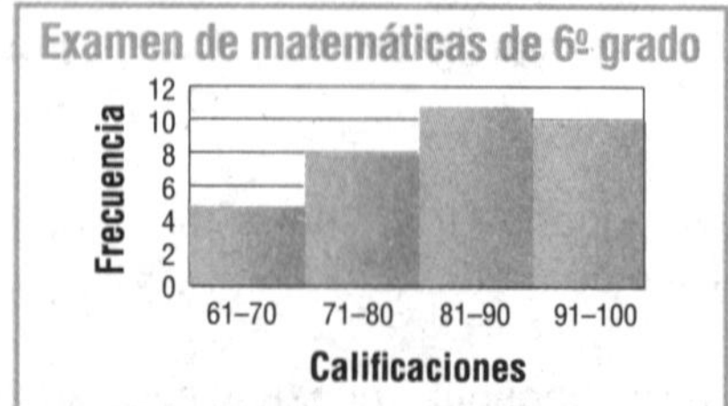

Identity Property of Zero The sum of an addend and zero is the addend. Example: $5 + 0 = 5$

independent events Two or more events in which the outcome of one event does not affect the outcome of the other event(s).

independent variable The variable in a function with a value that is subject to choice.

indirect measurement Finding a measurement using similar figures to find the length, width, or height of objects that are too difficult to measure directly.

inequality An open sentence that uses $<, >, \neq, \leq,$ or $\geq$ to compare two quantities.

integer Any number from the set $\{..., -4, -3, -2, -1, 0, 1, 2, 3, 4, ...\}$, where ... means continues without end.

interquartile range A measure of variation in a set of numerical data. It is the distance between first and third quartiles of the data set.

inverse variation A relationship where the product of x and y is a constant k. As x increases in value, y decreases in value, or as y decreases in value, x increases in value.

irrational number A number that cannot be expressed as the ratio of two integers.

isosceles triangle A triangle having at least two congruent sides.

propiedad de identidad del cero La suma de un sumando y cero es igual al sumando. Ejemplo: $5 + 0 = 5$

eventos independientes Dos o más eventos en los cuales el resultado de uno de ellos no afecta el resultado de los otros eventos.

variable independiente Variable en una función cuyo valor está sujeto a elección.

medición indirecta Hallar una medición usando figuras semejantes para calcular el largo, ancho o altura de objetos que son difíciles de medir directamente.

desigualdad Enunciado abierto que usa $<, >, \neq, \leq$ o $\geq$ para comparar dos cantidades.

entero Cualquier número del conjunto $\{..., -4, -3, -2, -1, 0, 1, 2, 3, 4, ...\}$, donde ... significa que continúa sin fin.

rango intercuartil Una medida de la variación en un conjunto de datos numéricos. Es la distancia entre el primer y el tercer cuartiles del conjunto de datos.

variación inversa Relación en la cual el producto de x y y es una constante k. A medida que aumenta el valor de x, disminuye el valor de y o a medida que disminuye el valor de y, aumenta el valor de x.

número irracional Número que no se puede expresar como el razón de dos enteros.

triángulo isósceles Triángulo que tiene por lo menos dos lados congruentes.

Kk

kilogram The base unit of mass in the metric system. One kilogram equals 1,000 grams.

kilogramo Unidad básica de masa del sistema métrico. Un kilogramo equivale a 1,000 gramos.

Ll

lateral face In a polyhedron, a face that is not a base.

cara lateral En un poliedro, las caras que no forman las bases.

lateral surface area The sum of the areas of all of the lateral faces of a solid.

área de superficie lateral Suma de las áreas de todas las caras de un sólido.

least common denominator (LCD) The least common multiple of the denominators of two or more fractions. You can use the LCD to compare fractions.

mínimo común denominador (mcd) El menor de los múltiplos de los denominadores de dos o más fracciones. Puedes usar el mínimo común denominador para comparar fracciones.

like fractions Fractions that have the same denominators.

fracciones semejantes Fracciones que tienen los mismos denominadores.

like terms Terms that contain the same variables raised to the same power. Example: 5x and 6x are like terms.

términos semejante Términos que contienen las mismas variables elevadas a la misma potencia. Ejemplo: 5*x* y 6*x* son *términos semejante*.

line graph A type of statistical graph using lines to show how values change over a period of time.

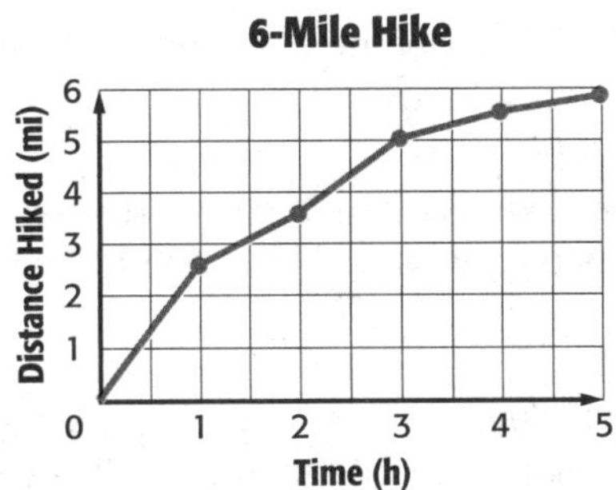

gráfica lineal Tipo de gráfica estadística que usa segmentos de recta para mostrar cómo cambian los valores durante un período de tiempo.

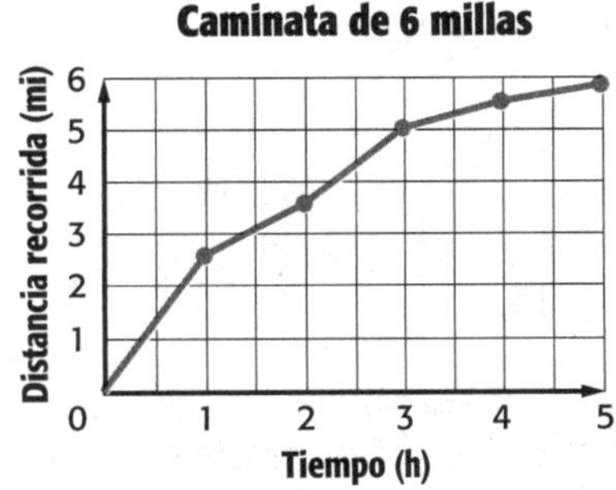

linear expression An algebraic expression in which the variable is raised to the first power.

expresión lineal Expresión algebraica en la cual la variable se eleva a la primera potencia.

linear function A function for which the graph is a straight line.

función lineal Función cuya gráfica es una recta.

linear relationship A relationship for which the graph is a straight line.

relación lineal Una relación para la cual la gráfica es una línea recta.

liter The base unit of capacity in the metric system. The amount of dry or liquid material an object can hold.

litro Unidad básica de capacidad del sistema métrico. La cantidad de materia líquida o sólida que puede contener un objeto.

markdown An amount by which the regular price of an item is reduced.

rebaja Una cantidad por la cual el precio regular de un artículo se reduce.

markup The amount the price of an item is increased above the price the store paid for the item.

margen de utilidad Cantidad de aumento en el precio de un artículo por encima del precio que paga la tienda por dicho artículo.

mean The sum of the data divided by the number of items in the data set.

media La suma de los datos dividida entre el número total de artículos en el conjunto de datos.

mean absolute deviation A measure of variation in a set of numerical data, computed by adding the distances between each data value and the mean, then dividing by the number of data values.

desviación media absoluta Una medida de variación en un conjunto de datos numéricos que se calcula sumando las distancias entre el valor de cada dato y la media, y luego dividiendo entre el número de valores.

measures of center Numbers that are used to describe the center of a set of data. These measures include the mean, median, and mode.

medidas del centro Números que se usan para describir el centro de un conjunto de datos. Estas medidas incluyen la media, la mediana y la moda.

measures of variation A measure used to describe the distribution of data.

medidas de variación Medida usada para describir la distribución de los datos.

median A measure of center in a set of numerical data. The median of a list of values is the value appearing at the center of a sorted version of the list—or the mean of the two central values, if the list contains an even number of values.

mediana Una medida del centro en un conjunto de dados númericos. La mediana de una lista de valores es el valor que aparace en el centro de una versíon ordenada de la lista, o la media de dos valores centrales si la lista contiene un número par de valores.

meter The base unit of length in the metric system.

metro Unidad fundamental de longitud del sistema métrico.

metric system A decimal system of measures. The prefixes commonly used in this system are kilo-, centi-, and milli-.

sistema métrico Sistema decimal de medidas. Los prefijos más comunes son kilo-, centi- y mili-.

mode The number or numbers that appear most often in a set of data. If there are two or more numbers that occur most often, all of them are modes.

moda El número o números que aparece con más frecuencia en un conjunto de datos. Si hay dos o más números que ocurren con más frecuencia, todosellos son modas.

monomial A number, variable, or product of a number and one or more variables.

monomio Número, variable o producto de un número y una o más variables.

Multiplication Property of Equality If you multiply each side of an equation by the same nonzero number, the two sides remain equal.

propiedad de multiplicación de la igualdad Si multiplicas ambos lados de una ecuación por el mismo número no nulo, lo lados permanecen iguales.

Multiplication Property of Inequality When you multiply each side of an inequality by a negative number, the inequality symbol must be reversed for the inequality to remain true.

propiedad de desigualdad en la multiplicación Cuando se multiplica cada lado de una desigualdad por un número negativo, el símbolo de desigualdad debe invertirse para que la desigualdad siga siendo verdadera.

Multiplicative Identity Property The product of any number and one is the number.

propiedad de identidad de la multiplicación El producto de cualquier número y uno es el mismo número.

Multiplicative Property of Zero The product of any number and zero is zero.

propiedad del cero en la multiplicación El producto de cualquier número y cero es cero.

multiplicative inverse Two numbers with a product of 1. For example, the multiplicative inverse of $\frac{2}{3}$ is $\frac{3}{2}$.

inverso multiplicativo Dos números cuyo producto es 1. Por ejemplo, el inverso multiplicativo de $\frac{2}{3}$ es $\frac{3}{2}$.

negative exponent Any nonzero number to the negative *n* power. It is the multiplicative inverse of its *n*th power.

exponente negativo Cualquier número que no sea cero a la potencia negative de *n*. Es el inverso multiplicativo de su *en*ésimo potencia.

negative integer An integer that is less than zero. Negative integers are written with a − sign.

entero negativo Número menor que cero. Se escriben con el signo −.

net A two-dimensional figure that can be used to build a three-dimensional figure.

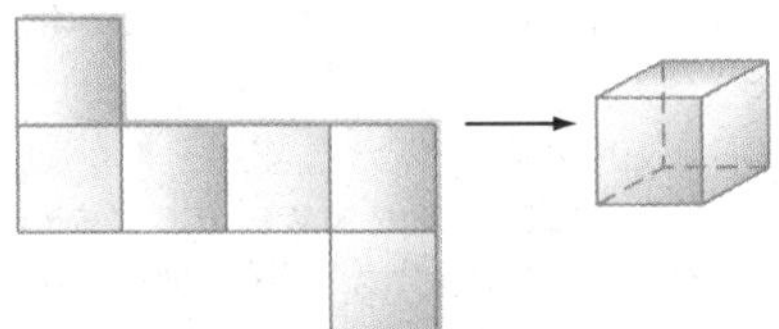

red Figura bidimensional que sirve para hacer una figura tridimensional.

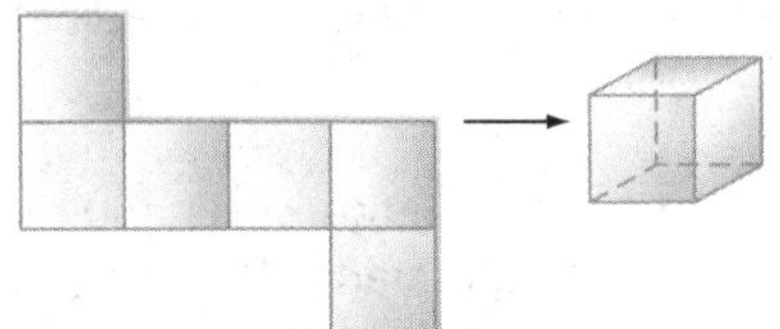

nonagon A polygon having nine sides.

enágono Polígono que tiene nueve lados.

nonlinear function A function for which the graph is *not* a straight line.

nonlinear function Función cuya gráfica *no* es una línea recta.

nonproportional The relationship between two ratios with a rate or ratio that is not constant.

no proporcional Relación entre dos razones cuya tasa o razón no es constante.

numerical expression A combination of numbers and operations.

expresión numérica Combinación de números y operaciones.

obtuse angle Any angle that measures greater than 90° but less than 180°.

ángulo obtuso Cualquier ángulo que mide más de 90° pero menos de 180°.

obtuse triangle A triangle having one obtuse angle.

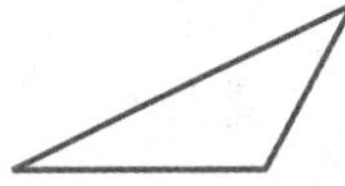

triángulo obtusángulo Triángulo que tiene un ángulo obtuso.

octagon A polygon having eight sides.

octágono Polígono que tiene ocho lados.

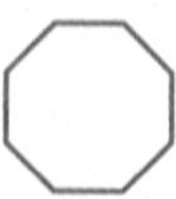

opposites Two integers are opposites if they are represented on the number line by points that are the same distance from zero, but on opposite sides of zero. The sum of two opposites is zero.

opuestos Dos enteros son opuestos si, en la recta numérica, están representados por puntos que equidistan de cero, pero en direcciones opuestas. La suma de dos opuestos es cero.

order of operations The rules to follow when more than one operation is used in a numerical expression.

1. Evaluate the expressions inside grouping symbols.
2. Evaluate all powers.
3. Multiply and divide in order from left to right.
4. Add and subtract in order from left to right.

orden de las operaciones Reglas a seguir cuando se usa más de una operación en una expresión numérica.

1. Primero, evalúa las expresiones dentro de los símbolos de agrupación.
2. Evalúa todas las potencias.
3. Multiplica y divide en orden de izquierda a derecha.
4. Suma y resta en orden de izquierda a derecha.

ordered pair A pair of numbers used to locate a point in the coordinate plane. An ordered pair is written in the form (*x*-coordinate, *y*-coordinate).

par ordenado Par de números que se utiliza para ubicar un punto en un plano de coordenadas. Se escribe de la siguiente forma: (coordenada *x*, coordenada *y*).

origin The point at which the *x*-axis and the *y*-axis intersect in a coordinate plane. The origin is at (0, 0).

origen Punto en que el eje *x* y el eje *y* se intersecan en un plano de coordenadas. El origen está ubicado en (0, 0).

outcome Any one of the possible results of an action. For example, 4 is an outcome when a number cube is rolled.

resultado Cualquiera de los resultados posibles de una acción. Por ejemplo, 4 puede ser un resultado al lanzar un cubo numerado.

outlier A data value that is either much *greater* or much *less* than the median.

valor atípico Valor de los datos que es mucho *mayor* o mucho *menor* que la mediana.

parallel lines Lines in a plane that never intersect.

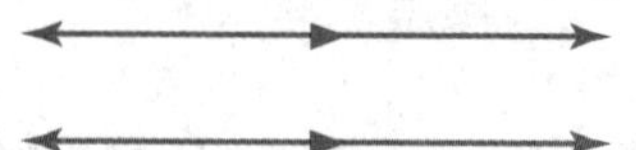

rectas paralelas Rectas en un plano que nunca se intersecan.

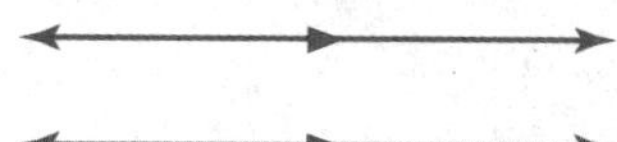

parallelogram A quadrilateral with opposite sides parallel and opposite sides congruent.

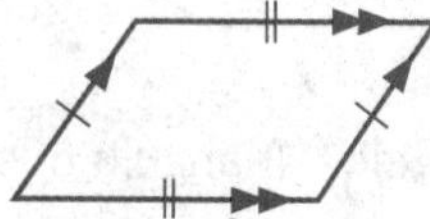

paralelogramo Cuadrilátero cuyos lados opuestos son paralelos y congruentes.

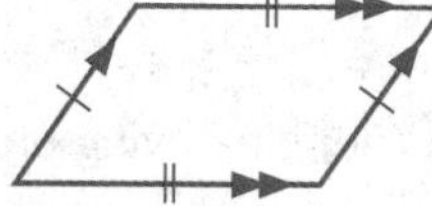

pentagon A polygon having five sides.

percent equation An equation that describes the relationship between the part, whole, and percent.

$$\text{part} = \text{percent} \cdot \text{whole}$$

percent error A ratio that compares the inaccuracy of an estimate (amount of error) to the actual amount.

percent of change A ratio that compares the change in a quantity to the original amount.

$$\text{percent of change} = \frac{\text{amount of change}}{\text{original amount}}$$

percent of decrease A negative percent of change.

percent of increase A positive percent of change.

percent proportion One ratio or fraction that compares part of a quantity to the whole quantity. The other ratio is the equivalent percent written as a fraction with a denominator of 100.

$$\frac{\text{part}}{\text{whole}} = \frac{\text{percent}}{100}$$

perfect squares Numbers with square roots that are whole numbers. 25 is a perfect square because the square root of 25 is 5.

permutation An arrangement, or listing, of objects in which order is important.

perpendicular lines Lines that meet or cross each other to form right angles.

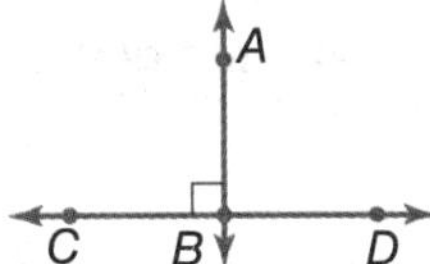

pi The ratio of the circumference of a circle to its diameter. The Greek letter π represents this number. The value of pi is 3.1415926. . . . Approximations for pi are 3.14 and $\frac{22}{7}$.

plane A two-dimensional flat surface that extends in all directions.

pentágono Polígono que tiene cinco lados.

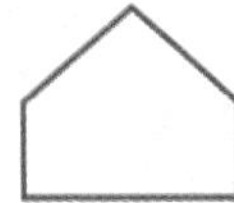

ecuación porcentual Ecuación que describe la relación entre la parte, el todo y el por ciento.

$$\text{parte} = \text{por ciento} \cdot \text{todo}$$

porcentaje de error Una razón que compara la inexactitud de una estimación (cantidad del error) con la cantidad real.

porcentaje de cambio Razón que compara el cambio en una cantidad a la cantidad original.

$$\text{porcentaje de cambio} = \frac{\text{cantidad del cambio}}{\text{cantidad original}}$$

porcentaje de disminución Porcentaje de cambio negativo.

porcentaje de aumento Porcentaje de cambio positivo.

proporción porcentual Razón o fracción que compara parte de una cantidad a toda la cantidad. La otra razón es el porcentaje equivalente escrito como fracción con 100 de denominador.

$$\frac{\text{parte}}{\text{todo}} = \frac{\text{porcentaje}}{100}$$

cuadrados perfectos Números cuya raíz cuadrada es un número entero. 25 es un cuadrado perfecto porque la raíz cuadrada de 25 es 5.

permutación Arreglo o lista de objetos en la cual el orden es importante.

rectas perpendiculares Rectas que al encontrarse o cruzarse forman ángulos rectos.

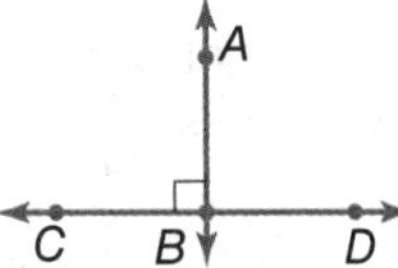

pi Relación entre la circunferencia de un círculo y su diámetro. La letra griega π representa este número. El valor de pi es 3.1415926. . . . Las aproximaciones de pi son 3.14 y $\frac{22}{7}$.

plano Superficie bidimensional que se extiende en todas direcciones.

polygon A simple closed figure formed by three or more straight line segments.

polígono Figura cerrada simple formada por tres o más segmentos de recta.

polyhedron A three-dimensional figure with faces that are polygons.

poliedro Una figura tridimensional con caras que son polígonos.

population The entire group of items or individuals from which the samples under consideration are taken.

población El grupo total de individuos o de artículos del cual se toman las muestras bajo estudio.

positive integer An integer that is greater than zero. They are written with or without a + sign.

entero positivo Entero que es mayor que cero; se escribe con o sin el signo +.

powers Numbers expressed using exponents. The power 3^2 is read *three to the second power,* or *three squared.*

potencias Números que se expresan usando exponentes. La potencia 3^2 se lee *tres a la segunda potencia o tres al cuadrado.*

precision The ability of a measurement to be consistently reproduced.

precisión Capacidad que tiene una medición de poder reproducirse consistentemente.

principal The amount of money deposited or borrowed.

capital Cantidad de dinero que se deposita o se toma prestada.

prism A polyhedron with two parallel congruent faces called bases.

prisma Un poliedro con dos caras congruentes paralelas llamadas bases.

probability The chance that some event will happen. It is the ratio of the number of favorable outcomes to the number of possible outcomes.

probabilidad La posibilidad de que suceda un evento. Es la razón del número de resultados favorables al número de resultados posibles.

probability model A model used to assign probabilities to outcomes of a chance process by examining the nature of the process.

modelo de probabilidad Un modelo usado para asignar probabilidades a resultados de un proceso aleatorio examinando la naturaleza del proceso.

properties Statements that are true for any number or variable.

propiedades Enunciados que son verdaderos para cualquier número o variable.

proportion An equation stating that two ratios or rates are equivalent.

proporción Ecuación que indica que dos razones o tasas son equivalentes.

proportional The relationship between two ratios with a constant rate or ratio.

proporcional Relación entre dos razones con una tasa o razón constante.

pyramid A polyhedron with one base that is a polygon and three or more triangular faces that meet at a common vertex.

pirámide Un poliedro con una base que es un polígono y tres o más caras triangulares que se encuentran en un vértice común.

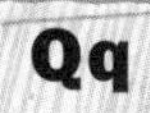

Qq

quadrant One of the four regions into which the two perpendicular number lines of the coordinate plane separate the plane.

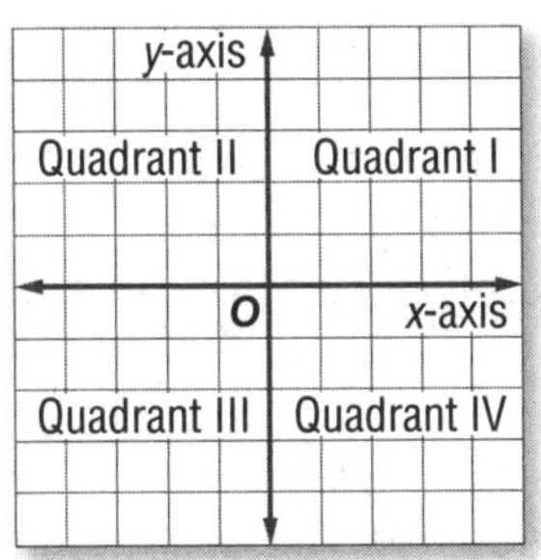

cuadrante Una de las cuatro regiones en que dos rectas numéricas perpendiculares dividen el plano de coordenadas.

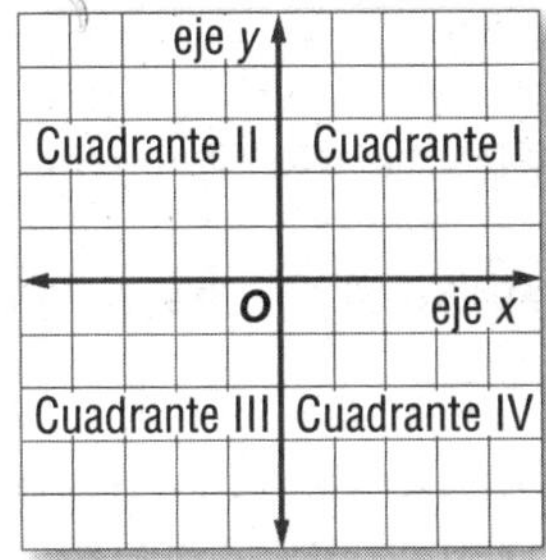

quadrilateral A closed figure having four sides and four angles.

cuadrilátero Figura cerrada que tiene cuatro lados y cuatro ángulos.

quartile A value that divides the data set into four equal parts.

cuartil Valor que divide el conjunto de datos en cuatro partes iguales.

Rr

radical sign The symbol used to indicate a nonnegative square root, $\sqrt{\ }$.

signo radical Símbolo que se usa para indicar una raíz cuadrada no negativa, $\sqrt{\ }$.

radius The distance from the center of a circle to any point on the circle.

radio Distancia desde el centro de un círculo hasta cualquiera de sus puntos.

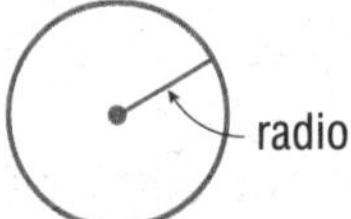

random Outcomes occur at random if each outcome occurs by chance. For example, rolling a number on a number cube occurs at random.

azar Los resultados ocurren aleatoriamente si cada resultado ocurre por casualidad. Por ejemplo, sacar un número en un cubo numerado ocurre al azar.

range The set of output values for a function.

rango Conjunto de valores de salida para una función.

range The difference between the greatest and least data value.

rango La diferencia entre el número mayor y el menor en un conjunto de datos.

rate A ratio that compares two quantities with different kinds of units.

tasa Razón que compara dos cantidades que tienen distintas unidades de medida.

rate of change A rate that describes how one quantity changes in relation to another. A rate of change is usually expressed as a unit rate.

tasa de cambio Tasa que describe cómo cambia una cantidad con respecto a otra. Por lo general, se expresa como tasa unitaria.

rational numbers The set of numbers that can be written in the form $\frac{a}{b}$, where a and b are integers and $b \neq 0$.
Examples: $1 = \frac{1}{1}, \frac{2}{9}, -2.3 = -2\frac{3}{10}$

real numbers A set made up of rational and irrational numbers.

reciprocal The multiplicative inverse of a number.

rectangle A parallelogram having four right angles.

rectangular prism A prism that has two parallel congruent bases that are rectangles.

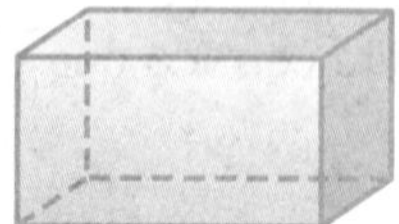

reduction An image smaller than the original.

regular polygon A polygon that has all sides congruent and all angles congruent.

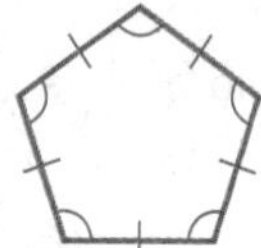

regular pyramid A pyramid whose base is a regular polygon and in which the segment from the vertex to the center of the base is the altitude.

relation Any set of ordered pairs.

relative frequency A ratio that compares the frequency of each category to the total.

repeating decimal The decimal form of a rational number.

rhombus A parallelogram having four congruent sides.

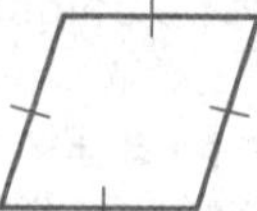

right angle An angle that measures exactly 90°.

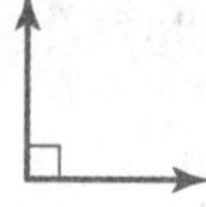

números racionales Conjunto de números que puede escribirse en la forma $\frac{a}{b}$ donde a y b son números enteros y $b \neq 0$.
Ejemplos: $1 = \frac{1}{1}, \frac{2}{9}, -2.3 = -2\frac{3}{10}$

números reales Conjunto de números racionales e irracionales.

recíproco El inverso multiplicativo de un número.

rectángulo Paralelogramo con cuatro ángulos rectos.

prisma rectangular Un prisma con dos bases paralelas congruentes que son rectángulos.

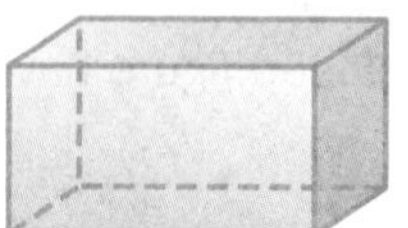

reducción Imagen más pequeña que la original.

polígono regular Polígono con todos los lados y todos los ángulos congruentes.

pirámide regular Pirámide cuya base es un polígono regular y en la cual el segmento desde el vértice hasta el centro de la base es la altura.

relación Cualquier conjunto de pares ordenados.

frecuencia relativa Razón que compara la frecuencia de cada categoría al total.

decimal periódico La forma decimal de un número racional.

rombo Paralelogramo que tiene cuatro lados congruentes.

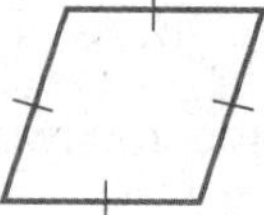

ángulo recto Ángulo que mide exactamente 90°.

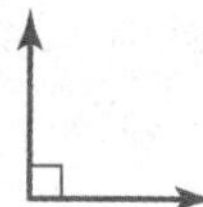

right triangle A triangle having one right angle.

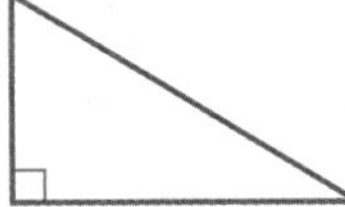

triángulo rectángulo Triángulo que tiene un ángulo recto.

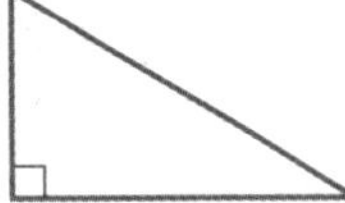

sales tax An additional amount of money charged on items that people buy.

impuesto sobre las ventas Cantidad de dinero adicional que se cobra por los artículos que se compran.

sample A randomly selected group chosen for the purpose of collecting data.

muestra Grupo escogido al azar o aleatoriamente que se usa con el propósito de recoger datos.

sample space The set of all possible outcomes of a probability experiment.

espacio muestral Conjunto de todos los resultados posibles de un experimento probabilístico.

scale The scale that gives the ratio that compares the measurements of a drawing or model to the measurements of the real object.

escala Razón que compara las medidas de un dibujo o modelo a las medidas del objeto real.

scale drawing A drawing that is used to represent objects that are too large or too small to be drawn at actual size.

dibujo a escala Dibujo que se usa para representar objetos que son demasiado grandes o demasiado pequeños como para dibujarlos de tamaño natural.

scale factor A scale written as a ratio without units in simplest form.

factor de escala Escala escrita como una razón sin unidades en forma simplificada.

scale model A model used to represent objects that are too large or too small to be built at actual size.

modelo a escala Réplica de un objeto real, el cual es demasiado grande o demasiado pequeño como para construirlo de tamaño natural.

scalene triangle A triangle having no congruent sides.

triángulo escaleno Triángulo sin lados congruentes.

scatter plot In a scatter plot, two sets of related data are plotted as ordered pairs on the same graph.

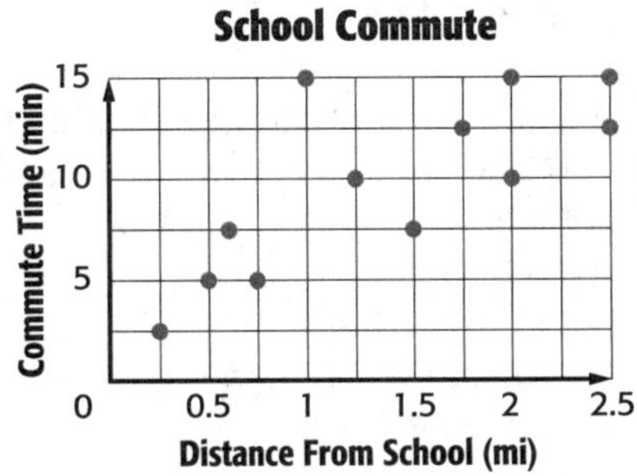

diagrama de dispersión Diagrama en que dos conjuntos de datos relacionados aparecen graficados como pares ordenados en la misma gráfica.

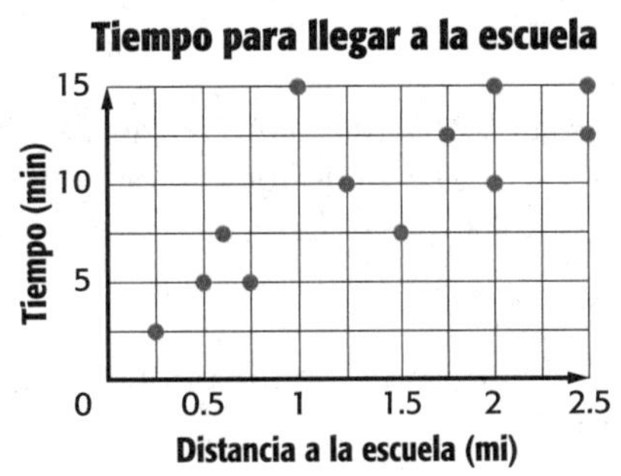

selling price The amount the customer pays for an item.

precio de venta Cantidad de dinero que paga un consumidor por un artículo.

semicircle Half of a circle. The formula for the area of a semicircle is $A = \frac{1}{2}\pi r^2$.

semicírculo Medio círculo La fórmula para el área de un semicírculo es $A = \frac{1}{2}\pi r^2$.

sequence An ordered list of numbers, such as 0, 1, 2, 3 or 2, 4, 6, 8.

sucesión Lista ordenada de números, como 0, 1, 2, 3 ó 2, 4, 6, 8.

similar figures Figures that have the same shape but not necessarily the same size.

figuras semejantes Figuras que tienen la misma forma, pero no necesariamente el mismo tamaño.

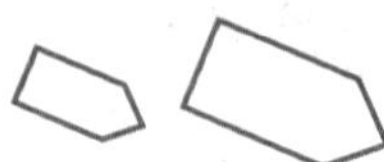

similar solids Solids with the same shape. Their corresponding linear measures are proportional.

sólidos semejantes Sólidos con la misma forma. Sus medidas lineales correspondientes son proporcionales.

simple event One outcome or a collection of outcomes.

eventos simples Un resultado o una colección de resultados.

simple interest The amount paid or earned for the use of money. The formula for simple interest is $I = prt$.

interés simple Cantidad que se paga o que se gana por el uso del dinero. La fórmula para calcular el interés simple es $I = prt$.

simple random sample An unbiased sample where each item or person in the population is as likely to be chosen as any other.

muestra aleatoria simple Muestra de una población que tiene la misma probabilidad de escogerse que cualquier otra.

simplest form An expression is in simplest form when it is replaced by an equivalent expression having no like terms or parentheses.

expresión mínima Expresión en su forma más simple cuando es reemplazada por una expresión equivalente que no tiene términos similares ni paréntesis.

simplify Write an expression in simplest form.

simplificar Escribir una expresión en su forma más simple.

simulation An experiment that is designed to model the action in a given situation.

simulación Un experimento diseñado para modelar la acción en una situación dada.

slant height The height of each lateral face.

altura oblicua Altura de cada cara lateral.

slope The rate of change between any two points on a line. It is the ratio of vertical change to horizontal change. The slope tells how steep the line is.

pendiente Razón de cambio entre cualquier par de puntos en una recta. Es la razón del cambio vertical al cambio horizontal. La pendiente indica el grado de inclinación de la recta.

solution A replacement value for the variable in an open sentence. A value for the variable that makes an equation true. Example: The *solution* of $12 = x + 7$ is 5.

solución Valor de reemplazo de la variable en un enunciado abierto. Valor de la variable que hace que una ecuación sea verdadera. Ejemplo: La *solución* de $12 = x + 7$ es 5.

square The product of a number and itself. 36 is the square of 6.

cuadrado Producto de un número por sí mismo. 36 es el cuadrado de 6.

square A parallelogram having four right angles and four congruent sides.

cuadrado Paralelogramo con cuatro ángulos rectos y cuatro lados congruentes.

square root The factors multiplied to form perfect squares.

al cuadrado Factores multiplicados para formar cuadrados perfectos.

squared The product of a number and itself. 36 is the square of 6.

raíz cuadrada El producto de un número por sí mismo. 36 es el cuadrado de 6.

standard form Numbers written without exponents.

forma estándar Números escritos sin exponentes.

statistics The study of collecting, organizing, and interpreting data.

estadística Estudio que consiste en recopilar, organizar e interpretar datos.

straight angle An angle that measures exactly 180°.

ángulo llano Ángulo que mide exactamente 180°.

Subtraction Property of Equality If you subtract the same number from each side of an equation, the two sides remain equal.

propiedad de sustracción de la igualdad Si restas el mismo número de ambos lados de una ecuación, los dos lados permanecen iguales.

Subtraction Property of Inequality If you subtract the same number from each side of an inequality, the inequality remains true.

propiedad de desigualdad en la resta Si se resta el mismo número a cada lado de una desigualdad, la desigualdad sigue siendo verdadera.

supplementary angles Two angles are supplementary if the sum of their measures is 180°.

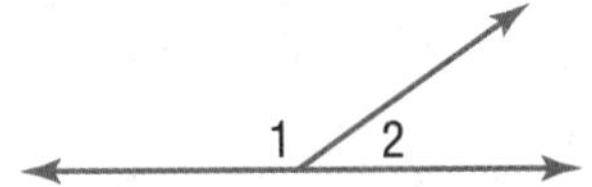

$\angle 1$ and $\angle 2$ are supplementary angles.

ángulos suplementarios Dos ángulos son suplementarios si la suma de sus medidas es 180°.

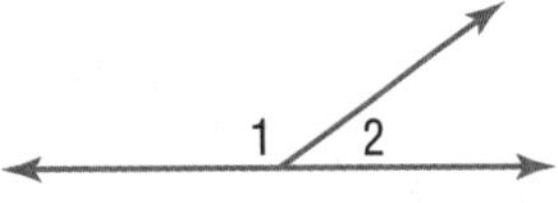

$\angle 1$ y $\angle 2$ son suplementarios.

surface area The sum of the areas of all the surfaces (faces) of a three-dimensional figure.

área de superficie La suma de las áreas de todas las superficies (caras) de una figura tridimensional.

survey A question or set of questions designed to collect data about a specific group of people, or population.

encuesta Pregunta o conjunto de preguntas diseñadas para recoger datos sobre un grupo específico de personas o población.

systematic random sample A sample where the items or people are selected according to a specific time or item interval.

muestra aleatoria sistemática Muestra en que los elementos o personas se eligen según un intervalo de tiempo o elemento específico.

term Each number in a sequence.

término Cada número en una sucesión.

term A number, a variable, or a product or quotient of numbers and variables.

término Número, variable, producto o cociente de números y de variables.

terminating decimal A repeating decimal which has a repeating digit of 0.

decimal finito Un decimal periódico que tiene un dígito que se repite que es 0.

theoretical probability The ratio of the number of ways an event can occur to the number of possible outcomes. It is based on what *should* happen when conducting a probability experiment.

three-dimensional figure A figure with length, width, and height.

third quartile For a data set with median M, the third quartile is the median of the data values greater than M.

tip Also known as a gratuity, it is a small amount of money in return for a service.

transversal The third line formed when two parallel lines are intersected.

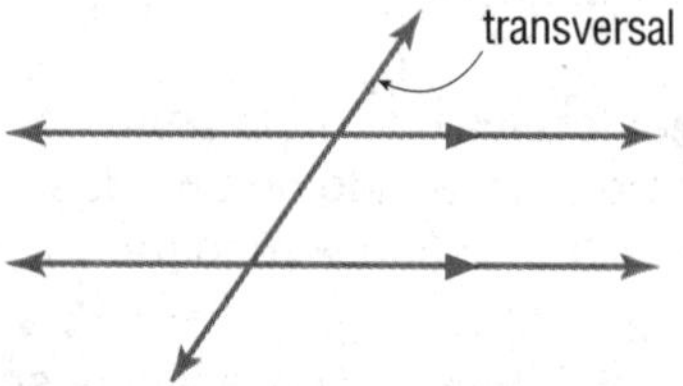

trapezoid A quadrilateral with one pair of parallel sides.

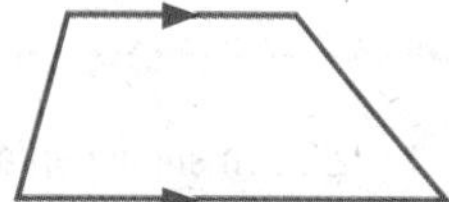

tree diagram A diagram used to show the sample space.

triangle A figure with three sides and three angles.

triangular prism A prism that has two parallel congruent bases that are triangles.

two-step equation An equation having two different operations.

two-step inequality An inequality than contains two operations.

probabilidad teórica Razón del número de maneras en que puede ocurrir un evento al número de resultados posibles. Se basa en lo que *debería* pasar cuando se conduce un experimento probabilístico.

figura tridimensional Figura que tiene largo, ancho y alto.

tercer cuartil Para un conjunto de datos con la mediana M, el tercer cuartil es la mediana de los valores mayores que M.

propina También conocida como gratificación; es una cantidad pequeña de dinero en recompensa por un servicio.

transversal Tercera recta que se forma cuando se intersecan dos rectas paralelas.

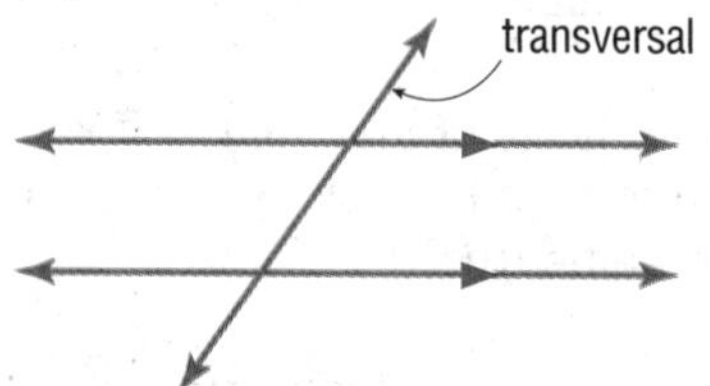

trapecio Cuadrilátero con un único par de lados paralelos.

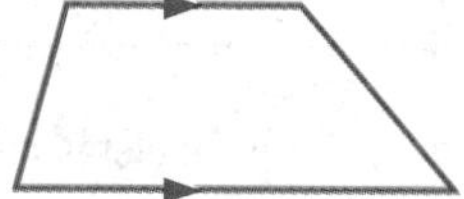

diagrama de árbol Diagrama que se usa para mostrar el espacio muestral.

triángulo Figura con tres lados y tres ángulos.

prisma triangular Un prisma que tiene dos bases congruentes paralelas que triángulos.

ecuación de dos pasos Ecuación que contiene dos operaciones distintas.

desigualdad de dos pasos Desigualdad que contiene dos operaciones.

Uu

unbiased sample A sample representative of the entire population.

unfair game A game where there is not a chance of each player being equally likely to win.

uniform probability model A probability model which assigns equal probability to all outcomes.

unit rate A rate that is simplified so that it has a denominator of 1 unit.

unit ratio A unit rate where the denominator is one unit.

unlike fractions Fractions with different denominators.

muestra no sesgada Muestra que se selecciona de modo que se representativa de la población entera.

juego injusto Juego donde cada jugador no tiene la misma posibilidad de ganar.

modelo de probabilidad uniforme Un modelo de probabilidad que asigna igual probabilidad a todos los resultados.

tasa unitaria Tasa simplificada para que tenga un denominador igual a 1.

razón unitaria Tasa unitaria en que el denominador es la unidad.

fracciones con distinto denominador Fracciones cuyos denominadores son diferentes.

Vv

variable A symbol, usually a letter, used to represent a number in mathematical expressions or sentences.

vertex A vertex of an angle is the common endpoint of the rays forming the angle.

vertex

vertex The point where three or more faces of a polyhedron intersect.

vertex The point at the tip of a cone.

vertical angles Opposite angles formed by the intersection of two lines. Vertical angles are congruent.

∠1 and ∠2 are vertical angles.

visual overlap A visual demonstration that compares the centers of two distributions with their variation, or spread.

variable Símbolo, por lo general una letra, que se usa para representar un número en expresiones o enunciados matemáticos.

vértice El vértice de un ángulo es el extremo común de los rayos que lo forman.

vértice El punto donde tres o más caras de un poliedro se cruzan.

vértice El punto en la punta de un cono.

ángulos opuestos por el vértice Ángulos opuestos formados por la intersección de dos rectas. Los ángulos opuestos por el vértice son congruentes.

∠1 y ∠2 son ángulos opuestos por el vértice.

superposición visual Una demostración visual que compara los centros de dos distribuciones con su variación, o magnitud.

volume The number of cubic units needed to fill the space occupied by a solid.

volumen Número de unidades cúbicas que se requieren para llenar el espacio que ocupa un sólido.

voluntary response sample A sample which involves only those who want to participate in the sampling.

muestra de respuesta voluntaria Muestra que involucra sólo aquellos que quieren participar en el muestreo.

***x*-axis** The horizontal number line in a coordinate plane.

eje *x* La recta numérica horizontal en el plano de coordenadas.

***x*-coordinate** The first number of an ordered pair. It corresponds to a number on the *x*-axis.

coordenada *x* El primer número de un par ordenado. Corresponde a un número en el eje *x*.

***y*-axis** The vertical number line in a coordinate plane.

eje *y* La recta numérica vertical en el plano de coordenadas.

***y*-coordinate** The second number of an ordered pair. It corresponds to a number on the *y*-axis.

coordenada *y* El segundo número de un par ordenado. Corresponde a un número en el eje *y*.

zero pair The result when one positive counter is paired with one negative counter. The value of a zero pair is 0.

par nulo Resultado de hacer coordinar una ficha positiva con una negativa. El valor de un par nulo es 0.

Go online for Step-by-Step Solutions. 

Chapter 1 Ratios and Proportional Reasoning

Page 6 Chapter 1 Are You Ready?

1. $\frac{2}{15}$ **3.** $\frac{1}{51}$ **5.** No; $\frac{12}{20} = \frac{3}{5}, \frac{15}{30} = \frac{1}{2}$

Pages 13–14 Lesson 1-1 Independent Practice

1. 60 mi/h **3.** 3.5 m/s **5.** Sample answer: about $0.50 per pair **7.** 510 words **9a.** 20.04 mi/h **b.** about 1.5 h **13.** Sometimes; a ratio that compares two measurements with different units is a rate, such as $\frac{2 \text{ miles}}{10 \text{ minutes}}$. **15.** C

Pages 15–16 Lesson 1-1 Extra Practice

17. 203.75 Calories per serving **19.** 32 mi/gal **21.** $108.75 ÷ 15 = $7.25, $7.25 × 18 = $130.50 **23.** C **25.** G **27.** $\frac{2}{7}$ **29.** $\frac{2}{3}$

Pages 21–22 Lesson 1-2 Independent Practice

1. $1\frac{1}{2}$ **3.** $\frac{4}{27}$ **5.** $\frac{2}{25}$ **7.** $6 per yard **9.** $\frac{5}{6}$ page **11.** $\frac{39}{250}$ **13.** $\frac{11}{200}$ **15.** Sample answer: If one of the numbers in the ratio is a fraction, then the ratio can be a complex fraction. **17.** $\frac{1}{2}$

Pages 23–24 Lesson 1-2 Extra Practice

19. 4 **21.** $\frac{1}{10}$ **23.** $\frac{1}{10}$ **25.** 8 costumes **27.** $\frac{3}{125}$ **29.** $\frac{1}{12}$ **31.** D **33.** C **35.** 24 **37.** 32 **39.** 1,000

Pages 29–30 Lesson 1-3 Independent Practice

1. 115 mi/h **3.** 322,000 m/h **5.** 6.1 mi/h **7.** 7,200 Mb/h **9.** Sample answer: Convert 42 miles per hour to miles per minute. **11.** 461.5 yd/h

Pages 31–32 Lesson 1-3 Extra Practice

13. 1,760 **15.** 66 **17.** 35.2 **19a.** 6.45 ft/s **19b.** 2,280 times **19c.** 0.11 mi **19d.** 900,000 times **21.** H **23.** no; Since the unit rates, $\frac{\$9}{1 \text{ baseball hat}}$ and $\frac{\$8}{1 \text{ baseball hat}}$ are not the same, the rates are not equivalent.

25.

Payment	$22	÷2×5	$55
Hours	2	÷2×5	5

Pages 37–38 Lesson 1-4 Independent Practice

1.

Time (days)	1	2	3	4
Water (L)	225	450	675	900

Yes; the time to water ratios are all equal to $\frac{1}{225}$.

3. The table for Desmond's Time shows a proportional relationship. The ratio between the time and the number of laps is always 73.

5a. yes; Sample answer:

Side Length (units)	1	2	3	4
Perimeter (units)	4	8	12	16

The side length to perimeter ratio for side lengths of 1, 2, 3, and 4 units is $\frac{1}{4}$, $\frac{2}{8}$ or $\frac{1}{4}$, $\frac{3}{12}$ or $\frac{1}{4}$, $\frac{4}{16}$ or $\frac{1}{4}$. Since these ratios are all equal to $\frac{1}{4}$, the measure of the side length of a square is proportional to the square's perimeter.
b. no; Sample answer:

Side Length (units)	1	2	3	4
Area (units^2)	1	4	9	16

The side length to area ratio for side lengths of 1, 2, 3, and 4 units is $\frac{1}{1}$ or 1, $\frac{2}{4}$ or $\frac{1}{2}$, $\frac{3}{9}$ or $\frac{1}{3}$, $\frac{4}{16}$ or $\frac{1}{4}$. Since these ratios are not equal, the measure of the side length of a square is not proportional to the square's area.
7. It is not proportional because the ratio of laps to time is not consistent; $\frac{4}{1} \neq \frac{6}{2} \neq \frac{8}{3} \neq \frac{10}{4}$. **9.** B

Pages 39–40 Lesson 1-4 Extra Practice

11.

Degrees Celsius	0	10	20	30
Degrees Fahrenheit	32	50	68	86

No; the degrees Celsius to degrees Fahrenheit ratios are not all equal. **13a.** No; the fee to ride tickets ratios are not equal. **13b.** no; Sample answer: The fee increase is inconsistent. The table shows an increase of $4.50 from 5 to 10 tickets, an increase of $4 from 10 to 15 tickets, and an increase of $2.50 from 15 to 20 tickets.

15.

n	30	60	120	173
p	90	180	360	519

17. 20 **19.** 12 **21.** 3

Page 43 Problem-Solving Investigation The Four-Step Plan

Case 3. $360 **Case 5.** Add 2 to the first term, 3 to the second, 4 to the third, and so on; 15, 21, 28.

Pages 49–50 Lesson 1-5 Independent Practice

1

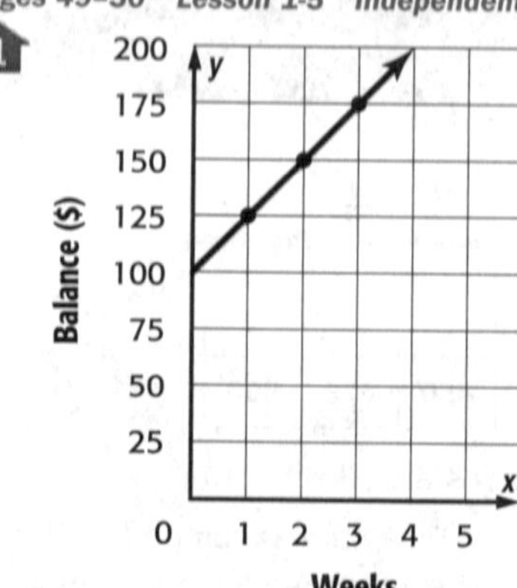

Not proportional; The graph does not pass through the origin. **3** Plant B; The graph is a straight line through the origin. **5.** Proportional; Sample answer: The ordered pairs would be (0, 0), (1, 35), (2, 70). This would be a straight line through the origin.

7.

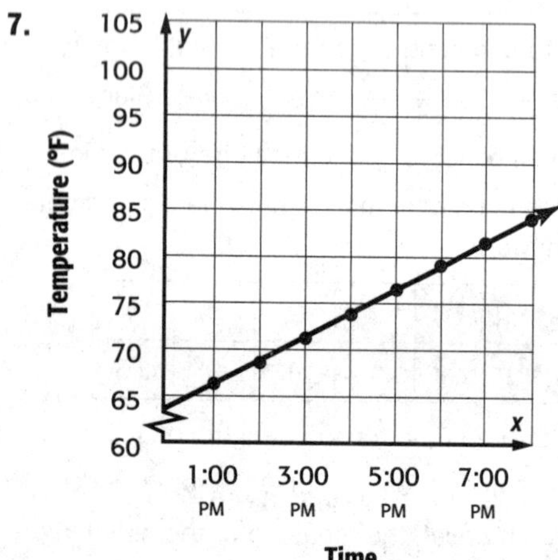

Not proportional; The graph does not pass through the origin.

Pages 51–52 Lesson 1-5 Extra Practice

9. Not proportional; The graph does not pass through the origin. **11.** Not proportional; The graph does not pass through the origin. **13.** The number of heartbeats is proportional to the number of seconds because the graph is a straight line through the origin. **15.** Samora's; The graph is a straight line through the origin. **17.** $\frac{5}{1}$ **19.** $\frac{1}{5}$

Pages 59–60 Lesson 1-6 Independent Practice

1 40 **3.** 3.5 **5.** $\frac{2}{5} = \frac{x}{20}$; 8 ounces **7** $c = 0.50p$; $4.00 **9.** $\frac{360}{3} = \frac{n}{7}$; 840 visitors **11.** 256 c; Sample answer: The ratio of cups of mix to cups of water is 1:8, which means that the proportion $\frac{1}{8} = \frac{32}{x}$ is true and can be solved. **13.** 18 **15.** B

Pages 61–62 Lesson 1-6 Extra Practice

17. 7.2 **19.** $\frac{6}{7} = \frac{c}{40}$; about 34 patients **21.** $s = 45w$; $360 **23.** B **25.** No, the ratios for each age and height are not equal. **27.** Yes; the unit rate is $\frac{15}{1}$ or $15 per hour. **29.** 500 kB/min

Pages 69–70 Lesson 1-7 Independent Practice

1 6 m per s **3** $9 per shirt; Sample answer: The point (0, 0) represents 0 T-shirts purchased and 0 dollars spent. The point (1, 9) represents 9 dollars spent for 1 T-shirt. **5.** 10 inches per hour

7. Sample answer:

Feet	Inches
3	18
6	36
9	54
12	72

9. C

Pages 71–72 Lesson 1-7 Extra Practice

11. $0.03 per minute **13.** Josh; sample answer: The unit rate for Ramona is $9 per hour. The unit rate for Josh is $10 per hour. **15.** A

17.

Input	Add 4	Output
1	1 + 4	**5**
2	2 + 4	**6**
3	3 + 4	**7**
4	4 + 4	**8**

19.

Input	Multiply by 2	Output
1	1 × 2	**2**
2	2 × 2	**4**
3	3 × 2	**6**
4	4 × 2	**8**

21.

Input	Add 6	Output
4	?	10
5	?	11
6	?	12
7	?	13

Pages 77–78 Lesson 1-8 Independent Practice

1 $\frac{50}{1}$ or 50; Adriano read 50 pages every hour.

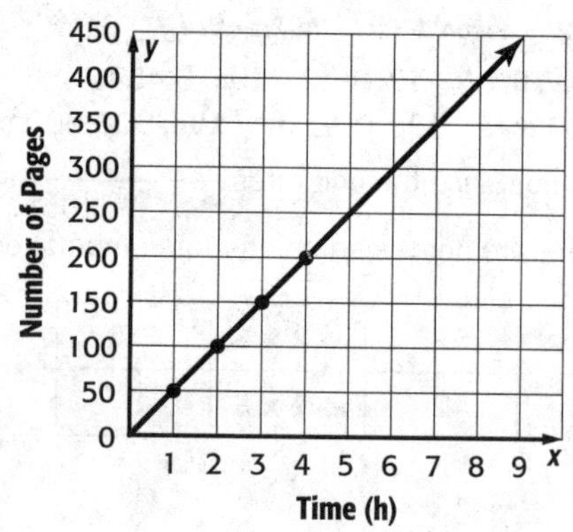

3a. It shows that car A travels 120 miles in 2 hours. **b.** It shows that car B travels 67.5 miles in 1.5 hours. **c.** the speed of each car at that point **d.** the average speed of the car **e.** Car A; the slope is steeper.
5. Marisol found $\frac{\text{run}}{\text{rise}}$. Her answer should be $\frac{3}{2}$. **7.** D

Pages 79–80 Lesson 1-8 Extra Practice

9a. It costs \$20 to rent a paddle boat from Water Wheels for 1 hour. **9b.** It costs \$50 to rent a paddle boat from Fun in the Sun for 2 hours.

11.

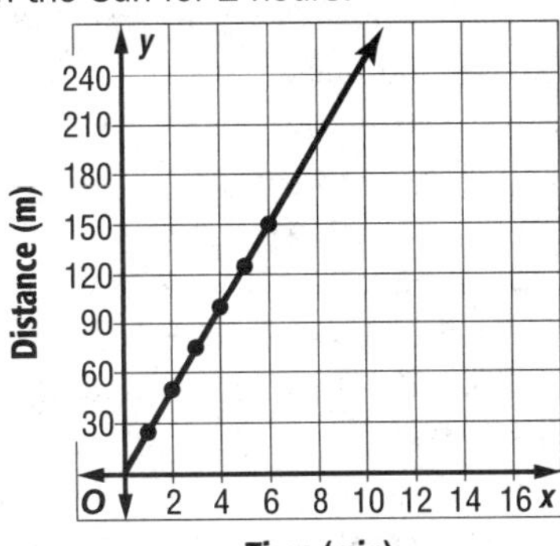

13.

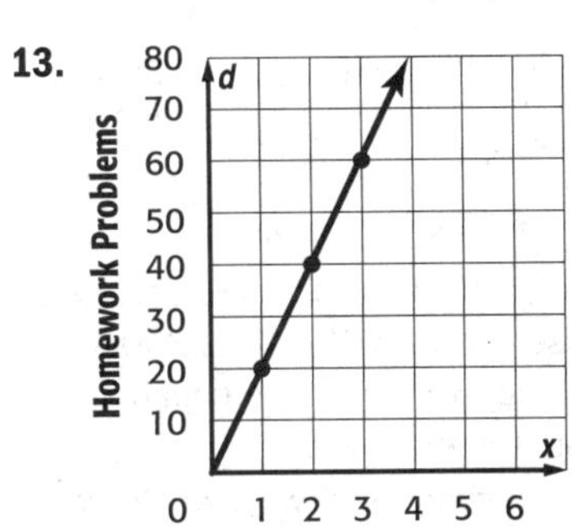

15. C **17.** No; sample answer: $\frac{3.50}{1} \neq \frac{4.50}{2}$ **19.** Yes; sample answer: $\frac{7.50}{1} = \frac{15}{2} = \frac{22.5}{3} = \frac{30}{4}$

Pages 85–86 Lesson 1-9 Independent Practice

1. 30 lb per bag

3.

Time (h)	1	2	3	4
Charge (\$)	75	100	125	150

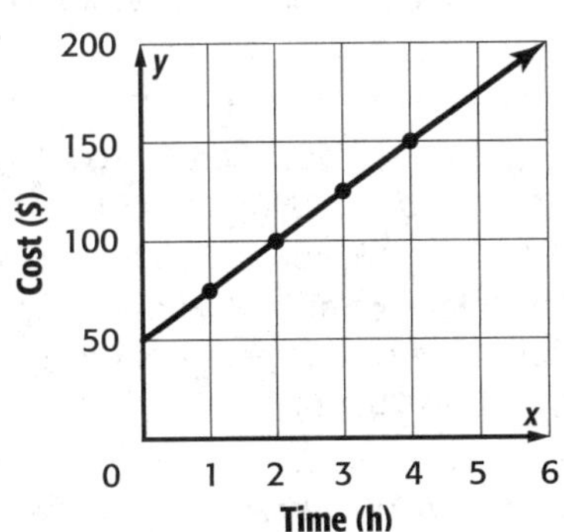

No; sample answer: $\frac{75}{1} \neq \frac{100}{2}$; Because there is no constant ratio and the line does not go through the origin, there is no direct variation. **5.** no **7.** no **9.** $y = \frac{7}{4}x$; 21 **11.** $y = \frac{1}{4}x$; 28 **13.** Sample answer: 9; $5\frac{1}{2}$; 36; 22 **15.** C

Pages 87–88 Lesson 1-9 Extra Practice

17. 7 c **19.** yes; 0.2 **21.** C **23.** yes; 36
25. $\frac{8}{1}$; Each ticket costs \$8.

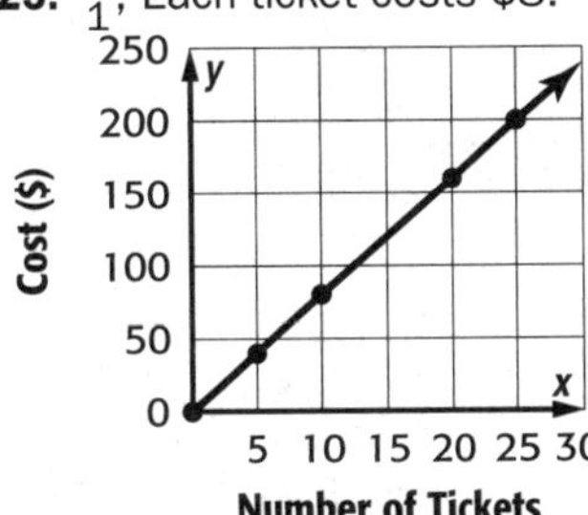

Page 91 Chapter Review Vocabulary Check

1. rate **3.** ordered **5.** complex **7.** slope **9.** proportion **11.** Dimensional

Page 92 Chapter Review Key Concept Check

1. denominator **3.** vertical change to horizontal change

Page 93 Chapter Review Problem Solving

1. the 16-ounce bottle **3.** No; Sample answer: The cost for 1 month of service is \$60, while the cost for 2 months is \$90; $\frac{60}{1} \neq \frac{90}{2}$ **5.** 721.8 lb

Chapter 2 Percents

Page 98 Chapter 2 Are You Ready?

1. 48 **3.** \$70 **5.** 72.5% **7.** 92%

Pages 107–108 Lesson 2-1 Independent Practice

1. 120.9 **3.** \$147.20 **5.** 17.5 **7.** 1.3 **9.** 30.1 **11.** \$7.19 at Pirate Bay, \$4.46 at Funtopia, \$9.62 at Zoomland **13.** 4 **15.** 0.61 **17.** 520 **19.** 158 **21.** 0.14 **23.** Sample answer: It is easiest to use a fraction when the denominator of the fraction is a multiple of the number. If this is not the case, a decimal may be easier to use.

Pages 109–110 Lesson 2-1 Extra Practice

25. 45.9 **27.** 14.7 **29.** \$54 **31.** 0.3 **33.** 2.25 **35.** \$19.95 **35.** D **37.** 91.8 **39.** 133.92 **41.** 160

Pages 115–116 Lesson 2-2 Independent Practice

1. 35

$$\frac{1}{2} \cdot 70 = 35$$
$$0.1 \cdot 70 = 7 \text{ and}$$
$$5 \cdot 7 = 35$$

3 18

$\frac{1}{5} \cdot 90 = 18$

$0.1 \cdot 90 = 9$ and

$2 \cdot 9 = 18$

5. 168

$\frac{7}{10} \cdot 240 = 168$

$0.1 \cdot 240 = 24$ and

$7 \cdot 24 = 168$

7. 720

$(2 \cdot 320) + \left(\frac{1}{4} \cdot 320\right) = 720$

9. 2

$0.01 \cdot 500 = 5$ and

$\frac{2}{5} \cdot 5 = 2$

11 about 96 mi; $0.01 \cdot 12{,}000 = 120$ and $\frac{4}{5} \cdot 120 = 96$

13. 6

$\frac{2}{3} \cdot 9 = 6$

15. 24

$\frac{1}{10} \cdot 240 = 24$

17a. Sample answer: about 260 canned foods; $200 + 0.3 \cdot 200$ **17b.** Sample answer: about 780 canned foods; $600 + 0.3 \cdot 600$ **19.** sometimes; Sample answer: one estimate for 37% of 60 is $\frac{2}{5} \cdot 60 = 24$.

Pages 117–118 Lesson 2-2 Extra Practice

21. 135

23. 90

$\frac{9}{10} \cdot 100 = 90$

$0.1 \cdot 100 = 10$ and

$9 \cdot 10 = 90$

25. 0.7

$0.01 \cdot 70 = 0.7$

27. about 12 muscles; $\frac{3}{10} \cdot 40 = 12$ **29a.** Sample answer: 420; $\frac{7}{10} \cdot 600 = 420$ **29b.** Greater; both the number of passes and the percent were rounded up. **29c.** Tony Romo; sample answer: 64% of 520 must be greater than 64% of 325. **31.** G **33.** 300 **35.** $\frac{1}{4}$

Pages 125–126 Lesson 2-3 Independent Practice

1. 25% **3** 75 **5.** 36 **7.** \$68 **9.** 80 **11** 0.2% **13a.** about 3.41% **13b.** about 24,795.62 km **13c.** about 6,378.16 km **15.** 20% of 500, 20% of 100, 5% of 100; If the percent is the same but the base is greater, then the part is greater. If the base is the same but the percent is greater, then the part is greater.

Pages 127–128 Lesson 2-3 Extra Practice

17. 45 **19.** 20 **21.** 20% **23.** 8 pencils; $0.25 \times 8 = 2$ **25.** 120% **27.** A **29.** 60% **31.** $\frac{3}{20}$ **33.** $\frac{8}{15}$ **35.** $\frac{7}{25}$

Pages 133–134 Lesson 2-4 Independent Practice

1 50%; $75 = n \cdot 150$ **3.** 63.7; $p = 0.65 \cdot 98$ **5.** 6; $p = 0.24 \cdot 25$ **7.** 50 books **9** **a.** 37% **b.** 31% **11.** 0.3; $p = 0.004 \cdot 82.1$ **13.** 115%; $230 = n \cdot 200$ **15.** Sample answer: If the percent is less than 100%, then the part is less than the whole; if the percent equals 100%, then the part equals the whole; if the percent is greater than 100%, then the part is greater than the whole.

Pages 135–136 Lesson 2-4 Extra Practice

17. 20% **19.** 25%; $98 = n \cdot 392$ **21.** 4.4; $1.45 = 0.33 \cdot w$ **23.** 42.5; $17 = 0.4 \cdot w$ **25.** $17 = n \cdot 27$; 63% **27.** \$6.15; $0.25 \cdot 6 = 0.15$ and $6 + 0.15 = 6.15$ **29.** F **31.** $<$ **33.** $<$ **35.** 140; There are 140 students that participate in fall sports.

Page 139 Problem-Solving Investigation Determine Reasonable Answers

Case 3. 70 families **Case 5.** 240 students; Sample answer: $0.6 \times 400 = 240$

Pages 147–148 Lesson 2-5 Independent Practice

1. 20%; increase **3** 25%; decrease **5.** 41%; decrease **7** 28% **9.** 38%; decrease **11a.** 100% **11b.** 300% **13.** about 4.2% **15.** He did not write a ratio comparing the change to the original amount. It should have had a denominator of \$52 and the percent of change would be about 140%.

Pages 149–150 Lesson 2-5 Extra Practice

17. 50%; decrease **19.** 33%; increase **21a.** about 3.8%; increase **21b.** about 2.9%; decrease **23.** 25% **25.** 6,500 comments **27.** 200% **29.** 45.93 **31.** 49,695.72 mi

Pages 155–156 Lesson 2-6 Independent Practice

1. \$69.60 **3** \$1,605 **5** \$35.79 **7.** \$334.80 **9.** \$10.29 **11.** 7% **13.** \$54, \$64.80; The percent of gratuity is 20%. All of the other pairs have a gratuity of 15%.

Pages 157–158 Lesson 2-6 Extra Practice

15. \$103.95 **17.** \$7.99 **19.** \$96.26 **21.** Yes; \$84 was earned. 5% × \$70 = \$3.50; \$70 + \$3.50 = \$73.50; 15% × \$70 = \$10.50; \$73.50 + \$10.50 = \$84 **23.** B **25.** Store B; The total cost of the boots at store A is \$58.19. The total cost of the boots at store B is \$56.98. **27.** 57.85 **29.** \$50

Pages 163–164 Lesson 2-7 Independent Practice

1. \$51.20 **3** \$6.35 **5** \$4.50 **7a.** \$28.76, \$25.29, \$28.87 **7b.** Funtopia **9.** \$9.00

11. Sample answers are given.

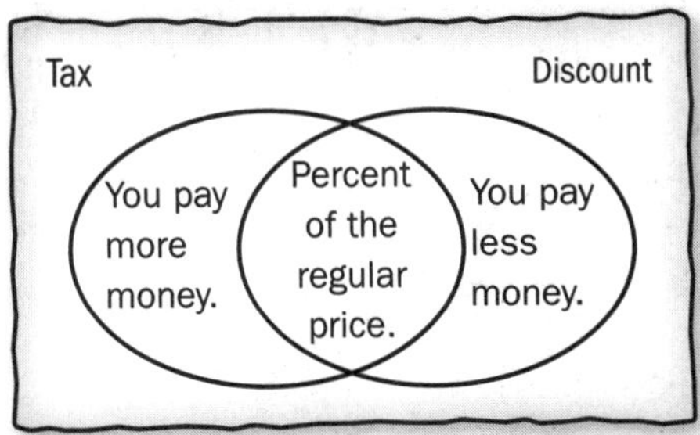

13. $25

Pages 165–166 Lesson 2-7 Extra Practice

15. $102.29 **17.** $169.15 **19.** Mr. Chang; $22.50 < $23.99 **21.** A **23.** 29%; increase **25.** 35%; decrease **27.** Carlos, 18 months; Karen, 16 months; Beng, 14 months

Pages 171–172 Lesson 2-8 Independent Practice

1. $38.40 **3.** $5.80 **5.** $1,417.50 **7.** $75.78 **9. a.** 5% **b.** Yes; he would have $5,208. **11.** Sample answer: If the rate is increased by 1%, then the interest earned is $60 more. If the time is increased by 1 year, then the interest earned is $36 more. **13.** C

Pages 173–174 Lesson 2-8 Extra Practice

15. $6.25 **17.** $123.75 **19.** $45.31 **21.** $14.06 **23.** C

25–28.

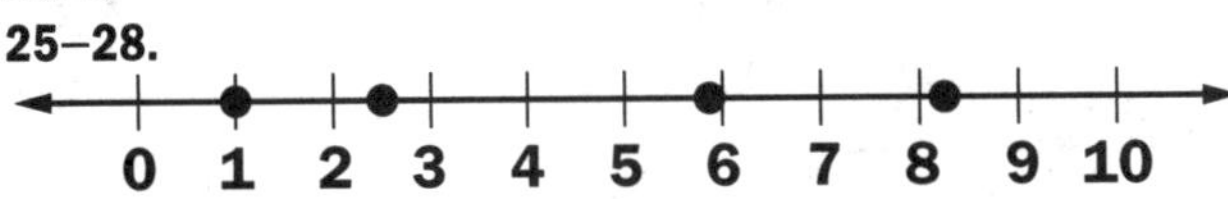

29. Belinda; Sample answer: Since 6 > 4, 5.6 > 5.4. So, Belinda walks a longer distance to school.

Page 179 Chapter Review Vocabulary Check

Down

1. increase **3.** markdown **5.** selling **7.** discount **9.** sales tax

Across

11. interest

Page 180 Chapter Review Key Concept Check

1. 300 **3.** 18 **5.** 12

Page 181 Chapter Review Problem Solving

1. 21 students; 12% = 0.12, 0.12(175) = 21 **3.** 5% **5.** $18

Chapter 3 Integers

Page 190 Chapter 3 Are You Ready?

1. 6 **3.** 24

4–9.

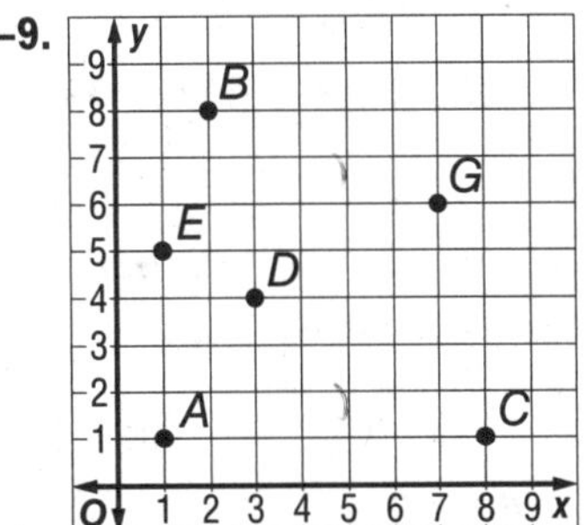

Pages 195–196 Lesson 3-1 Independent Practice

1. 9 **3.** –53

5.

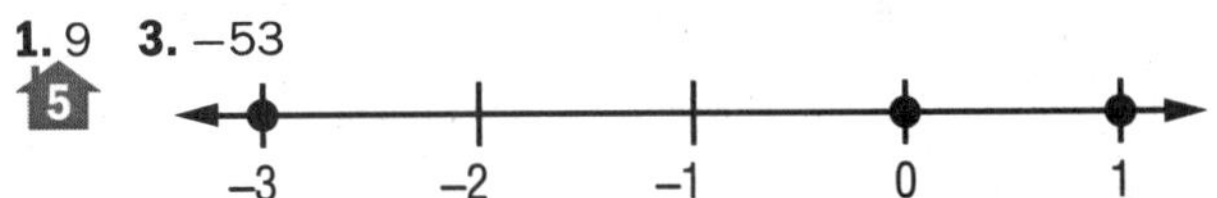

7. 10 **9.** 8 **11.** –7 **13.** $299.97; |–200| + |–40| + |–60| = 200 + 40 + 60 = 300 **15.** always; It is true if *A* and *B* are both positive or if *A* or *B* is negative, and if both *A* and *B* are negative. **17.** A

Pages 197–198 Lesson 3-1 Extra Practice

19. 12

21.

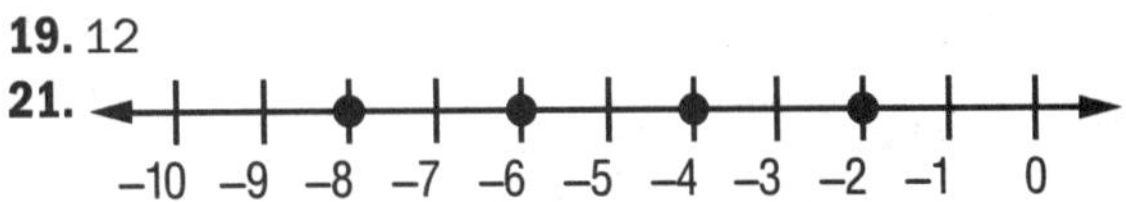

23. 11 **25.** 25 **27.** 5 **29.** C **31.** Wednesday **33.** (0, –2); *y*-axis **35.** (1, 1); I

36–39.

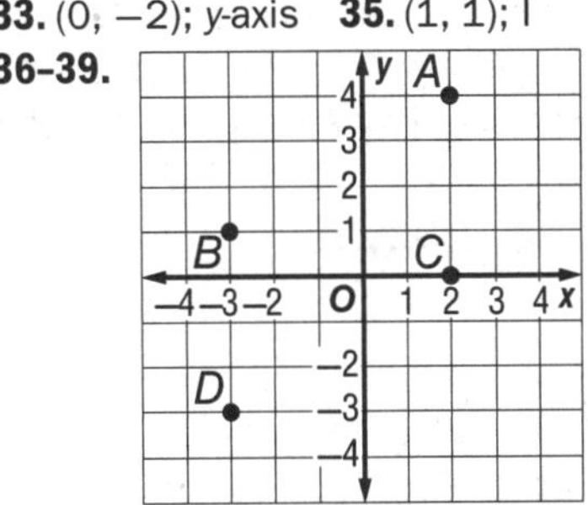

Pages 207–208 Lesson 3-2 Independent Practice

1. –38 **3.** 16 **5.** 0 **7.** 9 **9.** –4 **11.** green; profit of $1; white; profit of $3; black; profit of $3 **13.** Sample answer: In science, atoms may contain 2 positive charges and 2 negative charges. In business, a stock's value may fall 0.75 one day and rise 0.75 the next day. **15.** *a* **17.** *m* + (–15)

Pages 209–210 Lesson 3-2 Extra Practice

19. 13 **21.** –6 **23.** 15 **25.** 22 **27.** –19 **29.** –5 + (–15) + 12; The team has lost a total of 8 yards. **31.** D **33.** –8 + (–3) = –11 **35.** –8 **37.** 4 **39.** 5

Pages 219–220 Lesson 3-3 Independent Practice

1. –10 **3.** –12 **5.** –30 **7.** 23 **9.** 104 **11.** 0 **13. a.** 2,415 ft **b.** 3,124 ft **c.** 627 ft **d.** 8 ft **15.** 16 **17.** Sample answer: –5 – 11 = –5 + (–11) = –16; Add 5 and 11 and keep the negative sign. **19.** He did not find the additive inverse of –18. –15 – (–18) = –15 + 18 or 3. The correct answer is 3. **21.** D

Pages 221–222 Lesson 3-3 Extra Practice

23. 35 **25.** −14 **27.** 6 **29.** 15 **31.** 11 **33.** 1 **35.** A **37.** 10 − 12 **39.** 195 **41.** 12 **43.** 2

Page 227 Problem-Solving Investigation Look for a Pattern

Case 3. Add the previous 2 terms; 89, 144
Case 5. 13 toothpicks

Pages 237–238 Lesson 3-4 Independent Practice

1. −96 **3.** 36 **5** −64 **7** 5(−650); −3,250; Ethan burns 3,250 Calories each week. **9.** 5 black T-shirts

11.

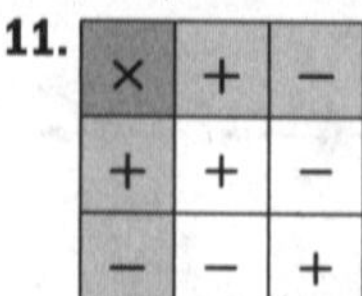

×	+	−
+	+	−
−	−	+

Sample answer: When you multiply a negative and a positive integer, the product is negative. When you multiply two negative integers the product is positive. **13.** Sample answer: Evaluate −7 + 7 first. Since −7 + 7 = 0, and any number times 0 is 0, the value of the expression is 0.
15. D

Pages 239–240 Lesson 3-4 Extra Practice

17. 160 **19.** −64 **21.** −45 **23.** 12(−4); −48; Lily's gift card has $48 less than its starting amount. **25.** 16 **27.** −12 **29.** 648 **31.** −243 **33.** Sample answer: The answer should be −24. A negative multiplied by a negative will be positive. Then, if it is multiplied by a negative it will be negative. **35.** 8(−15); −120 **37.** A **39.** < **41.** >

43.

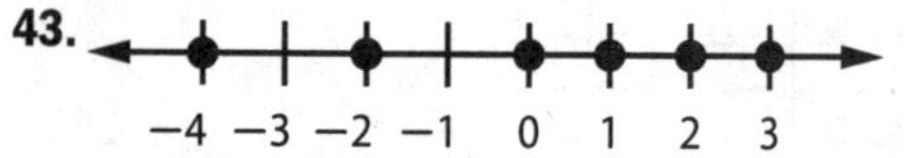

Pages 247–248 Lesson 3-5 Independent Practice

1. −10 **3** 5 **5.** −11 **7.** −2 **9.** −3 **11.** −6 **13** −60 miles per hour **15.** 4 **17.** 16 **19.** No; Sample answer: 9 ÷ 3 ≠ 3 ÷ 9 **21.** −2 **23.** B

Pages 249–250 Lesson 3-5 Extra Practice

25. 9 **27.** 4 **29.** 9 **31.** −12 **33.** 2 **35.** −10°F; The boiling point decreases 10°F at an altitude of 5,000 ft. **37.** B **39.** 4; Sample answer: Christopher answered 6 questions incorrectly. If each question is worth the same, each incorrect answer is worth −24 ÷ 6 or −4 points. So, Nythia answered −16 ÷ (−4) or 4 questions incorrectly. **41.** −9 **43.** 5 **45.** III

Page 253 Chapter Review Vocabulary Check

1. additive **3.** integers **5.** opposites

Page 254 Chapter Review Key Concept Check

1. not correct; |−5| + |2| = 5 + 2 or 7 **3.** not correct; −24 ÷ |−2| = −24 ÷ 2 = −12

Page 255 Chapter Review Problem Solving

1.

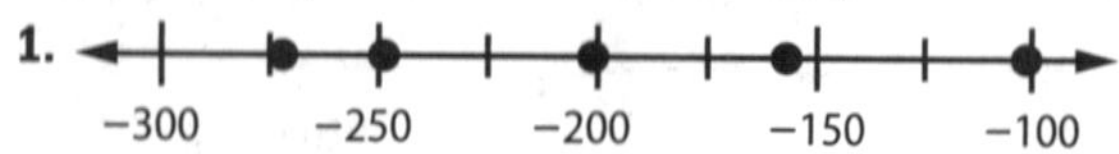

3. 100°C **5.** 4(−2); $33

Chapter 4 Rational Numbers

Page 260 Chapter 4 Are You Ready?

1. $\frac{2}{3}$ **3.** $\frac{8}{11}$

4–7.

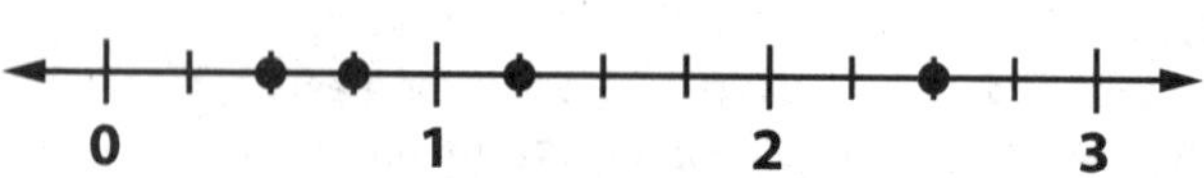

Pages 267–268 Lesson 4-1 Independent Practice

1. 0.5 **3** 0.125 **5.** −0.66 **7.** 5.875 **9.** $-0.\overline{8}$ **11.** $-0.\overline{72}$ **13.** $-\frac{1}{5}$ **15.** $5\frac{24}{25}$ **17** $10\frac{1}{2}$ **19.** Sample answer: $\frac{3}{5}$ **21.** Sample answer: $3\frac{1}{7} \approx 3.14286$ and $3\frac{10}{71} \approx 3.14085$; Since 3.1415926... is between $3\frac{1}{7}$ and $3\frac{10}{71}$, Archimedes was correct.

Pages 269–270 Lesson 4-1 Extra Practice

23. 0.8 **25.** $-0.\overline{4}$ **27.** 0.75 **29.** $\frac{17}{50}$ **31.** $-\frac{13}{1}$ **33.** $-\frac{16}{5}$
35. D **37.** B **39.** 0.1

41–43.

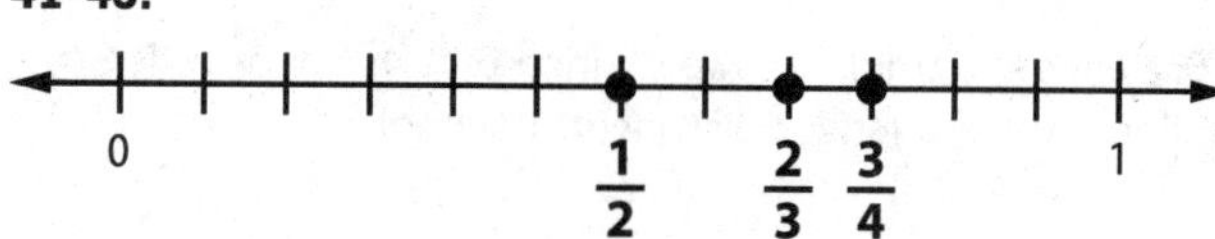

Pages 275–276 Lesson 4-2 Independent Practice

1. >

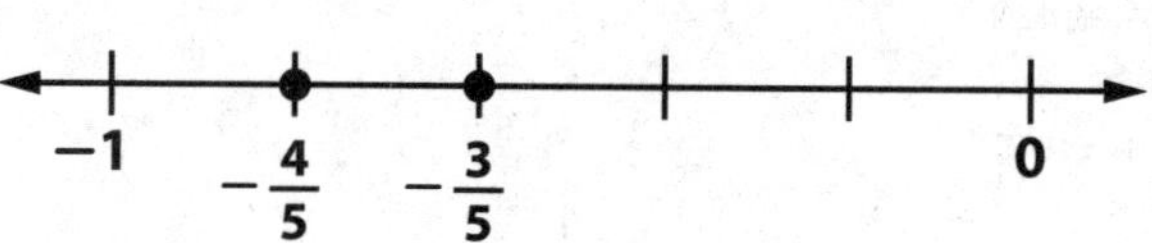

3. > **5** first quiz **7.** $-\frac{5}{8}$, −0.62, −0.615 **9** < **11.** Yes; $69\frac{1}{8} < 69\frac{6}{8}$. **13.** Sample answer: $\frac{63}{32}$ is closest to 2 because the difference of $\frac{63}{32}$ and 2 is the least.

Pages 277–278 Lesson 4-2 Extra Practice

15. < **17.** < **19.** Jim; $\frac{10}{16} > \frac{4}{15}$ **21.** −1.4, −1.25, $-1\frac{1}{25}$ **23.** C **25.** D **27.** > **29.** > **31.** >

Selected Answers

Pages 287–288 Lesson 4-3 Independent Practice

1. $1\frac{4}{7}$ **3.** $-\frac{2}{3}$ **5** $-1\frac{1}{2}$ **7** $\frac{3}{14}$ **9a.** $\frac{33}{100}$ **9b.** $\frac{67}{100}$ **9c.** $\frac{41}{100}$ **11.** Sample answer: $\frac{11}{18}$ and $\frac{5}{18}$; $\frac{11}{18} - \frac{5}{18} = \frac{6}{18}$, which simplifies to $\frac{1}{3}$. **13.** C

Pages 289–290 Lesson 4-3 Extra Practice

15. $-1\frac{2}{3}$ **17.** $\frac{1}{4}$ **19.** $\frac{1}{9}$ **21.** $1\frac{47}{100}$ **23.** $\frac{1}{2}$ c **25.** D **27.** 4 **29.** < **31.** < **33.** 28 **35.** 60

Pages 295–296 Lesson 4-4 Independent Practice

1 $\frac{13}{24}$ **3.** $1\frac{2}{5}$ **5.** $\frac{4}{9}$ **7.** $-\frac{26}{45}$ **9.** $1\frac{11}{18}$
11 Subtraction; Sample answer: To find how much time remained, subtract $\left(\frac{1}{6} + \frac{1}{4}\right)$ from $\frac{2}{3}$; $\frac{1}{4}$ h

13.

Homework	Fraction of Time	
	Pepita	Francisco
Math	$\frac{1}{6}$	$\frac{1}{2}$
English	$\frac{2}{3}$	$\frac{1}{8}$
Science	$\frac{1}{6}$	$\frac{3}{8}$

15. Sample answer: Let $\frac{1}{a}$ and $\frac{1}{b}$ represent the unit fractions, where a and b are not zero. Multiply the first numerator by b and the second numerator by a. Write the product over the denominator ab. Write in simplest form. **17.** C

Pages 297–298 Lesson 4-4 Extra Practice

19. $\frac{19}{30}$ **21.** $\frac{11}{20}$ **23.** $-\frac{13}{24}$ **25.** Subtraction; Sample answer: To find how much more turkey Makalaya bought, subtract $\frac{1}{4}$ from $\frac{5}{8}$; $\frac{3}{8}$ lb **27.** Theresa did not rename the fractions using the LCD. $\frac{5}{20} + \frac{12}{20} = \frac{17}{20}$ **29.** I **31.** $1\frac{2}{5}$ **33.** $1\frac{1}{100}$ **35.** $7\frac{7}{10}$ **37.** 26 **39.** 27

Pages 303–304 Lesson 4-5 Independent Practice

1. $9\frac{5}{9}$ **3.** $8\frac{3}{5}$ **5** $7\frac{5}{12}$ **7.** $4\frac{14}{15}$ **9.** $4\frac{1}{3}$
11 Subtraction; the width is shorter than the length; $1\frac{3}{4}$ ft
13. −5 **15.** $13\frac{5}{9}$ **17.** Sample answer: A board with a length of $3\frac{7}{8}$ ft needs to be cut from a $5\frac{1}{2}$ -foot existing board. How much wood will be left after the cut is made?; $1\frac{5}{8}$ ft **19.** B

Pages 305–306 Lesson 4-5 Extra Practice

21. $18\frac{17}{24}$ **23.** $7\frac{5}{7}$ **25.** $5\frac{7}{8}$ **27.** Subtraction twice; the amount of flour is less than the original amount; $2\frac{2}{3}$ c

29. $7\frac{1}{8}$ yd **31.** D **33.** 5; 8; 40 **35.** 5; 11; 55 **37.** 14 mi; Sample answer: $6\frac{4}{5} \approx 7$ and $1\frac{3}{4} \approx 2$; $7 \times 2 = 14$

Page 309 Problem-Solving Investigation Draw a Diagram

Case 3. $\frac{3}{8}$ **Case 5.** $\frac{1}{4}$ mi

Pages 315–316 Lesson 4-6 Independent Practice

1. $\frac{3}{32}$ **3.** $-4\frac{1}{2}$ **5.** $\frac{1}{6}$ **7** $\frac{3}{8}$ **9.** −1 **11** $\frac{1}{16}$
13. $\frac{1}{3} \times \left(\frac{11}{16}\right) = \frac{11}{48}$ **15.** Sample answer: Three fourths of the students at Walnut Middle School were on the honor roll. Of that group, only $\frac{1}{8}$ of them received all As. What fraction of the students received all As? **17.** A

Pages 317–318 Lesson 4-6 Extra Practice

19. $\frac{1}{9}$ **21.** $\frac{1}{4}$ **23.** $2\frac{1}{6}$ **25.** $\frac{3}{16}$ **27.** $-\frac{8}{27}$ **29.** broccoli: $1\frac{7}{8}$ c, pasta: $5\frac{5}{8}$ c, salad dressing: 1 c, cheese: 2 c; Multiply each amount by $1\frac{1}{2}$. **31.** B **33.** < **35.** $\frac{1}{18} \div \frac{1}{3} = \frac{1}{6}$; $\frac{1}{18} \div \frac{1}{6} = \frac{1}{3}$ **37.** $6\frac{3}{4} \div 1\frac{1}{5} = 5\frac{5}{8}$; $6\frac{3}{4} \div 5\frac{5}{8} = 1\frac{1}{5}$
39. $5\frac{1}{4}$ pints

Pages 323–324 Lesson 4-7 Independent Practice

1. 12.7 **3** 128.17 **5.** 0.04 **7.** 15.75 **9.** 1.5 **11.** 887.21 mL **13** 1.5 lb **15.** 1,000 mL or 1 L **17.** 0.031 m, 0.1 ft, 0.6 in., 1.2 cm **19.** 0.7 gal, 950 mL, 0.4 L, $1\frac{1}{4}$ c **21.** C

Pages 325–326 Lesson 4-7 Extra Practice

23. 158.76 **25.** 121.28 **27.** 41.89 **29.** 2 L **31.** 3 gal **33.** 4 mi **35.** B **37.** 5.7 **39.** 15,840 **41.** 1 **43.** 5 **45.** 1

Pages 331–332 Lesson 4-8 Independent Practice

1. $\frac{7}{16}$ **3** $\frac{1}{15}$ **5.** $\frac{2}{9}$ **7** 84 movies
9. $1\frac{1}{4}$

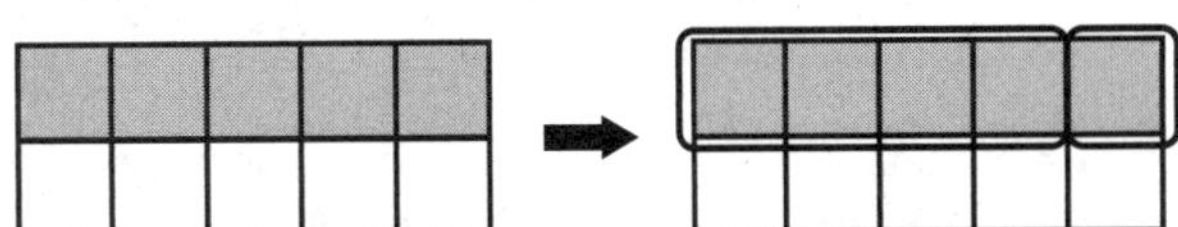

Sample answer: The model on the left shows that one half of a rectangle with ten sections is five sections. Two fifths of ten sections is four sections. The model on the right shows the five sections divided into $1\frac{1}{4}$ groups of four sections. **11.** $\frac{1}{6}$ of a dozen; 2 folders **13.** $\frac{10}{3}$

Pages 333–334 Lesson 4-8 Extra Practice

15. $\frac{2}{3}$ **17.** $-7\frac{4}{5}$ **19.** 11 servings **21.** $\frac{1}{2}$ **23.** C **25.** $\frac{9}{20}$ **27.** $\frac{46}{63}$ **29.** $\frac{3}{4}$ ft **31a.** $\frac{5}{8}$ mi **31b.** $\frac{13}{16}$ mi

Page 337 Chapter Review Vocabulary Check

1. bar notation **3.** common denominator **5.** terminating

Page 338 Chapter Review Key Concept Check

1. $\frac{3}{5}$ **3.** denominator **5.** multiply

Page 339 Chapter Review Problem Solving

1. $5.1\overline{3}$ min **3.** $1\frac{5}{8}$ c **5.** $\frac{7}{30}$; Sample answer: The product of $\frac{7}{20}$ and $\frac{2}{3}$ is $\frac{14}{60}$ or $\frac{7}{30}$. **7.** 140 oz

Index

Aa

Bb

Cc

Ff

Gg

Hh

Ii

Kk

Ll

Mm

Nn

Oo

Ss

Tt

Uu

Zz

Name ______________________________

=

Name ______________________________

Work Mats

Name ____

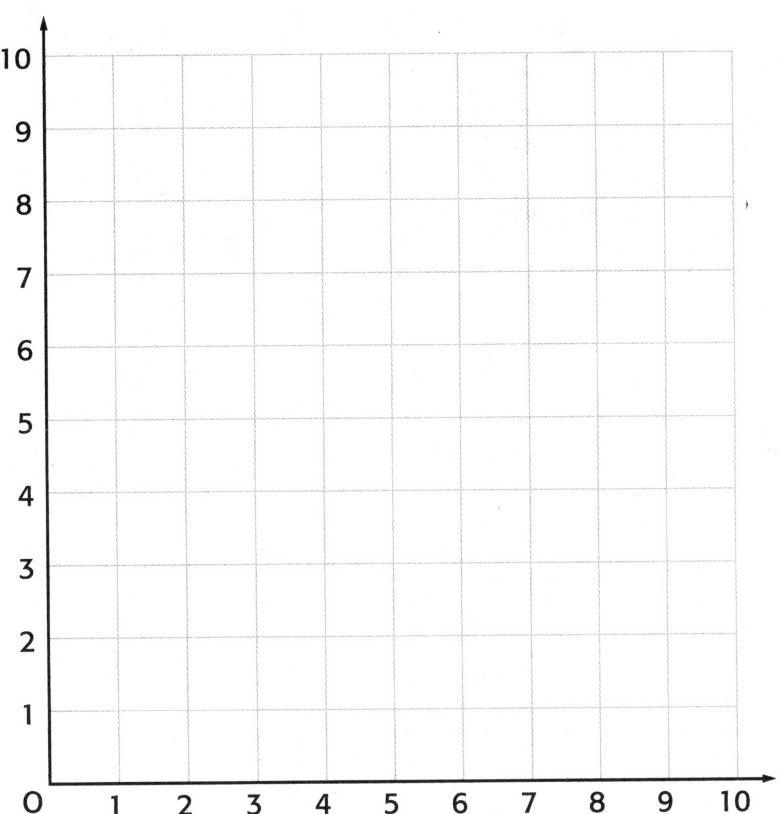

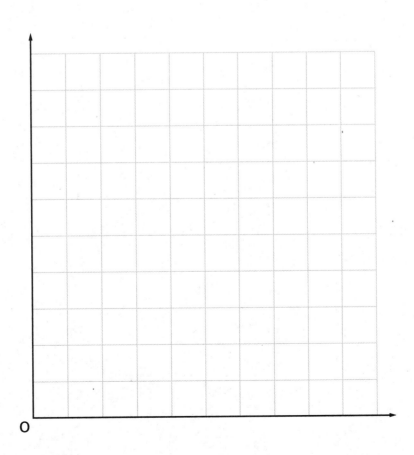

Name ______________________________

0 1 2 3 4 5 6 7 8 9

–11 –10 –9 –8 –7 –6 –5 –4 –3 –2 –1 0 1 2 3 4 5 6 7 8 9 10 11

What Are Foldables and How Do I Create Them?

Foldables are three-dimensional graphic organizers that help you create study guides for each chapter in your book.

Step 1 Go to the back of your book to find the Foldable for the chapter you are currently studying. Follow the cutting and assembly instructions at the top of the page.

Step 2 Go to the Key Concept Check at the end of the chapter you are currently studying. Match up the tabs and attach your Foldable to this page. Dotted tabs show where to place your Foldable. Striped tabs indicate where to tape the Foldable.

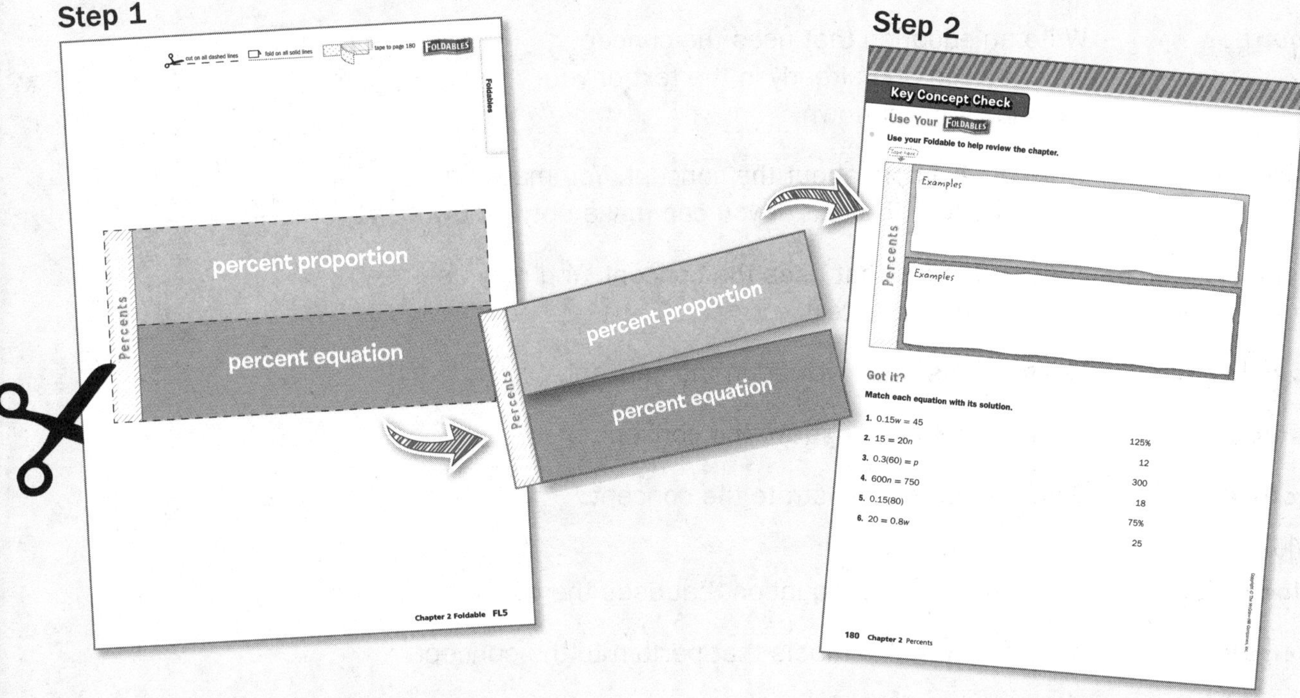

How Will I Know When to Use My Foldable?

When it's time to work on your Foldable, you will see a Foldables logo at the bottom of the **Rate Yourself!** box on the Guided Practice pages. This lets you know that it is time to update it with concepts from that lesson. Once you've completed your Foldable, use it to study for the chapter test.

How Do I Complete My Foldable?

No two Foldables in your book will look alike. However, some will ask you to fill in similar information. Below are some of the instructions you'll see as you complete your Foldable. **HAVE FUN** learning math using Foldables!

Instructions and what they mean

Best Used to...	Complete the sentence explaining when the concept should be used.
Definition	Write a definition in your own words.
Description	Describe the concept using words.
Equation	Write an equation that uses the concept. You may use one already in the text or you can make up your own.
Example	Write an example about the concept. You may use one already in the text or you can make up your own.
Formulas	Write a formula that uses the concept. You may use one already in the text.
How do I ...?	Explain the steps involved in the concept.
Models	Draw a model to illustrate the concept.
Picture	Draw a picture to illustrate the concept.
Solve Algebraically	Write and solve an equation that uses the concept.
Symbols	Write or use the symbols that pertain to the concept.
Write About It	Write a definition or description in your own words.
Words	Write the words that pertain to the concept.

Meet Foldables Author Dinah Zike

Dinah Zike is known for designing hands-on manipulatives that are used nationally and internationally by teachers and parents. Dinah is an explosion of energy and ideas. Her excitement and joy for learning inspires everyone she touches.

cut on all dashed lines fold on all solid lines tape to page 92 FOLDABLES

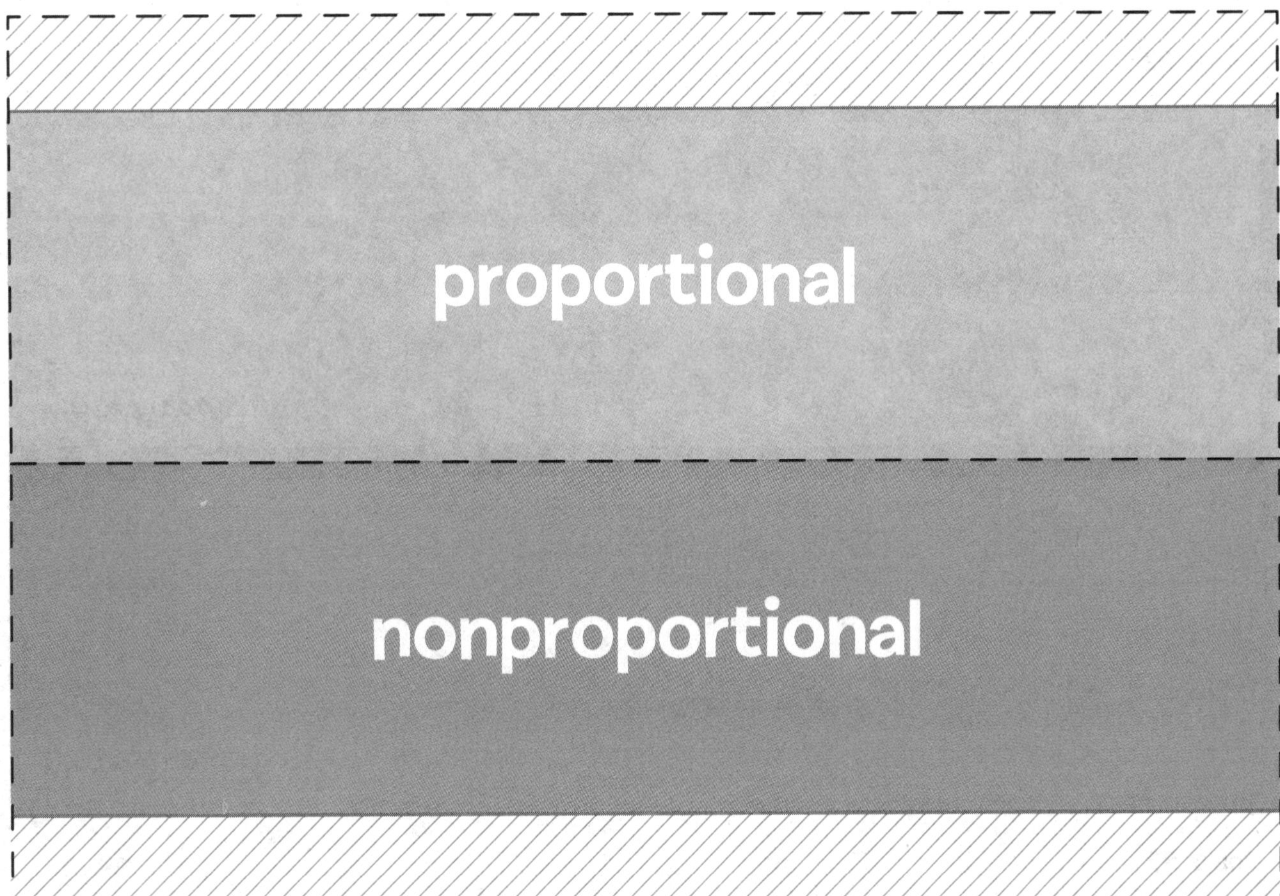

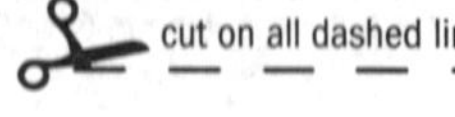

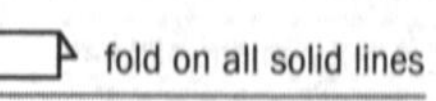

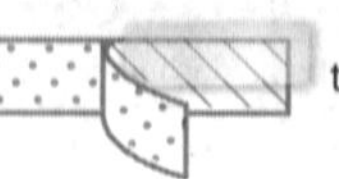

page 92

Tab 1

Write About It

Write About It

page 92

Tab 2

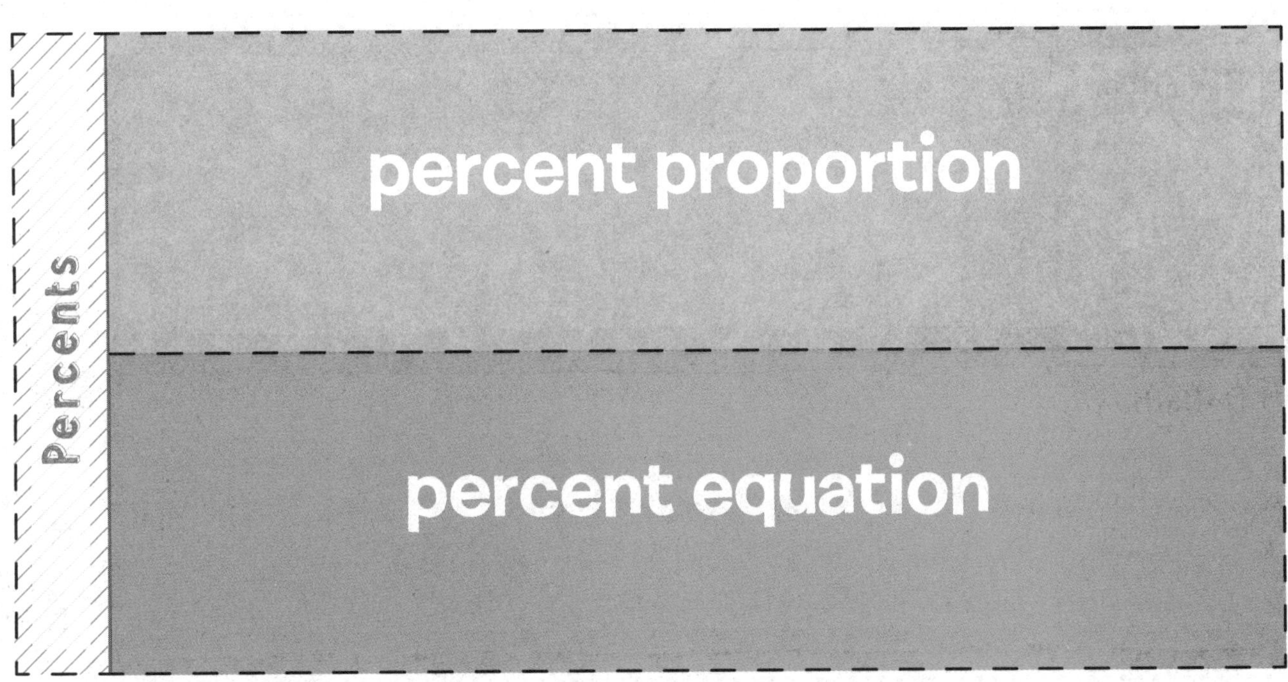
Percents
percent proportion
percent equation

cut on all dashed lines — fold on all solid lines — tape to page 180

FOLDABLES

Definition

Definition

page 180

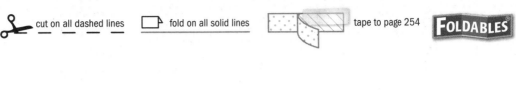

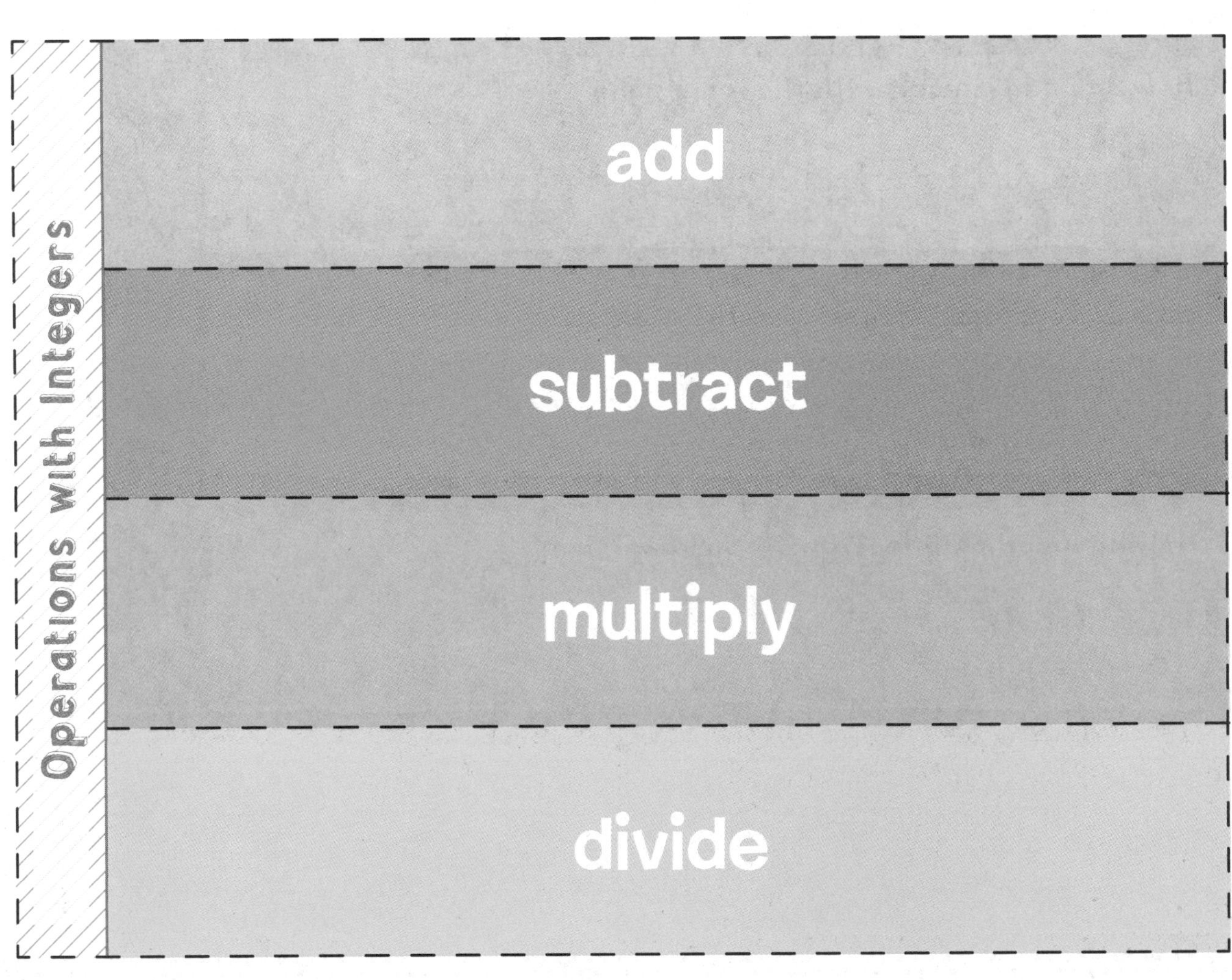
Operations with Integers
add
subtract
multiply
divide

cut on all dashed lines   fold on all solid lines   tape to page 254   FOLDABLES

How do I add integers with the same sign?

+

How do I subtract integers with the same sign?

−

How do I multiply integers with the same sign?

×

How do I divide integers with the same sign?

÷

page 254

cut on all dashed lines   fold on all solid lines   tape to page 338

FOLDABLES®

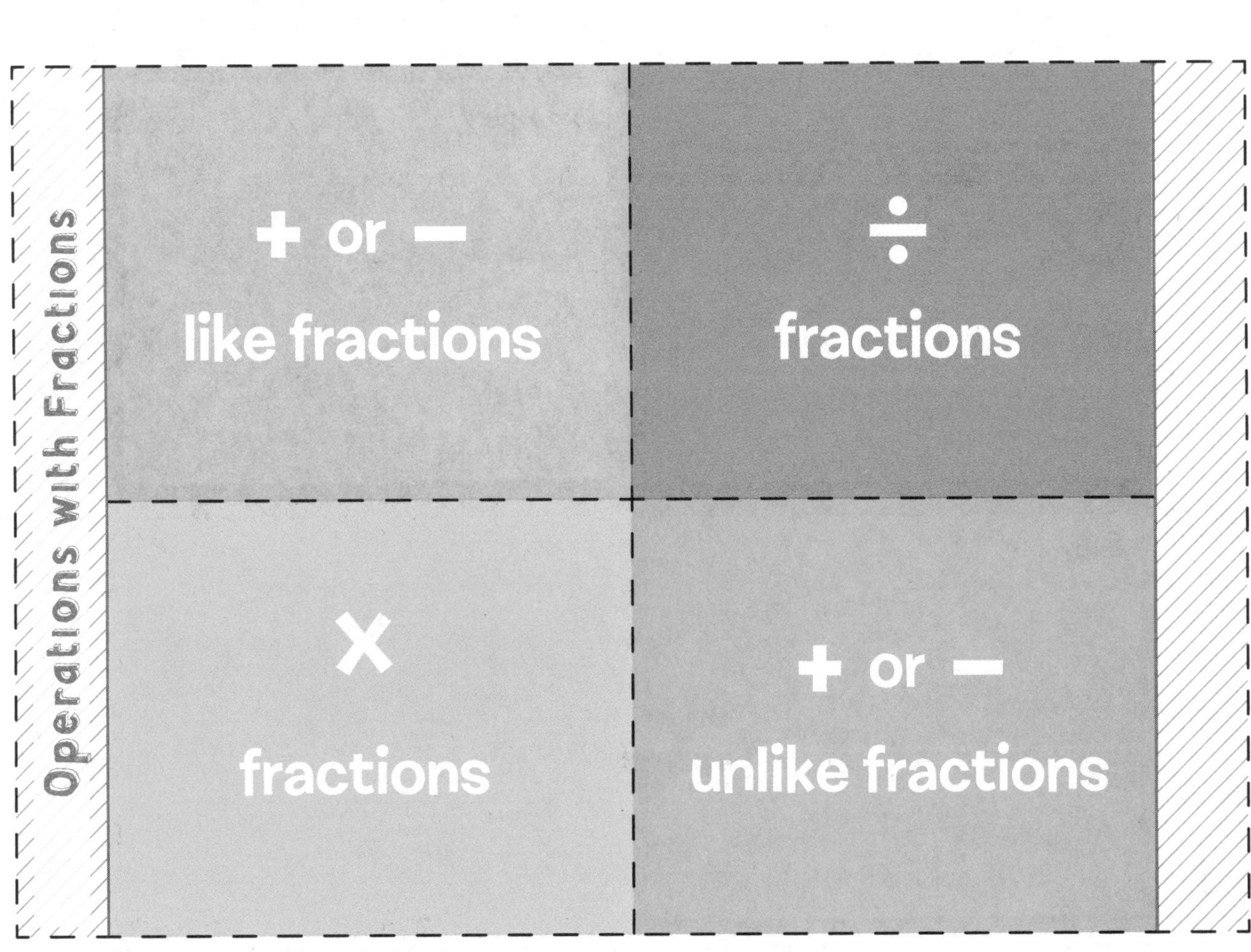

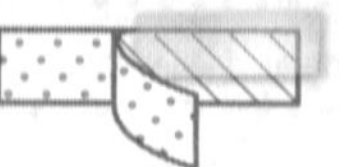
cut on all dashed lines
fold on all solid lines
tape to page 338

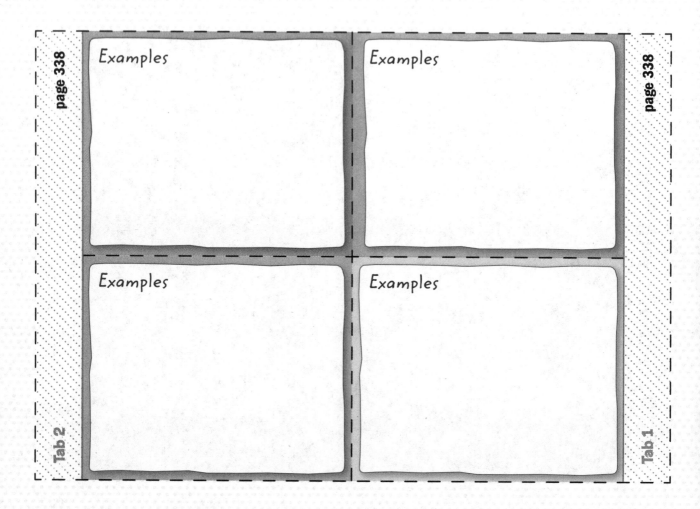

page 338

Examples

Examples

page 338

Examples

Examples

Tab 2

Tab 1